大数据技术精品系列教材

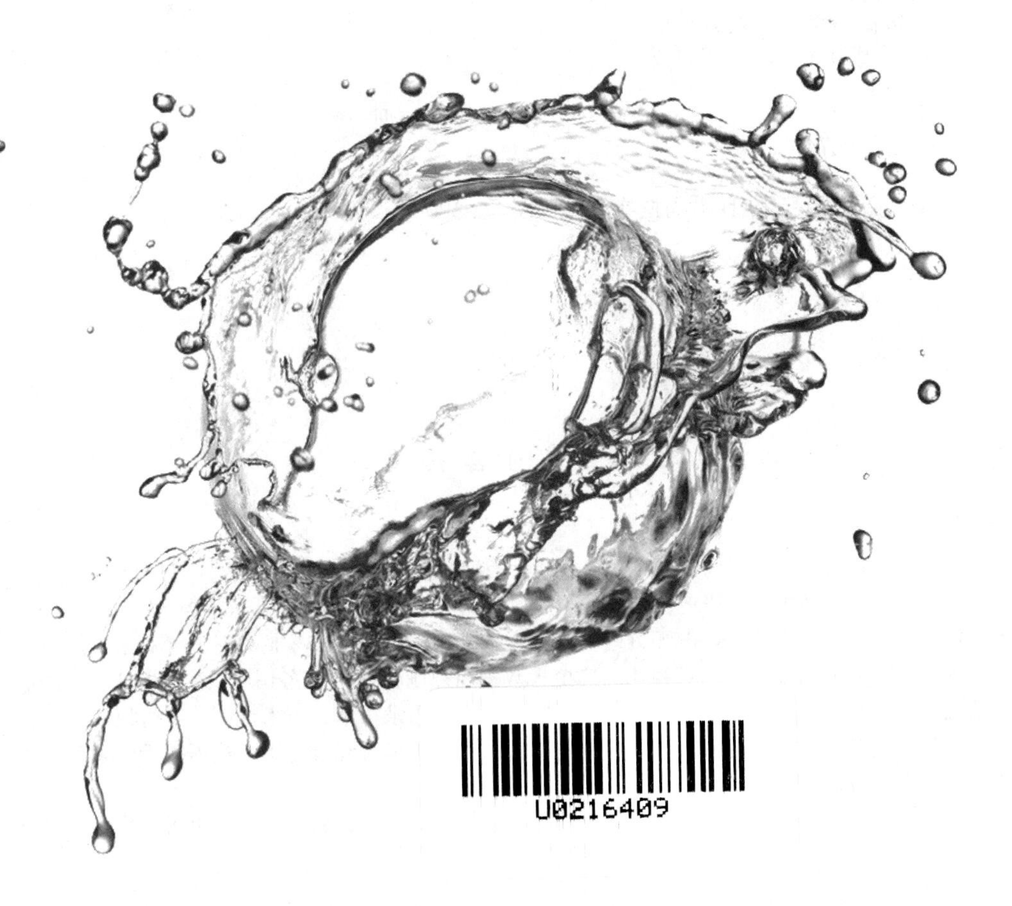

大数据开发项目实战

Hands-on Big Bata Development

祝锡永 张良均◉主编
陈小伟 胡军浩 王爱国◉副主编

人民邮电出版社
北京

图书在版编目（CIP）数据

大数据开发项目实战 / 祝锡永，张良均主编. -- 北京：人民邮电出版社，2022.9（2023.8重印）
大数据技术精品系列教材
ISBN 978-7-115-59804-2

Ⅰ. ①大… Ⅱ. ①祝… ②张… Ⅲ. ①数据处理－教材 Ⅳ. ①TP274

中国版本图书馆CIP数据核字(2022)第140326号

内 容 提 要

本书以项目案例为导向，贯穿讲解一个大数据的实战项目：广电大数据用户画像。全书共 8 章，具体内容包括大数据项目概述、Hadoop 生态组件基础、广电大数据用户画像——需求分析、广电大数据用户画像——数据采集与预处理、广电大数据用户画像——实时统计订单信息、广电大数据用户画像——用户标签计算与可视化、广电大数据用户画像——任务调度实现、基于 TipDM 大数据挖掘建模平台实现广电大数据用户画像。本书在项目需求探索、技术选型、架构设计、集群安装部署与集成开发以及项目开发方面进行实战讲解，有助于读者综合运用大数据技术知识和各种工具软件，实现大数据项目开发全流程操作。

本书可以作为高校大数据技术类专业的大数据项目实训课程的教材，也可以作为大数据技术爱好者的自学用书。

◆ 主　　编　祝锡永　张良均
　副 主 编　陈小伟　胡军浩　王爱国
　责任编辑　初美呈
　责任印制　王　郁　焦志炜

◆ 人民邮电出版社出版发行　　北京市丰台区成寿寺路 11 号
　邮编　100164　　电子邮件　315@ptpress.com.cn
　网址　https://www.ptpress.com.cn
　北京隆昌伟业印刷有限公司印刷

◆ 开本：787×1092　1/16
　印张：16.25　　2022 年 9 月第 1 版
　字数：372 千字　　2023 年 8 月北京第 3 次印刷

定价：59.80 元

读者服务热线：(010)81055256　印装质量热线：(010)81055316
反盗版热线：(010)81055315
广告经营许可证：京东市监广登字 20170147 号

大数据技术精品系列教材

专家委员会

序 FOREWORD

随着大数据时代的到来，移动互联网和智能手机迅速普及，多种形态的移动互联网应用蓬勃发展，电子商务、云计算、互联网金融、物联网、虚拟现实、智能机器人等不断渗透并重塑传统产业，而与此同时，大数据当之无愧地成为新的产业革命核心。

2019 年 8 月，联合国教科文组织以联合国 6 种官方语言正式发布《北京共识——人工智能与教育》。其中提出，“通过人工智能与教育的系统融合，全面创新教育、教学和学习方式，并利用人工智能加快建设开放灵活的教育体系，确保全民享有公平、适合每个人且优质的终身学习机会”。这表明基于大数据的人工智能和教育均进入了新的阶段。

高等教育是教育系统中的重要组成部分，高等院校作为人才培养的重要载体，肩负着为社会培育人才的重要使命。2018 年 6 月 21 日的新时代全国高等学校本科教育工作会议首次提出了“金课”的概念。“金专”“金课”“金师”迅速成为新时代高等教育的热词。如何建设具有中国特色的大数据相关专业，以及如何打造世界水平的“金专”“金课”“金师”“金教材”是当代教育教学改革的难点和热点。

实践教学是在一定的理论指导下，通过实践引导，使学习者获得实践知识、掌握实践技能、锻炼实践能力、提高综合素质的教学活动。实践教学在高校人才培养中有着重要的地位，是巩固和加深理论知识的有效途径。目前，高校大数据相关专业的教学体系设置过多地偏向理论教学，课程设置冗余或缺漏，知识体系不健全，且与企业实际应用契合度不高，学生无法把理论转化为实践应用技能。为了有效解决该问题，“泰迪杯”数据挖掘挑战赛组委会与人民邮电出版社共同策划了“大数据技术精品系列教材”，这恰与 2019 年 10 月 24 日教育部发布的《教育部关于一流本科课程建设的实施意见》（教高〔2019〕8 号）中提出的“坚持分类建设”“坚持扶强扶特”“提升高阶性”“突出创新性”“增加挑战度”原则完全契合。

“泰迪杯”数据挖掘挑战赛自 2013 年创办以来，一直致力于推广高校数据挖掘实践教学，培养学生数据挖掘的应用和创新能力。挑战赛的赛题均为经过适当简化和加工的实际问题，来源于各企业、管理机构和科研院所等，非常贴近现实热点需求。赛题中的数据只做必要的脱敏处理，力求保持原始状态。挑战赛围绕数据挖掘的整个流程，从数据采集、数据迁移、数据存储、数据分析与挖掘，到数据可视化，涵盖了企业应用中的各个环节，与目前大数据专业人才培养目标高度一致。“泰迪杯”数据挖掘挑战赛不依赖于数学建模，甚至不依赖传统模型的竞赛形式，使得“泰迪杯”数据挖掘挑战赛在全国各大高校反响热烈，且得到了全国各界专家学者的认可与支持。2018年，“泰迪杯”数据挖掘挑战赛增加了子赛项——数据分析技能赛，为应用型本科、高职和中职技能型人才培养提供理论、技术和资源方面的支持。截至2021年，全国共有超1000所高校，约2万名研究生、9万名本科生、2万名高职生参加了“泰迪杯”数据挖掘挑战赛和数据分析技能赛。

本系列教材的第一大特点是注重学生的实践能力培养，针对高校实践教学中的痛点，首次提出“鱼骨教学法”的概念。以企业真实需求为导向，学生学习技能时紧紧围绕企业实际应用需求，将学生需掌握的理论知识通过企业案例的形式进行衔接，达到知行合一、以用促学的目的。第二大特点是以大数据技术应用为核心，紧紧围绕大数据应用闭环的流程进行教学。本系列教材涵盖了企业大数据应用中的各个环节，符合企业大数据应用的真实场景，使学生从宏观上理解大数据技术在企业中的具体应用场景及应用方法。

在教育部全面实施“六卓越一拔尖”计划2.0的背景下，对如何促进我国高等教育人才培养体制机制的综合改革，以及如何重新定位和全面提升我国高等教育质量，本系列教材将起到抛砖引玉的作用，从而加快推进以新工科、新医科、新农科、新文科为代表的一流本科专业的“双万计划”建设；落实“让学生忙起来，管理严起来和教学活起来”措施，让大数据相关专业的人才培养质量有一个质的提升；借助数据科学的引导，在文、理、农、工、医等方面全方位发力，培养各个行业的卓越人才及未来的领军人才。同时本系列教材将根据读者的反馈意见和建议及时改进、完善，努力成为大数据时代的新型“编写、使用、反馈”螺旋式上升的系列教材样板。

汕头大学校长
教育部高校大学数学课程教学指导委员会副主任委员
“泰迪杯”数据挖掘挑战赛组织委员会主任
“泰迪杯”数据分析技能赛组织委员会主任

2021年7月于粤港澳大湾区

前 言 PREFACE

进入 21 世纪以后，互联网技术发展快速，人类产生的数据正处于“大爆炸”的阶段。不断增长的数据量和多种多样的数据类型迫使各行各业从传统的数据技术应用转向大数据技术应用。为了解决海量数据的存储和处理问题，各行各业涌现了大量基于 Hadoop 生态圈等技术的大数据解决方案。目前，大部分互联网和金融行业搭建了基于 Hadoop、Spark、Hive、Kafka 等技术的技术栈，用于海量数据的离线分析和实时分析，以满足数据采集、数据存储、数据处理、数据可视化等业务要求。

可以预见，未来各行业所产生的数据，形式必将更加丰富、体量必将更加庞大、处理过程必将更加高效，大数据技术的应用范围也必将更加广泛。这不仅要求现有的 IT 行业人员需要学习大数据技术，也要求高校建立大数据技术课程体系，为社会输送具备大数据专业素养的人才，从而满足社会对大数据人才日益增长的需求。

本书特色

目前市面上的大数据图书种类繁多，其中部分偏向理论知识，往往缺少实战指导，部分结合技术讲解与实战，但总体综合性较低，以实际企业项目为基础的实战型高价值图书并不多。本书全面贯彻党的二十大精神，以社会主义核心价值观、新时代中国特色社会主义思想为引领，加强基础研究、发扬斗争精神，为建成教育强国、科技强国、人才强国、文化强国添砖加瓦。本书以企业实际项目为基础，先介绍企业大数据项目数据处理流程、架构分析、人员安排、实战环境和涉及技术；再介绍 Hadoop 常用生态组件，然后依据项目开发流程，详细介绍从项目数据获取到最后项目部署上线的完整流程。在项目实战过程中，本书再现实际项目开发流程，从需求分析入手，引导读者进行思考，注重项目过程中思路的启发，分析每一步实现的原因，解释每一个实现结果的意义，使读者对项目开发的流程有更加深刻的体会。本书的项目综合性强，采用时下企业常用的大数据技术，包括数据采集、数据分析、数据存储、数据可视化等，综合考量离线分析技术的应用和实时分析技术的应用。本书力求使读者对大数据技术与大数据开发有准确的理解和更深的掌握。

本书适用对象

- 开设大数据相关课程的高校的教师和学生。

目前，国内很多高校的计算机、数学、大数据等专业均开设了与大数据技术相关的课程，但大数据方向的教材相对较少，讲解大数据项目开发的教材较为少见；学生学习基础知识后缺乏实战训练，对于大数据知识的运用不够灵活。本书通过实战项目，引导读者进行思考，使读者充分发挥主观能动性和创造性。本书将理论与实际相结合，便于读者把大数据技术掌握得更加牢固。

● 想从事大数据行业的 IT 技术人员、大数据技术爱好者。

对想从事与大数据相关的软件开发和管理人员来说，本书是一本非常好的参考书。读者仅仅掌握大数据的技术，还不足以很好地完成大数据的相关开发工作，通过本书了解大数据开发的技术应用，积攒大数据项目开发经验，是进入大数据行业的重要一步。

代码下载及问题反馈

为了帮助读者更好地使用本书，本书配套原始数据文件、程序代码，以及 PPT 课件、教学大纲、教学进度表和教案等教学资源，读者可以从泰迪云教材网站免费下载，也可登录人邮教育社区（www.ryjiaoyu.com）下载。同时欢迎教师加入 QQ 交流群“人邮大数据教师服务群”（669819871）进行交流探讨。

本书的编写和出版得到了浙江省普通高校“十三五”新形态教材建设项目（2020-55）的资助。由于编者水平有限，书中难免出现一些疏漏和不足之处。如果读者有更多的宝贵意见，欢迎在泰迪学社微信公众号（TipDataMining）回复“图书反馈”进行反馈。更多本系列图书的信息可以在泰迪云教材网站查阅。

编　者

2023 年 5 月

泰迪云教材

目录 CONTENTS

第 1 章 大数据项目概述

IT 企业对大数据人才的需求越来越高，而大数据行业的高需求、高待遇也促使很多人学习大数据技术甚至转行至大数据行业，促进我国更快更好地进入高质量发展阶段，加快建设制造强国、质量强国、航天强国、交通强国、网络强国、数字中国。大数据技术用于对大规模数据进行专业化处理，且用好大数据技术确实能显著提升企业的运营效率。做好大数据项目，对企业的发展将大有裨益。互联网的从业人员更是需要高度关注与大数据相关的技术及应用。本章将从 IT 企业项目数据处理流程入手，对 IT 企业的偏软件交付类项目进行介绍，引入企业大数据项目架构及如何将大数据架构应用到企业项目处理流程中的内容，并对本书项目实战中使用的软硬件环境进行详细说明，读者可以参考本章内容部署相关环境。

学习目标

（1）了解企业大数据项目的数据处理流程与架构设计。

（2）了解企业大数据项目各个阶段的人员安排。

（3）熟悉本书项目所需的实战环境和涉及的技术。

（4）熟悉 Spark 任务调用的实现和 Spark 任务提交到集群并进行任务监控的过程。

1.1 企业大数据项目简介

一个成功的大数据项目需要有一个良好的开端。进行企业大数据项目核心开发工作之前，需要先明确企业大数据项目数据处理的流程、分析项目的架构，再了解项目人员的安排。本章内容将为读者后续进行企业真实大数据项目实战打下基础。

1.1.1 数据处理流程

企业项目，如开发一款软件、一个信息管理系统、部署实施一套网络环境等，涉及多个方面的多个环节。而在本书中，企业项目仅限定为 IT 企业的偏软件交付类项目。在此类项目中，一般涉及两个公司：一个公司发布需求（甲方招标，即提出软件、系统或硬件的需求），而另一个公司满足需求（乙方投标，即完成对需求的响应）。本书更多的是描述乙方的工作内容。一般情况下，偏软件交付类项目开发可以分为可行性分析、需求分析、软件设计（概要设计、详细设计）、编码（软件设计实现）、测试、运行维护 6 个阶段。有时，因为甲方会提前进行可行性分析，所以乙方的工作更加偏向后面几个阶段。

企业偏软件交付类项目包含的项目种类比较多，常见的项目如客户关系管理（Customer Relationship Management，CRM）系统、用户个性化推荐系统等。本书的项目更加偏向于后者，也就是数据处理类项目。数据处理类项目的处理流程如图 1-1 所示。

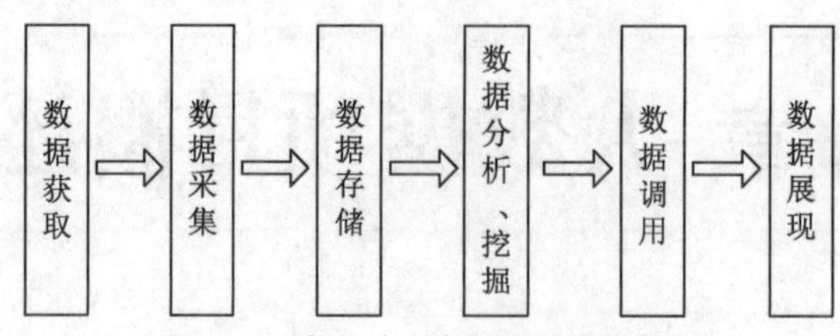

图 1-1　企业大数据项目处理流程

图 1-1 所示的企业大数据项目处理流程，也称大数据功能模块，说明如下。

（1）数据获取：数据获取可能是一个文本文件、一个数据库、一个网络端口、一个爬虫程序等，其主要功能是提供数据。

（2）数据采集：数据采集即数据传输（本书指数据系统狭义的概念），如通过文件传送协议（File Transfer Protocol，FTP）下载文件就是一种数据传输的过程。数据采集更广义的概念是指数据经过抽取-转换-加载（Extract-Transform-Load，ETL）过程整合至数据存储平台中，以供更长时间的存储或供计算引擎分析与处理。

（3）数据存储：数据采集后，需要一个地方来存储这些数据，以供备份或数据分析、挖掘使用。

（4）数据分析、挖掘：此层主要完成一些和业务相关的计算、分析、挖掘任务。

（5）数据调用：数据调用是针对数据分析、挖掘的结果提供某种获取结果的接口，供第三方（或本系统）访问调用。

（6）数据展现：数据展现负责数据的最终结果的展现。展现方式有多种，如表格、趋势图等，需要和具体业务挂钩。

1.1.2　架构分析

1.1.1 小节中介绍了大数据项目的数据处理流程，读者需要知道作为乙方，要如何完成一个数据处理类项目。在项目开发的过程中，读者按照图 1-1 所示的流程来完成项目目标即可，只是对于不同的项目，其使用的技术可能不一样。本小节则介绍如何将大数据技术应用于企业项目大数据处理流程，使读者对整个流程有一个清晰的认识，为学习后面章节的内容打下基础。

把大数据技术和企业大数据项目数据处理流程结合起来即可形成企业大数据项目一般架构，如图 1-2 所示。

从图 1-2 中可以发现在企业项目数据处理流程的过程中，各层均使用了大数据技术来实现。对图 1-2 中各层使用的技术介绍如下。

1. 数据获取层

数据的来源多种多样，如文本文件、端口数据、移动设备数据、互联网数据。一般情况下，如果项目是科技项目或论文结题项目，那么数据源会直接给出数据，如 CSV 文件。而项目是企业级项目时，如建立一个推荐系统，企业的数据一般是放在数据库中的，如放在 Oracle 或 MySQL 中。还有一些项目需要通过爬虫程序去爬取互联网中的数据，此时数据源就是一个爬虫程序。

图 1-2 企业大数据项目一般架构

2. 数据采集层

数据采集的框架有很多，但是在大数据项目中使用比较广泛的是 Flume、Sqoop、Kafka。

Flume 是 Apache 软件基金会的一个数据采集框架，它是一个分布式的、可靠的、高可用的，可以从多种不同的数据源收集、聚集、移动大量日志数据至集中数据存储层的框架。Flume 可以匹配多种输入、输出源，使得不同输入、输出源的连接配置简单化。

Apache Sqoop（简称 Sqoop）项目旨在协助关系数据库管理系统（Relational Database Management System，RDBMS）与 Hadoop 系统（Hadoop 生态环境）进行高效的大数据传输。用户可以在 Sqoop 的帮助下，轻松地把关系数据库的数据导入 Hadoop 和其他相关的系统（如 HBase 和 Hive），同时可以把数据从 Hadoop 系统中抽取并导出到关系数据库中。Sqoop 可以建立一个连接 RDBMS 和 Hadoop 系统的数据传输通道。

Kafka 是由 Apache 软件基金会开发的一个开源流处理平台，其目标是为处理实时数据提供一个统一、高吞吐量、低延迟的平台。Kafka 的持久化层本质上是一个"按照分布式事务日志架构的大规模发布/订阅消息队列"，因此 Kafka 更适合用于需要进行实时采集大批量数据的数据源（如端口数据源）。

3. 数据存储层

Hadoop 分布式文件系统（Hadoop Distributed File System，HDFS）、HBase、Hive 都是 Hadoop 技术流的数据存储框架。HDFS 是 HBase、Hive 的底层存储技术。HBase 使用键值对的存储结构，适用于针对特定键的搜索场景。Hive 则是一个大数据仓库，它可以针对 HDFS 中的数据建立元数据，并对元数据进行各种查询操作。

Elasticsearch（简称 ES）是一个基于 Lucene 的开源搜索引擎，它不但稳定、可靠、快速，而且具有良好的水平扩展能力，是专门为分布式环境设计的。因为 ES 是面向文档型数据库的，所以它存储的是整个对象或者文档。它还会为存储的数据建立索引，因此可以在 ES 中高效地搜索、排序和过滤文档。

MongoDB 是 NoSQL 数据库，它是一个高扩展、高性能和高可用的数据库。MongoDB 是一种面向文档的数据库，以 JavaScript 对象表示法（JavaScript Object Notation，JSON）的形式进行数据存储。和 ES 一样，MongoDB 也支持全文搜索，MongoDB 和 ES 的区别主要体现在使用场景上，可根据不同的使用场景（建议读者根据自己的实际环境进行判断）有针对性地选择这些产品。

需要明确一点，在大数据项目架构的数据存储层也可能会有传统数据库的身影，如 MySQL、Oracle、DB2 等，传统数据库的作用是存储结果。举个简单的例子，某超市现在要统计一年中所有产品的销售数据。如果这是一个大型超市，消费者及其购买的商品特别多，数据总量则会特别庞大，使用大数据技术计算数据后，最终得到的数据量是比较小的，因此基于效率考量，可以把这类数据存入传统数据库。

4. 数据分析、挖掘层

在数据分析、挖掘层中，常用的计算引擎有内存计算引擎（Spark）、离线计算引擎（MapReduce）、流处理引擎（Streaming）、搜索引擎（ES）。这几种引擎都有各自的应用场景，在实际设计时，需要结合具体问题具体分析。总的来说，这些计算引擎都用于完成数据统计、数据分析或机器学习任务，为具体业务中的任务提供匹配的计算能力。

5. 数据调用层

数据调用层要完成的任务就是为第三方提供可方便调用的接口。举个简单的例子，现在有一个应用系统，需要基于计算层的计算结果进行展现，如果当前的应用系统的数据调用层没有提供接口，那么只有当前的应用系统能获取得到这些计算结果。如果现在要换一个应用系统，则需要另编写代码来获取这些计算结果。数据调用层为获取数据分析、挖掘层的计算结果提供了一种标准化的接口，使得如果各个应用都按照设计的标准来获取数据，那么各个应用都可以通过标准化的接口来获取这些数据，而不需要额外进行编码。

6. 数据展现层

简单理解，数据展现层指的是系统最终对外提供的服务。例如，淘宝 App、京东 App 对外其实提供了买东西的服务（对于客户来说）和卖东西的服务（对于销售者来说）。又如，一个音乐 App 除了提供歌曲的播放、搜索等基本功能，可能还提供推荐歌曲的服务等。此外，与大数据结合比较紧密的大屏应用可视化在公安、电力、园区管理、网络、航天等信息化程度相对较高的领域发挥了巨大作用，可以帮助行业从业务管理、事前预警、事中指挥调度、事后分析研判等多个方面提升智能化决策能力。例如，大屏应用可视化在电力领域可以满足常态下电网信息的实时监测监管、应急态下协同处置指挥调度的需要；在园区管理方面可以实现园区建设规划、管网运行、能耗监测、园区交通管理、安防管理、园区资源管理等多个维度的日常运行监测与协调管理。因此，一个可视化应用也可以作为一个业务来理解。

一个成熟的企业大数据项目架构除了上面所述的 6 层，还需要一些额外的模块来支撑整个系统，如集群服务模块，提供基本的集群权限、集群监控、资源调度、元数据管理等服务；系统管理模块，提供基本的应用管理、数据管理、运维管理、调度管理等服务；以及基于角色、资源来进行权限控制的模块，提供用户管理、服务治理等服务。这些模块与大数据技术模块共同协作，一起保证了整个架构的实现、系统的稳定运行。

1.1.3 人员安排

1.1.1 小节中给出了项目开发的 6 个阶段：可行性分析、需求分析、软件设计（概要设计、详细设计）、编码（软件设计实现）、测试、运行维护。本小节将结合这些阶段尽量简化如下问题：什么人、什么阶段、做什么事。

销售（售前）：项目前期，乙方（完成项目的一方）的售前人员需要和甲方沟通具体需求、签订合同等。双方需要明确需求，并将其书面化、文档化。这一操作体现在类似需求说明书等文件中，确保双方对于需求的理解是一致的。

数据分析师/架构师：拿到需求说明书后，数据分析师针对这些需求提出预研方案（或模型），并进行预研探索（包括一些基本的数据处理、模型构建等）；架构师需要结合需求以及数据分析师的预研结果来提出需要实现的工程系统架构及方案，同时需要通过对提出的一个或多个架构、方案进行分析，根据其可行性、适合性来确定最终的工程系统架构及系统实现方案。

程序开发人员：架构师设计出整个系统的架构后，程序开发人员会根据系统架构搭建系统框架，并和项目组其他成员共同制订各种开发计划、细则、要求等，而项目组成员负责整个系统的技术实现及各自的单元测试部分。

测试人员：整个系统实现后，需要有专门的测试人员对系统实现的各个模块进行一系列的集成测试、系统测试，并协助完成最后的验收测试。

实施/运维人员：乙方在测试环境中部署、测试整个系统后，需要在甲方提供的实际环境中再次部署，因此需要由实施人员到甲方现场部署系统。项目后期，系统运行过程中出现的各种问题，都需要通过运维人员来解决；如果涉及程序漏洞等，那么可以协调相关人员解决。一般来说，项目进行到运维阶段基本上就算结束了，也意味着甲、乙双方合同关系终止。合同终止一般是在运维结束后，也有可能是在运维结束前，如果合同终止在运维阶段前，那么运维阶段一般会再签署补充的运维协议。

1.2 大数据项目实战基础

在实际的项目开发中，开发人员需要从多方面进行分析考虑，搭建出符合项目需求的环境，同时明确项目涉及的技术及需要掌握的能力。

1.2.1 实战环境

本小节会对本书项目所需的实战环境做详细介绍，读者可以参考，以搭建符合自身需求的项目实战环境。下面将从硬件环境、软件环境、开发环境 3 个方面进行说明。

1. 硬件环境

本书项目使用的硬件环境主要包括 2 个方面：CDH 集群、客户端开发设备。

本书项目的 CDH 集群硬件配置如表 1-1 所示。

表 1-1　本书项目的 CDH 集群硬件配置

设备名	IP 地址	CPU 核数	内存大小/GB	磁盘大小/GB
node1	192.168.111.75	24	24	1200
node2	192.168.111.76	24	24	1200
node3	192.168.111.77	24	24	1200
node4	192.168.111.78	24	24	1200
server1	192.168.111.73	16	16	700
server2	192.168.111.74	16	16	700
server3	192.168.111.240	16	16	300

客户端开发设备是一台普通的笔记本计算机，其配置为 8GB 内存、Intel Core i5 双核 CPU、190GB SSD。

2. 软件环境

因为本书项目使用的实战环境资源不足，所以把传统数据库 MySQL 部署在 node1 节点上、应用服务器 Tomcat 部署在 server2 节点上，具体可参考各章说明。

本书项目软件环境包含的各个软件的版本描述如表 1-2 所示。

表 1-2　软件环境包含的各个软件的版本

软件	版本	备注
Cloudera Manager	5.7.3	与之配套的 Hadoop 相关软件受 Cloudera Manager 大版本的影响
Hadoop	2.6.0	2.6.0-cdh5.7.3
Spark	1.6.0	1.6.0-cdh5.7.3
Hive	1.1.0	1.1.0-cdh5.7.3
HBase	1.2.0	1.2.0-cdh5.7.3
ZooKeeper	—	HBase 自带
Sqoop	1.4.6	1.4.6-cdh5.7.3
Kafka	0.10.2-kafka2.2.0	—
Elasticsearch	6.3.2	—
JDK	1.8	开发环境和集群环境 JDK 版本保持一致
IntelliJ IDEA	2016 及以上	2016～2018 版本均可
Maven	3.3.1	3 及以上版本均可
Scala	2.10.6	Scala 插件和 IntelliJ IDEA 匹配即可

在本书项目的实战环境中，CDH 集群各服务部署情况如图 1-3 所示。

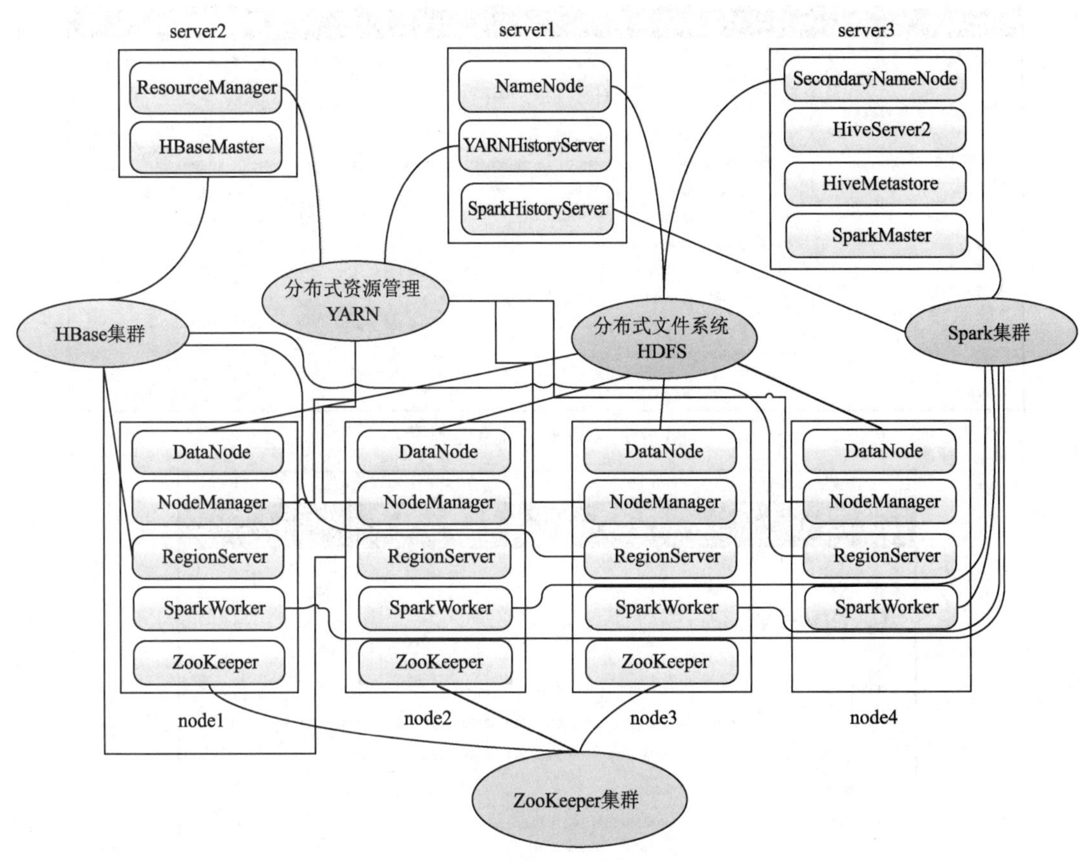

图 1-3 CDH 集群各服务部署情况

读者可以参考 Cloudera 官网相关说明进行 CDH 集群的安装。需要注意的是，图 1-3 所示的 CDH 集群各服务部署情况只是针对表 1-1 所示的硬件配置，读者应该根据实际的硬件环境来进行调整。

本书项目直接使用部署好的 CDH 集群，其部分监控界面截图分别如图 1-4～图 1-10 所示。

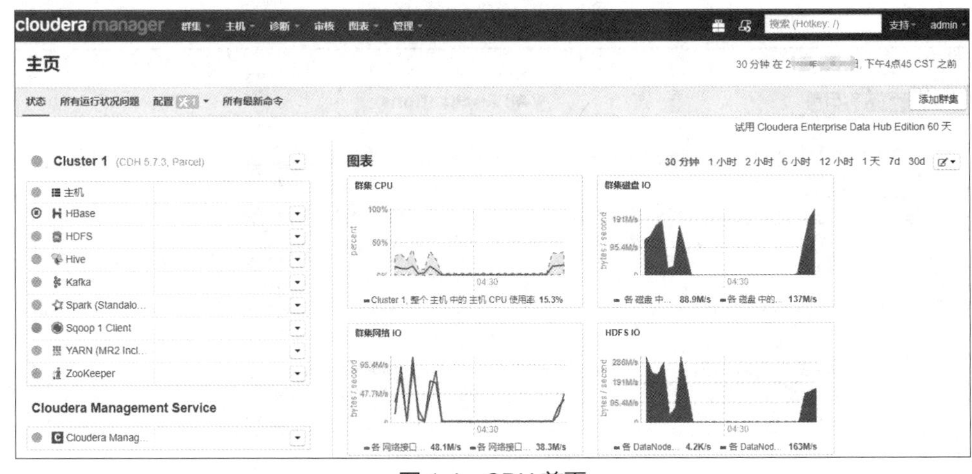

图 1-4 CDH 首页

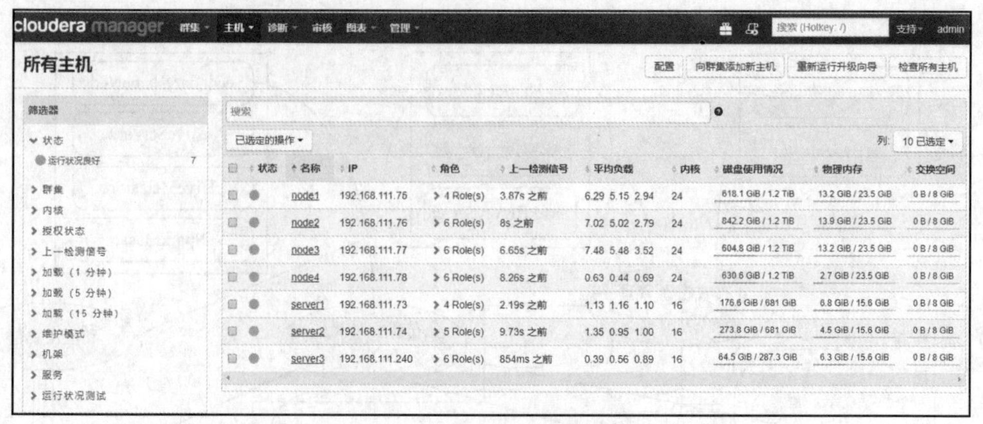

图 1-5　CDH 主机监控界面

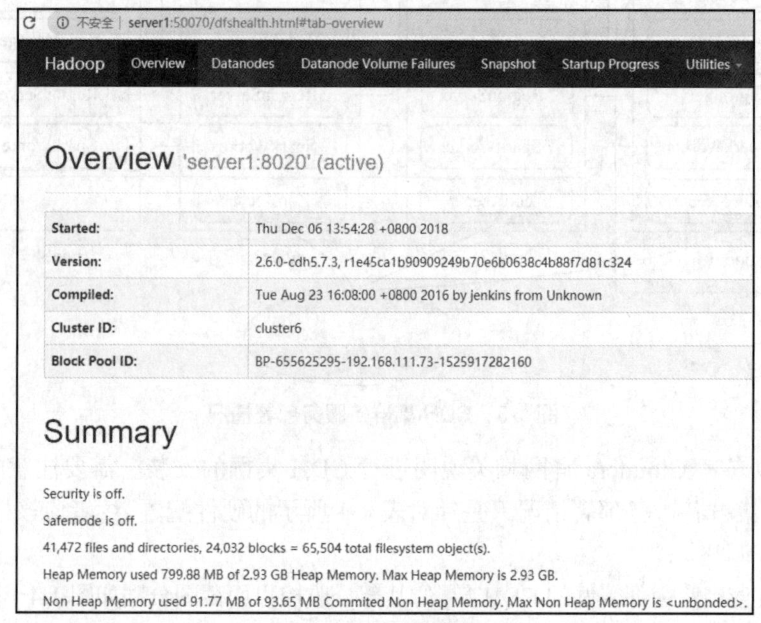

图 1-6　HDFS 监控界面

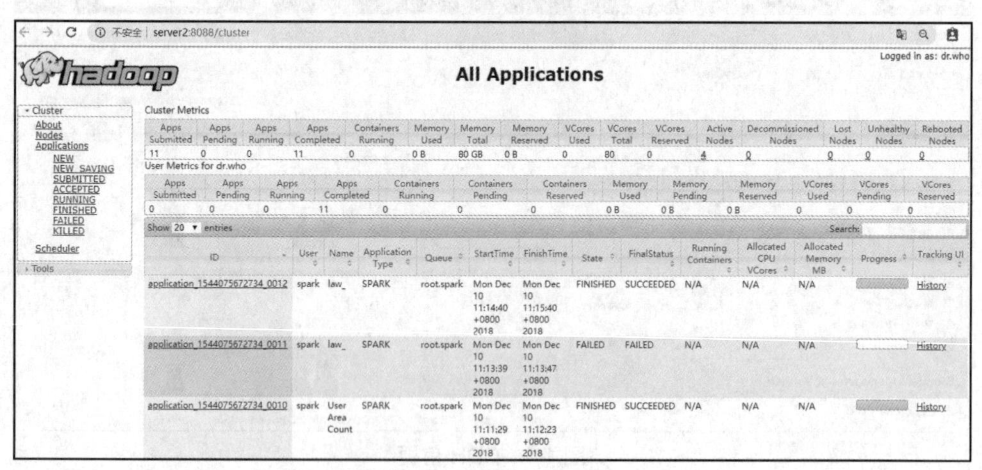

图 1-7　YARN 监控界面

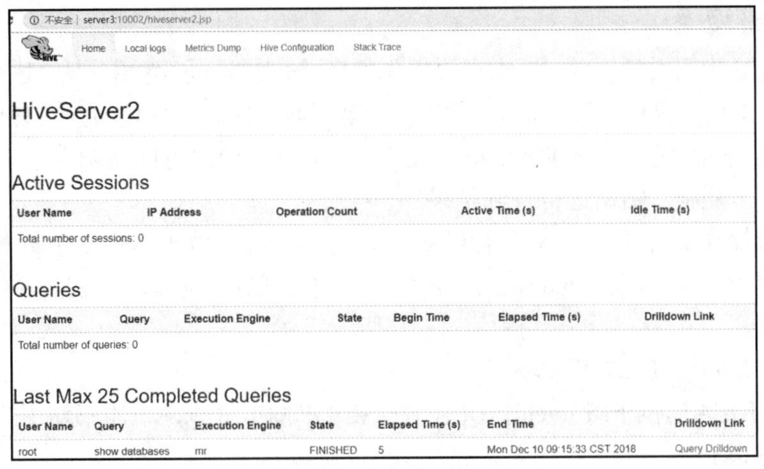

图 1-8　Hive 监控界面

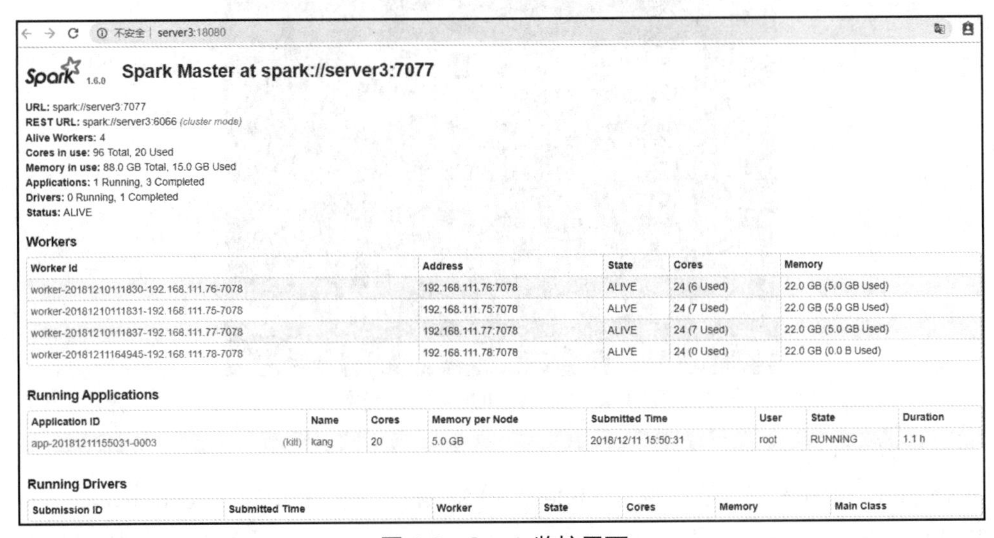

图 1-9　Spark 监控界面

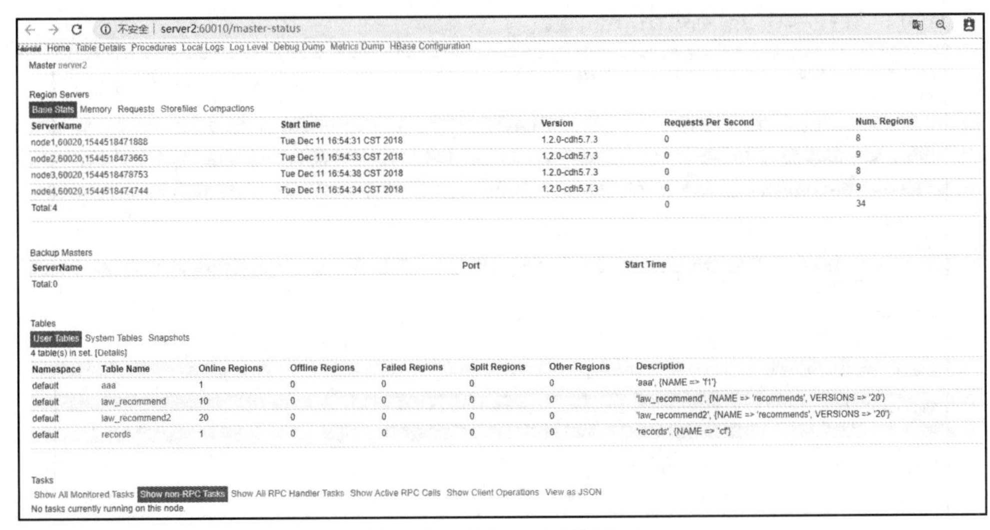

图 1-10　HBase 监控界面

3. 开发环境

开发环境分为两大类，第一类指直接操作类环境，如 Linux 终端（直接执行 Shell 命令）或 Spark Shell（直接执行一段 Spark 代码）；而第二类是代码工程化的开发环境，如前面已经执行过一段 Spark 代码，现在需要把这段代码工程化，使之可以在调度中运行得到结果，而不是通过人工复制、粘贴运行得到结果。

针对直接操作类环境，后文会有说明，此处不详述。由于本项目的开发工程需用到 Windows 10 系统，所以以 Windows 10 的系统环境为基础，配置项目实战中需要用到的第二类开发环境。

（1）安装 IntelliJ IDEA 开发环境

在 IntelliJ IDEA 官网下载 IntelliJ IDEA 并安装，如图 1-11 所示。本书使用的 IntelliJ IDEA 版本为 2017.3.6。

图 1-11 IntelliJ IDEA 安装版本

（2）配置 Maven 插件

打开 IntelliJ IDEA 后，依次选择“File”→“Settings”→“Build, Execution, Deployment”→“Build Tools”→“Maven”选项，并在“Maven home directory”下拉列表框中配置安装好的 Maven 插件，如图 1-12 所示。

Build, Execution, Deployment › Build Tools › Maven For current project
☐ Work offline
☐ Use plugin registry
☑ Execute goals recursively
☐ Print exception stack traces
☐ Always update snapshots
Output level: Info
Checksum policy: No Global Policy
Multiproject build fail policy: Default
Plugin update policy: Default ignored by Maven 3+
Threads (-T option):
Maven home directory: C:/Program Files/apache-maven-3.6.0-bin/apache-maven-3.3.9
(Version: 3.3.9)
User settings file: C:\Users\fansy\.m2\settings.xml ☐ Override
Local repository: C:\Users\fansy\.m2\repository ☐ Override

图 1-12 配置 Maven 插件

（3）配置 Scala 插件

依次选择“File”→“Settings”→“Plugins”→“Install JetBrains Plugins”选项，在弹出的搜索框中输入“scala”，即可看到 Scala 插件（双击即可进行安装），如图 1-13 所示。

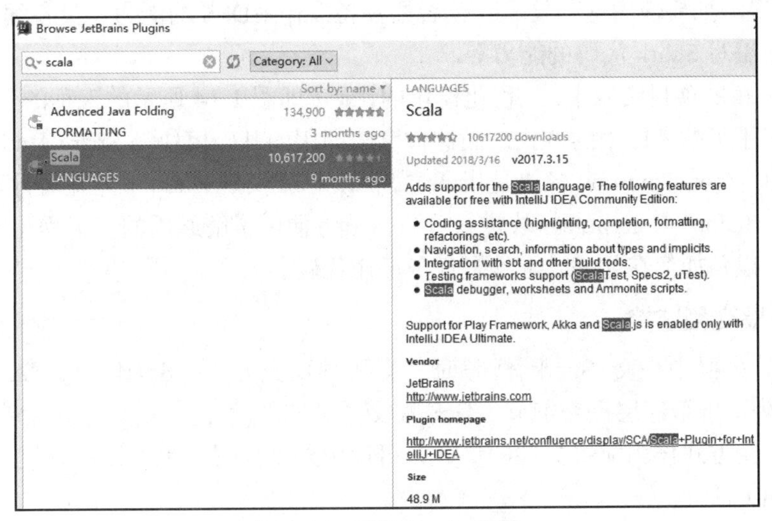

图 1-13 配置 Scala 插件

（4）配置随书附带的代码资源

下载并解压缩随书附带的代码资源 big_data_case_study.zip，解压缩后得到所有代码工程文件。使用 IntelliJ IDEA 导入代码工程，导入后的代码工程如图 1-14 所示。

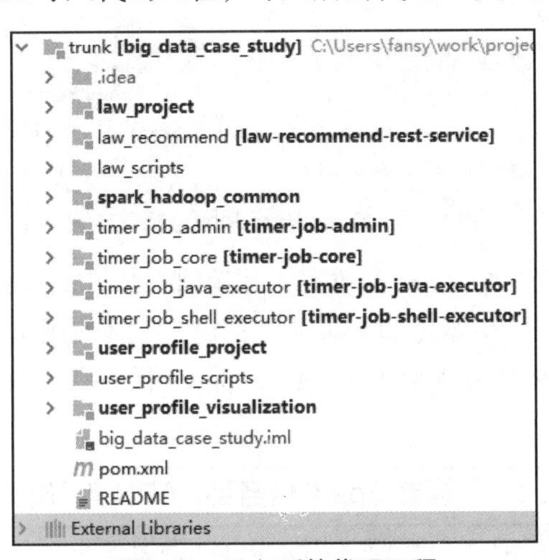

图 1-14 导入后的代码工程

1.2.2 涉及的技术及需掌握的能力

本书项目涉及的技术与能力包括但不限于以下两类。

（1）基础开发类：包括基本的 Linux 操作技能，如熟练使用 Linux 命令、编辑 Linux 配置文件等；基本的 Java 开发能力，能使用 Spring Boot 完成一个简单应用的开发实现；基本的 Scala 程序编写能力，如能熟练使用 Scala 的函数式编程；基本的 SQL（Structure Query Language，结构查询语言）增加、删除、修改、查询（简称增、删、改、查）能力，如能使用SQL完成复杂的统计分析脚本的编写。

（2）大数据技术类：包括分布式文件系统HDFS、分布式计算框架MapReduce、分布式资源管理框架YARN、内存计算框架Spark、大数据仓库Hive/HBase、数据传输工具Sqoop/Kafka等。如要使用这些技术，读者要具备操作HDFS的能力，具备编写Hive SQL的能力，具备编写Spark代码的能力等。

在第2、3章的项目实战中，工程化模块的代码（即图1-14所示的工程的代码）对Spark任务的调用采用了拿来即用的方式，而这种拿来即用的基础模块是Spark Hadoop Common模块。Spark Hadoop Common模块是指通过Spark框架实现Hadoop常用程序和库的模块，下面对Spark Hadoop Common模块进行分析，读者便可了解通用的、工程化的Spark任务调用的实现，以及如何在此基础上编写任务调用代码。

1. 任务提交和监控

在Spark集群中提交Spark任务时，有两种提交方式：Spark On YARN和Spark Standalone。为了屏蔽底层提交细节，任务提交做了如下设计。

（1）把提交方式作为配置文件的参数，可以由外部配置，也可以由开发人员在提交时指定，以增加灵活性。

（2）在提交时，只需要指定必要的参数即可，其他参数不用进行指定，使用配置文件中的默认设置即可。

基于以上两个规则，为实现Spark On YARN的任务提交方法设计了一个工具类SparkYarnJob，其有3个静态（关键字为static）方法，如图1-15所示。

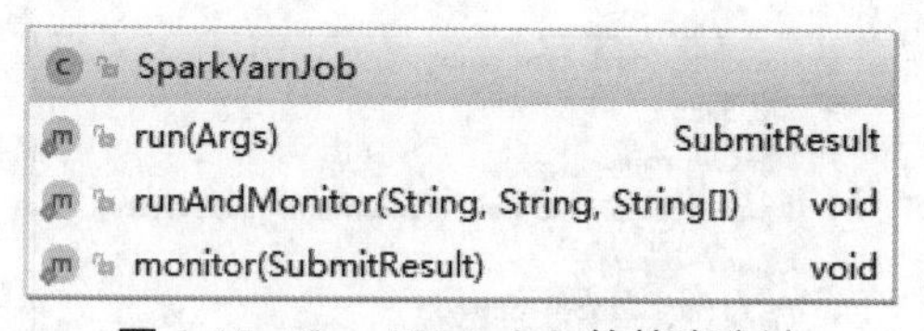

图1-15 SparkYarnJob的静态方法

在run()方法中，接收一个Args参数类，并返回一个SubmitResult返回值类。在run()方法中，主要根据Args中的引擎（如果有动态指定，那么使用指定的引擎，否则使用默认的引擎）选择不同的Spark任务提交方案。提交Spark任务到YARN集群的代码如代码1-1所示。

代码1-1 提交Spark任务到YARN集群的代码

```
System.setProperty("SPARK_YARN_MODE", "true");
SparkConf sparkConf = getSparkConf(EngineType.YARN);
ClientArguments cArgs = new ClientArguments(args.argsForYarn(), sparkConf);
Client client = new Client(cArgs, getConf(), sparkConf);
ApplicationId appId = client.submitApplication();
return SubmitResult.getSubmitResult(appId.toString(),args.getEngineType());
```

在代码1-1中，需要先设置系统属性SPARK_YARN_MODE为"true"，初始化SparkConf，使用Args的argsForYarn函数初始化ClientArguments，并使用SparkConf、ClientArguments和Hadoop Configuration（getConf函数，该函数会在后面说明）来初始化Client，通过Client

即可向 YARN 集群提交 Spark 任务。代码 1-1 的重点在于 Args 的 argsForYarn 函数，这个函数中包含了参数的具体构造（在参数类中会有说明）。

提交任务到 Spark 集群的代码如代码 1-2 所示。

代码 1-2 提交任务到 Spark 集群的代码

```
String jobId = SparkEngine.run(args.getAppName(), args.getMainClass(),
args. getArgs());
return SubmitResult.getSubmitResult(jobId, args.getEngineType());
// SparkEngine.run method
public static String run(String appName,String mainClass,String[] args){
        log.info("run args:\n"+ Arrays.toString(args));
        SparkConf sparkConf = getSparkConf(EngineType.SPARK);
        sparkConf.setAppName(appName +" "+ System.currentTimeMillis());
        Map<String,String> env = filterSystemEnvironment(System.getenv());
        CreateSubmissionResponse response = null;
        try {
                response = (CreateSubmissionResponse)
                                RestSubmissionClient.run(getValue("spark.appResource"),
mainClass, args, sparkConf,new scala.collection.immutable.HashMap());
        }catch (Exception e){
                e.printStackTrace();
                return null;
        }
        return response.submissionId();
}
```

在代码 1-2 中，代码分为两部分。其中一部分是调用 SparkEngine 的代码，同时针对返回值进行了构造，SparkEngine 的代码核心是 run()方法。run()方法的核心是使用了 RestSubmissionClient 类中的静态方法 run()，其他都是参考 Spark 源码（流程化的代码）实现的。需要注意的是，SparkEngine 需要放到 org.apache. spark.deploy.rest 包中，因为在 SparkEngine 的代码中使用了 RestSubmissionClient 类，而 RestSubmissionClient 类只可在 org.apache.spark.deploy.rest 包中调用。

在任务调用阶段返回 SubmitResult，根据 SubmitResult 即可进行监控。监控同样分为两种：提交到 YARN 集群的任务监控和提交到 Spark 集群的任务监控。其中，提交到 YARN 集群的任务监控的核心代码如代码 1-3 所示。

代码 1-3 提交到 YARN 集群的任务监控的核心代码

```
while (!finished) {
        try {
                logger.info("Checking Job {} , running...", jobInfo.getJobId() );
                Thread.sleep(getRandomInterval());
                FinalApplicationStatus applicationStatus = getFinalStatus(jobInfo.
getJobId());
                switch (applicationStatus) {
                        case SUCCEEDED:
                                logger.info("=== {} 成功运行并完成!",jobInfo.getJobId());
                                finished = true;
```

```
                        break;
                case FAILED:
                        logger.warn("=== {} 运行异常!",jobInfo.getJobId());
                        cleanupStagingDir(jobInfo.getJobId());
                        finished = true;
                        break;
                case KILLED:
                        logger.warn("=== {} 任务被杀死!", jobInfo.getJobId());
                        cleanupStagingDir(jobInfo.getJobId());
                        finished = true;
                        break;
                case UNDEFINED: // 继续检查
                        break;
                default:
                        logger.error("=== {} 任务状态获取异常!", jobInfo.
getJobId());
                        cleanupStagingDir(jobInfo.getJobId());
                        finished = true;
                        break;
            }
        } catch (InterruptedException | YarnException | IOException e) {
            e.printStackTrace();
        }
}
// SparkUtils.getFinalStatus method
public static FinalApplicationStatus getFinalStatus(String jobIdStr) throws
IOException, YarnException {
        ApplicationId jobId = ConverterUtils.toApplicationId(jobIdStr);
        ApplicationReport appReport = null;
        try {
            appReport = getClient().getApplicationReport(jobId);
            return appReport.getFinalApplicationStatus();
        } catch (YarnException | IOException e) {
            e.printStackTrace();
            throw e;
        }
}
// getClient method
public static YarnClient getClient() {
        if (client == null) {
            client = YarnClient.createYarnClient();
            client.init(getConf());
            client.start();
        }
        return client;
}
```

在代码 1-3 中，监控的核心流程就是根据任务 ID（jobId）获取当前任务的状态，并进行判断，如果任务运行完成（运行成功、运行异常或被杀死），那么退出循环；否则，休眠一定时间后，再次获取任务状态。从代码 1-3 可以看出，任务状态的获取主要通过 YarnClient

的 getApplicationReport()方法实现，而 YarnClient 的获取则需要根据 Hadoop Configuration（getConf 函数）来指定使用的集群参数。

提交到 Spark 集群的任务监控的核心代码如代码 1-4 所示。

代码 1-4 提交到 Spark 集群的任务监控的核心代码

```
SubmissionStatusResponse response = null;
while(!finished) {
        response = (SubmissionStatusResponse)
        getRestSubmissionClient().requestSubmissionStatus(jobInfo.getJobId(), true);
        logger.info("DriverState :{}",response.driverState());
        if("FINISHED".equals(response.driverState())){
                finished = true;
        }
        if( "ERROR".equals(response.driverState())){
                finished = true;
        }
        if( "FAILED".equals(response.driverState())){
                finished = true;
        }
        try {
                Thread.sleep(getRandomInterval());
        }catch (InterruptedException e){
                e.printStackTrace();
        }
}
// getRestSubmissionClient method
public static RestSubmissionClient getRestSubmissionClient(){
        if(restSubmissionClient == null){
                restSubmissionClient =  new RestSubmissionClient(getValue("spark.
master"));
        }
        return restSubmissionClient;
}
```

Spark 集群的任务监控通过 RestSubmissionClient 的 requestSubmissionStatus()方法来实现，其中 RestSubmissionClient 的初始化需要指定 Spark 集群的 URL。此外，Spark 集群任务监控的其他流程和 YARN 集群的任务监控流程一样。

2. 参数和返回值设计

参数类 Args 的方法如图 1-16 所示。

参数类 Args 采用了构造函数私有化的设计方法，对外提供 getArgs()的方法调用来实例化 Args。实例化 Args 需要提供 3 个参数，即 String appName、String mainClass、String[] args。其中，appName 指任务名称，mainClass 指调用的 Scala 类，args 则指 mainClass 对应的参数。另外，还有一个重载函数，其参数多了一个 Engine Type（引擎类型，Spark 或 YARN），可以动态指定或使用配置文件默认值。

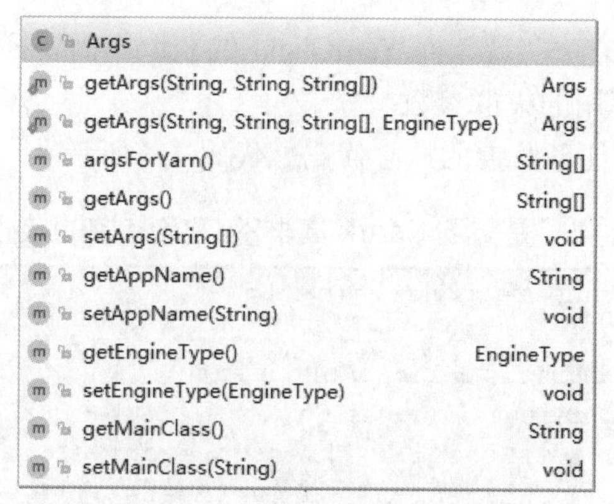

图 1-16　参数类 Args 的方法

参数类 Args 中除了对外提供获取 appName、mainClass、args、Engine Type 等参数，还提供 argsForYarn 函数，以实现提交任务到 YARN 集群做额外的参数配置。argsForYarn 函数实现额外的参数配置代码如代码 1-5 所示。

代码 1-5　参数类 Args 构造提交任务到 YARN 集群的参数

```
public String[] argsForYarn(){
        int defaultSize = 16;
        int defaultClassArgsLength = args.length * 2;
        List<String> argsList = new ArrayList<String>() {
                add("--name");
                add(appName);
                add("--class");
                add(mainClass);
                add("--driver-memory");
                add(getValue("yarn.spark.driver.memory"));
                add("--num-executors");
                add(getValue("yarn.spark.num.executors"));
                add("--executor-memory");
                add(getValue("yarn.spark.executor.memory"));
                add("--executor-cores");
                add(getValue("yarn.spark.executor.cores"));
                add("--jar");
                add(getValue("yarn.spark.algorithm.jar"));
                add("--files");
                add(getValue("yarn.spark.files"));
        };
        int j = defaultSize;
        for (int i = 0; i < defaultClassArgsLength / 2; i++) {
                argsList.add("--arg");
                argsList.add(args[i]);
        }
        return argsList.toArray(new String[0]);
}
```

返回值类 SubmitResult 同样采用了私有化构造方法的设计方法，对外提供 getSubmitResult 的静态函数来进行实例化。SubmitResult 类中主要提供两个参数：引擎、任务 ID。

3. 获取 SparkConf

提交任务到不同的集群时，所需的参数是不一样的，对于 SparkConf 参数的获取，有两种构造方法，如代码 1-6 所示。

代码 1-6 获取 SparkConf 参数

```
public static SparkConf getSparkConf(EngineType engineType){
        SparkConf sparkConf = new SparkConf();
        switch (engineType){
                case YARN:
                        sparkConf.set("spark.yarn.jar",  getValue("yarn.spark.
assemble.jar"));
                        sparkConf.set("spark.yarn.scheduler.heartbeat.interval-ms",
                          getValue("yarn.spark.yarn.scheduler.heartbeat.interval-ms"));
                        sparkConf.set("spark.yarn.appMasterEnv.SPARK_DIST_CLASSPATH",
                        getValue("yarn.spark.yarn.appMasterEnv.SPARK_DIST_CLASSPATH"));
                        sparkConf.set("spark.driver.extraJavaOptions", getValue
("yarn.spark.driver.extraJavaOptions"));
                        break;
                case SPARK:
                        sparkConf.set("spark.master", getValue("spark.master"));
                        sparkConf.set("spark.executor.memory", getValue("spark.
executor.memory"));
                        sparkConf.set("spark.cores.max", getValue("spark.cores.
max"));
                        sparkConf.set("spark.executor.cores",  getValue("spark.
executor.cores"));
                        sparkConf.set("spark.executor.extraClassPath", getValue
("spark.executor.extraClassPath"));
                        sparkConf.set("spark.driver.extraClassPath",getValue
("spark.driver.extraClassPath"));
                        //以下代码不能更改
                        sparkConf.set("spark.submit.deployMode","cluster");
                        sparkConf.set("spark.driver.supervise","false");
                        sparkConf.set("spark.files", getValue("spark.files"));
                        break;
                default:
                        logger.warn("Not support type:{}",engineType.name());
        }
        return sparkConf;
}
```

4. 获取 Hadoop Configuration

Hadoop Configuration 的获取可以通过静态方法 getConf()来实现，如代码 1-7 所示。

代码 1-7 获取 Hadoop Configuration

```
public static Configuration getConf() {
    try {
        if (conf == null) {
            conf = new Configuration();
            conf.set("mapreduce.app-submission.cross-platform",
                    getValue("mapreduce.app-submission.cross-platform"));
```

```
                    // 配置使用跨平台提交任务
            File file = getConfigurationFileParent();
            if (!file.exists() || !file.isDirectory()) {
                logger.error("路径{}不是目录或不存在！",
                file.getAbsolutePath());
                return conf;
            }
            for (File f : file.listFiles()) {
                if (f.getAbsolutePath().lastIndexOf("xml") != -1) {
                    logger.info("添加{}资源！", f.getName());
                    conf.addResource(f.toURI().toURL());
                } else {
                    logger.debug("资源{}不是以 xml 结尾的！",
                    f.getAbsolutePath());
                }
            }
            /**
             * CDH 集群远程提交 Spark 任务到 YARN 集群，出现
             * java.lang.NoClassDefFoundError: org/apache/hadoop/conf/
Configuration
             * 异常，需要设置 mapreduce.application.classpath 参数或
             * yarn.application.classpath 参数
             */
            conf.set("yarn.application.classpath", getValue("yarn.spark.yarn.
appMasterEnv.SPARK_DIST_CLASSPATH"));
        }
    } catch (Exception e) {
        e.printStackTrace();
    }
    return conf;
}
```

在代码 1-7 中，参考懒汉式单例模式，对外提供统一的 Configuration 实例。而此 Configuration 的实例化采用了默认的构造方法和构造方法中参数的配置。参数的配置主要使用了 Configuration 类提供的 addResource()方法加载集群的配置文件的方式。

5. 开发示例程序——提交任务到集群并进行任务监控

熟悉 Spark 任务调用实现的工程代码后，即可编写 Scala 任务程序。下面编写一个简单的 Scala 单词计数程序，如代码 1-8 所示。

代码 1-8　简单的 Scala 单词计数程序

```
package com.tipdm
import org.apache.spark.{SparkConf, SparkContext}
/**
  * WordCount Demo Program
  */
object WordCount {
  def main(args: Array[String]): Unit = {
    if(args.length !=2){
      println("com.tipdm.WordCount <in> <out>")
      System.exit(1)
```

```
    }
    val in = args(0)
    val out = args(1)
    val conf = new SparkConf().setAppName("Word Count")
    val sc = new SparkContext(conf)
    sc.textFile(in).flatMap(x => x.split(" ")).map(x => (x,1)).reduceByKey
((x,y) => x+y).saveAsTextFile(out)
    sc.stop()
  }
}
```

代码编写完成后，使用IntelliJ IDEA工具将其打包成wordcount.jar的JAR包，并将其上传至HDFS集群中（如/user/root目录），同时修改配置文件中的资源文件，如代码1-9所示。

代码1-9 任务资源文件配置

```
# YARN Engine Configuration
yarn.spark.algorithm.jar=hdfs://namenode:8020/user/root/wordcount.jar
# Spark Engine Configuration
spark.appResource=hdfs://namenode:8020/user/root/wordcount.jar
```

编写调用代码即可在IntelliJ IDEA中直接提交并监控任务，如代码1-10所示。

代码1-10 提交并监控任务

```
String appName ="WordCount";
String mainClass="com.tipdm.WordCount";
String[] myArgs= new String[]{"/user/root/words.txt","/user/root/tmp/wordcount_output"};
// Args innerArgs = Args.getArgs(appName,mainClass,myArgs);
// use default engineArgs innerArgs = Args.getArgs(appName,mainClass,myArgs,
EngineType.SPARK);
// use specific engine SubmitResult submitResult = SparkYarnJob.run(innerArgs);
SparkYarnJob.monitor(submitResult);
```

WordCount示例任务调用运行的部分截图如图1-17所示。

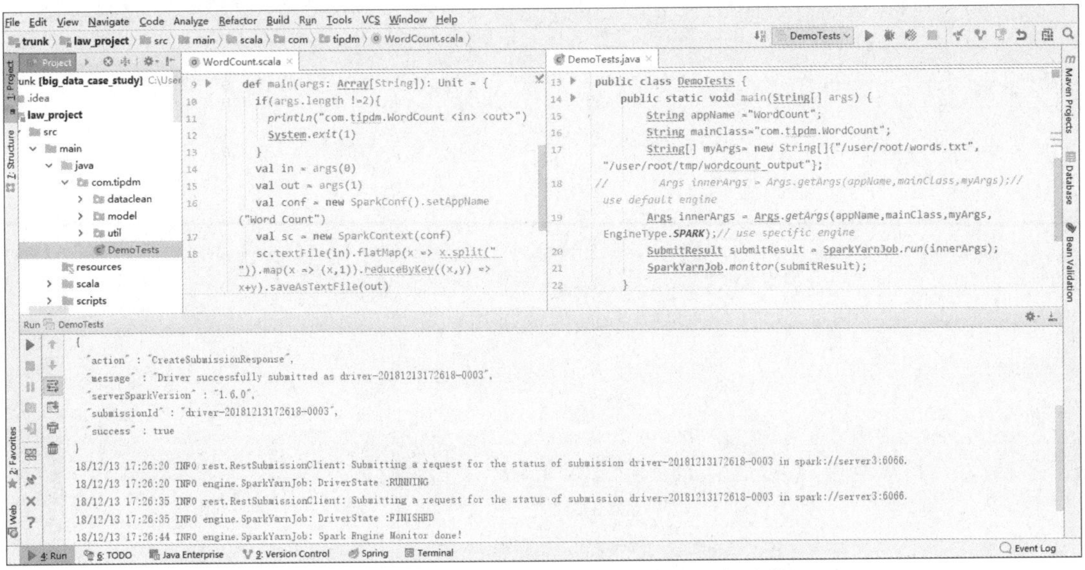

图1-17 WordCount示例任务调用运行的部分截图

以上是针对Spark Hadoop Common模块以及使用该模块来开发一个简单的任务程序的过程,读者可以在此基础上来开发应用程序。此外,本书的第2、3章中也是使用Spark Hadoop Common模块来进行开发的。

小结

本章以企业大数据类项目开发流程为例，介绍了企业大数据项目开发的一般流程与架构设计的分析依据，并对参与项目开发的人员安排进行了分析。此外，本章根据项目开发所需的实战环境和技术，介绍了 CDH 集群的资源配置情况、集群中配置的大数据技术、IntelliJ IDEA 的安装及开发环境的配置等，并演示了提交 Spark 任务到集群的过程以及对 Spark 任务的监控，以帮助读者为后续项目开发奠定基础。

第 2 章 Hadoop 生态组件基础

大数据时代，Hadoop 作为处理大数据的分布式存储和计算框架，在国内外大、中、小型企业得到了广泛应用，学习 Hadoop 技术几乎是从事大数据工作必不可少的一步。本章将介绍 Hadoop 框架的理论知识，包括 Hadoop 概述、集群安装与配置、框架组成和应用实践，并介绍 Hadoop 生态系统组件中的 Hive 组件和 Spark 组件的基础知识，加强基础研究。

学习目标

（1）了解 Hadoop 框架的发展历程、优点和生态系统常用组件。

（2）熟悉 Hadoop 集群、Hive 和 Spark 的安装与配置。

（3）熟悉 Hadoop 框架、Hive 和 Spark 的架构组成。

（4）掌握使用 Hadoop、Hive、Spark 进行简单的数据分析及存储操作的方法。

2.1 Hadoop 基础

Hadoop 是为大数据而生的，它用于设计分布式系统基础架构，以进行大数据的处理。对于大数据而言，Hadoop 可以使由大量普通机器组成的集群执行大规模运算，包括大规模计算和大规模存储。

2.1.1 Hadoop 概述

本小节将介绍 Hadoop 的概念、Hadoop 的发展历程、Hadoop 的优点和 Hadoop 的生态系统，并对 Hadoop 生态系统中的常用组件进行简单介绍。

1. Hadoop 的概念

随着互联网的快速发展和移动设备的广泛使用，数据的增量和存量快速增加，硬件发展却赶不上数据发展，单机设备很多时候已经无法处理数据规模达到 TB 甚至 PB 级别的数据。如果一头牛拉不动货物，显然找几头牛一起拉会比培育一头更强壮的牛更容易。同理，对于单机无法解决的问题，综合利用多台普通计算机来解决问题要比打造一台超级计算机的做法更加可行，这就是 Hadoop 的设计思想。

Hadoop 是由 Apache 软件基金会开发的一个可靠的、可扩展的、用于分布式计算的分布式系统基础架构和开源软件。Hadoop 软件库是一个框架，允许使用简单的编程模型在计算机集群中对大规模数据集进行分布式处理。它的目的是当单一的服务器扩展到成千上万台机器时，将集群部署在多台机器上，每台机器提供本地计算和存储服务，并将存储的数据备份在多个节点上，由此提升集群的可用性，而不是通过机器的硬件提升集群的可用性。当一台机器宕机时，其他节点依然可以提供备份数据和计算服务。

Hadoop 框架核心的设计是 HDFS 和 MapReduce。HDFS 是可扩展的、高容错的、高性能的分布式文件系统，负责数据的分布式存储和备份，文件写入后只能读取不能修改。MapReduce 是分布式计算框架，其处理过程包含 Map（映射）和 Reduce（规约）两个过程。

2. Hadoop 的发展历程

Hadoop 是由 Apache Lucence 创始人道格·卡廷创建的，Lucence 是一个应用广泛的文本搜索系统库。Hadoop 起源于开源的网络搜索引擎——Apache Nutch，Nutch 本身也是 Lucence 项目的一部分。Hadoop 的发展历程如图 2-1 所示。

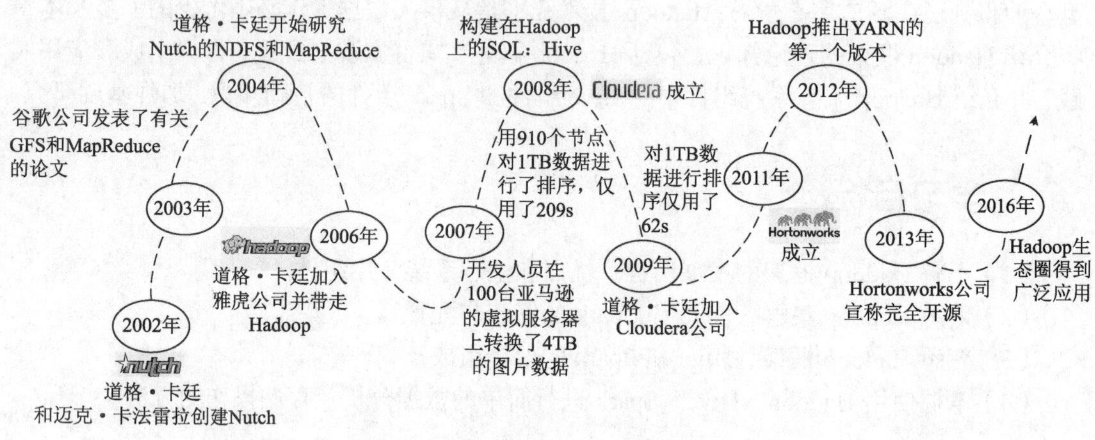

图 2-1 Hadoop 的发展历程

Nutch 项目开始于 2002 年，当时，道格·卡廷与其好友迈克·卡法雷拉认为网络搜索引擎由一个互联网公司垄断十分可怕，这个公司将掌握信息的入口。因此两位开发者决定自己开发一个可以代替当时主流搜索产品的开源搜索引擎项目，并将该项目命名为 Nutch。Nutch 致力于提供开源搜索引擎所需的全部工具集。后来两位开发者发现这一架构的灵活性不够，其只能支持几亿张网页数据的抓取、索引和搜索，不足以解决数十亿张网页的搜索问题。而谷歌公司在 2003 年发表的论文 *The Google File System* 和 2004 年发表的论文 *MapReduce:Simplified Data Processing on Large Clusters* 为该问题提供了可行的解决方案。

2004 年，道格·卡廷借鉴谷歌公司的新技术——谷歌文件系统（Google File System，GFS）和谷歌公司的 MapReduce 开发出了 Nutch 分布式文件系统（Nutch Distributed File System，NDFS），并模仿谷歌公司的 MapReduce 框架的设计思路，用 Java 设计实现了一套新的 MapReduce 并行处理软件系统，并在 Nutch 上开发了一个可工作的 MapReduce 框架。

2006 年 2 月，开发人员将 NDFS 和 MapReduce 移出 Nutch，形成 Lucence 的子项目，并将其命名为 Hadoop，据说这个名称来源于道格·卡廷儿子的一只玩具象。随后，道格·卡廷加入了雅虎公司，雅虎公司为此组织了专门的团队和资源，致力于将 Hadoop 发展成能够处理海量 Web 数据的分布式系统。

道格·卡廷加入雅虎公司后，Hadoop 项目逐渐发展并成熟起来。2007 年，开发人员在 100 台亚马逊的虚拟服务器上使用 Hadoop 转换了 4TB 的图片数据，加深了人们对 Hadoop 的印象。

2008 年，一位谷歌公司的工程师与几位朋友成立了一个专门商业化 Hadoop 的公司——Cloudera。同年，一个团队在 Hadoop 上开发了一个名为 Hive 的数据仓库工具，用于将 SQL 转换为 Hadoop 的 MapReduce 程序。2008 年 1 月，Hadoop 已经成为 Apache 的顶级项目。同年 4 月，Hadoop 打破世界纪录，成为最快的 TB 级数据排序系统。在一个有 910 个节点的集群中，Hadoop 用 209s 完成了对 1TB 数据的排序，刷新了 2007 年用时 297s 的纪录。

2009 年，道格·卡廷加入 Cloudera 公司。同年 5 月，一个团队使用 Hadoop 对 1TB 数据进行排序只用了 62s。

2011 年，雅虎公司将 Hadoop 团队独立出来，成立了子公司 Hortonworks，专门提供 Hadoop 的相关服务。

2012 年，Hortonworks 公司在 Hadoop 的基础上推出了与原框架有很大不同的 YARN 框架的第一个版本，从此 Hadoop 的研究迈进了一个新的层面。

2013 年，大型 IT 公司，如易安信（EMC）、微软（Microsoft）、英特尔（Intel）、天睿（Teradata）、思科（Cisco）等，都明显增加了 Hadoop 方面的投入，Hortonworks 公司宣称要 100%开源软件，Hadoop 2.0 的转型基本上无可阻挡。

2016 年，Hadoop 及其生态圈（包括 Spark 等）在各行各业“落地”并得到广泛应用，YARN 框架也在持续发展以支持更多应用。

3. Hadoop 的优点

Hadoop 是一个能够让用户轻松使用的分布式计算平台，用户可以轻松地在 Hadoop 上开发和运行处理海量数据的应用程序，其优点主要有以下 7 个。

（1）高可靠性

存储的数据有多个备份，设置在不同节点上，可以防止一个节点宕机造成集群损坏。当数据处理请求失败后，Hadoop 会自动重新部署计算任务。Hadoop 框架中有备份机制和检验模式，Hadoop 会对出现问题的部分进行修复，并通过设置快照的方式在集群出现问题时将系统恢复到之前的一个时间点。

（2）高扩展性

Hadoop 是在可用的计算机集群间分配数据并完成计算任务的。为集群添加新的节点并不复杂，所以集群可以很容易扩展节点。

（3）高效性

Hadoop 能够在节点之间动态地移动数据，在数据所在节点进行并发处理，并保证各个节点的动态平衡，因此处理速度非常快。

（4）高容错性

HDFS 在存储文件时会在多个节点或多台机器上存储文件的备份，当读取文档出错或者某台机器宕机后，系统会调用其他机器上的备份文件，保证程序顺利运行。如果启动的任务失败，则 Hadoop 会重新运行该任务或启用其他任务完成这个任务没有完成的部分。

（5）低成本

Hadoop 是开源的，即不需要支付任何费用即可下载并安装使用 Hadoop，节省了购买软件的成本。

（6）可构建在廉价机器上

Hadoop 不要求机器的配置达到极高的水准，大部分普通商用服务器即可满足要求，它通过提供多个副本和容错机制提高集群的可靠性。

（7）Hadoop 框架使用 Java 语言编写

Hadoop 是使用 Java 语言编写的框架，因此运行在 Linux 生产平台上是非常理想的。Hadoop 上的应用程序也可以使用其他语言编写，如 C++等。

4. Hadoop 的生态系统

Hadoop 面世之后快速发展，相继开发出了很多组件，这些组件各有特点，共同提供服务给 Hadoop 相关的工程，并逐渐形成了系列化的组件系统，这个系统通常被称为 Hadoop 的生态系统。因为大部分组件的 Logo 选用了动物图形，所以 Hadoop 的生态系统就像一群动物在狂欢，如图 2-2 所示。

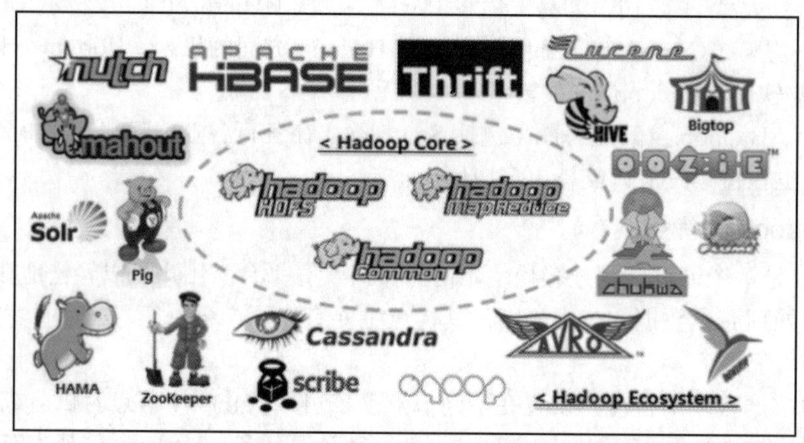

图 2-2　Hadoop 的生态系统

不同的组件用于提供特定的服务，下面对 Hadoop 生态系统中的常用组件进行介绍。

（1）HBase

HBase 是一个针对非结构化数据的可伸缩、高可靠、高性能、分布式和面向列的动态模式数据库。HBase 提供了对大规模数据的随机、实时读写访问。同时，HBase 中保存的数据可以使用 MapReduce 进行处理，MapReduce 将数据存储和并行计算完美地结合在一起。

HBase 适用于大数据量（TB/100s 级数据）且有快速随机访问的需求（如淘宝交易记录），能及时响应用户发出的访问请求；业务场景简单，不需要关系数据库中的很多特性（如交叉、连接）；需要扩充数据库结构等场景。

（2）Hive

Hive 是建立在 Hadoop 上的数据仓库基础构架，它提供了一系列的工具，用于存储数据、查询和分析存储在 Hadoop 中的大规模数据。Hive 定义了一种类 SQL 的语言——Hive 查询语言（Hive Query Language，HQL），通过简单的 HQL 可以将数据操作转换为复杂的 MapReduce 运行在 Hadoop 大数据平台上。

Hive 是一款分析历史数据的利器，但是 Hive 只有在结构化数据的情况下才能大显“神威”。Hive 处理有延迟，较为合适的使用场景是大数据集的批处理作业，如网络日志分析。

（3）Sqoop

Sqoop 是一款开源的工具，主要用于在 Hadoop（如 Hive 组件）与传统的数据库（如 MySQL、PostgreSQL）之间进行数据的传递。Sqoop 可以将一个关系数据库中的数据导入 HDFS，也可以将 HDFS 的数据导出到关系数据库中。

Sqoop 适用于结构化数据库，它可以将结构化数据库中的数据并行批量存储至 HDFS 中。

（4）Flume

Flume 是 Cloudera 提供的一个高可用的、高可靠的、分布式的海量日志采集、聚合和传输的系统，用于日志文件的采集。Flume 支持在日志系统中定制各类数据发送方，用于收集数据。同时，Flume 提供对数据进行简单处理，并写到各种数据接收方（可定制）的功能。

如果数据来源很多、数据流向很多，那么使用 Flume 采集数据是一个很好的选择。

（5）ZooKeeper

ZooKeeper 可以解决分布式环境下的数据管理问题，如统一命名、状态同步、集群管理、配置同步等。

使用 ZooKeeper 主要是为了保证集群各项功能的正常进行，在集群出现异常时能够及时通知管理员进行处理，保持数据的一致性，ZooKeeper 的作用是对整个集群进行监控。

（6）Mahout

Mahout 的主要目标是创建一些可扩展的机器学习领域经典算法，旨在帮助开发人员更加方便、快捷地创建智能应用程序。Mahout 现在已经包含了聚类、分类、推荐引擎（协同过滤）和频繁项集挖掘等广泛使用的数据挖掘方法。除了算法，Mahout 还包含数据的输入/输出工具、与其他存储系统（如数据库、MongoDB 或 Cassandra）集成等数据挖掘支持架构。

Mahout 的主要应用是通过提供机器算法包，使用户在使用的过程中能够通过调用算法包减少编程时间，同时减少用户的复杂算法程序对资源的消耗（用户的特殊需求除外）。

（7）Spark

Spark 是一个基于内存计算的大数据并行计算框架，可用于构建大型的、低延迟的数据分析应用程序。Spark 能弥补 MapReduce 计算模型延迟过高的缺陷，满足实时、快速计算的需求。而且 Spark 的中间数据计算结果可以保存在内存中，大大减少了对 HDFS 进行读写操作的次数，因此 Spark 可以更好地适用于数据挖掘与机器学习中需要迭代的算法。

大数据应用场景的普遍特点是数据计算量大、效率高，Spark 计算框架刚好可以满足这些计算要求。

2.1.2 Hadoop 集群安装与配置

为了更好地学习 Hadoop，需要学习搭建 Hadoop 集群环境。Hadoop 集群环境可以分为单机版环境、伪分布式环境和完全分布式环境。单机版环境是指在一台计算机上运行 Hadoop，没有分布式文件系统，而是直接读取本地操作系统中的文件；伪分布式环境可以看作在一台计算机上模拟组建的多节点集群；而完全分布式环境是在多台计算机上组建的分布式集群。

CDH 是对 Hadoop 集成环境的封装，可以使用 Cloudera Manager 进行自动化安装，支持大多数 Hadoop 组件，包括 HDFS、MapReduce、Hive、Pig、HBase、ZooKeeper、Sqoop 等，简化了大数据平台的安装。为了让读者更好地了解 Hadoop 生态系统组件的安装配置过程，本书直接演示使用 VMware 在个人计算机上模拟搭建 Hadoop 完全分布式环境的过程。

为了保证能较顺畅地运行 Hadoop 完全分布式集群，并可以进行基本的大数据开发调试，建议个人计算机硬件的配置如下：内存容量至少 8GB，硬盘可用容量至少 100GB，CPU 为 Intel Core i3 以上的处理器。在搭建 Hadoop 完全分布式集群之前，还需要准备好必要的软件，Hadoop 相关软件及版本如表 2-1 所示。

表 2-1　Hadoop 相关软件及版本

软件	版本	安装包	备注
Linux OS	CentOS 6.8	CentOS-6.8-x86_64-bin-DVD1.iso	64 位
JDK	1.7 以上	jdk-7u80-linux-x64.rpm	64 位
VMware	10	vmware-workstation-full-10.0.15255226.zip	—
Hadoop	2.6.4	hadoop-2.6.4.tar.gz	已编译好的安装包
Eclipse	4.5.1	eclipse-jee-mars-1-win32-x86_64.zip	64 位
Eclipse Hadoop 插件	2.6.0	hadoop-eclipse-plugin-2.6.0	—
SSH 连接工具	5	Xme5.exe	—

Hadoop 完全分布式集群是典型的主从架构，一般需要使用多台服务器来组建。本书中使用的集群拓扑结构如图 2-3 所示。请注意各个服务器的 IP 地址与名称，在后续的配置工作中将会经常被使用。

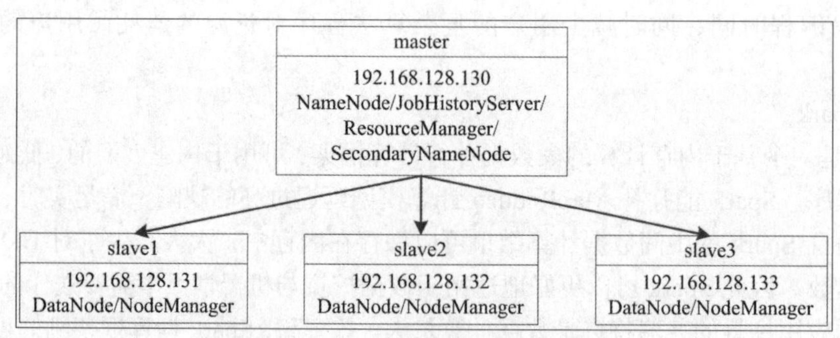

图 2-3　Hadoop 的集群拓扑结构

1. 搭建 Hadoop 完全分布式集群

在搭建 Hadoop 完全分布式集群（以下简称 Hadoop 集群）之前，需要先使用安装包 vmware-workstation-full-10.0.15255226.zip 安装 VMware，并在 VMware 中创建图 2-3 所示的 4 个虚拟机，设置好每个节点的固定 IP 地址和 IP 地址与主机名的映射，在每个节点上安装好 Java 环境，并通过 YUM 源安装软件 vim、zip、openssh-server 和 openssh-clients，如表 2-2 所示。

表 2-2 Hadoop 集群节点 IP 地址及配置软件

节点	IP 地址	语言环境	配置软件	备注
master	192.168.128.130	Java	vim、zip、openssh-server、openssh-clients	主节点
slave1	192.168.128.131	Java	vim、zip、openssh-server、openssh-clients	从节点
slave2	192.168.128.132	Java	vim、zip、openssh-server、openssh-clients	从节点
slave3	192.168.128.133	Java	vim、zip、openssh-server、openssh-clients	从节点

（1）修改配置文件

在虚拟机 master 上进行 Hadoop 的相关配置，需要先把 Hadoop 安装包 hadoop-2.6.4.tar.gz 上传至虚拟机 master 的/opt 目录下，再执行命令“tar -zxf hadoop-2.6.4.tar.gz -C /usr/local/”，将 Hadoop 安装包解压到虚拟机 master 的/usr/local 目录下。

配置 Hadoop 涉及的文件有 7 个：core-site.xml、hadoop-env.sh、yarn-env.sh、mapred-site.xml、yarn-site.xml、slaves、hdfs-site.xml。这些文件都在/usr/local/hadoop-2.6.4/etc/hadoop/目录下，进入该目录，并依次修改这些配置文件。

① 修改 core-site.xml 文件。core-site.xml 是 Hadoop 的核心配置文件，修改该文件，如代码 2-1 所示。在该文件中需要配置两个属性：fs.defaultFS 和 hadoop.tmp.dir。其中，fs.defaultFS 配置了 Hadoop 的 HDFS 的名称，其位置为主机的 8020 端口，需要注意的是，文件中的 hdfs://master:8020 中的 master 需要重命名为 NameNode 所在虚拟机的主机名；hadoop.tmp.dir 配置了 Hadoop 的临时文件的存放位置。

代码 2-1 修改 core-site.xml 文件

```
<configuration>
   <property>
   <name>fs.defaultFS</name>
     <value>hdfs://master:8020</value>
     </property>
   <property>
     <name>hadoop.tmp.dir</name>
     <value>/var/log/hadoop/tmp</value>
   </property>
</configuration>
```

② 修改 hadoop-env.sh 文件。hadoop-env.sh 文件是 Hadoop 运行环境的配置文件，需要在其中修改 JDK 所在位置。因此，将该文件中 JAVA_HOME 的值修改为 JDK 在本机的安装目录，如代码 2-2 所示。

代码 2-2 修改 hadoop-env.sh 文件

```
export JAVA_HOME=/usr/java/jdk1.7.0_80
```

③ 修改 yarn-env.sh 文件。yarn-env.sh 文件是 YARN 框架运行环境的配置文件，同样需要在其中修改 JDK 所在位置。将该文件中 JAVA_HOME 的值修改为 JDK 在本机的安装目录，如代码 2-3 所示。

代码 2-3　修改 yarn-env.sh 文件

```
# export JAVA_HOME=/home/y/libexec/jdk1.6.0/
export JAVA_HOME=/usr/java/jdk1.7.0_80
```

④ 修改 mapred-site.xml 文件。mapred-site.xml 文件是 MapReduce 的相关配置文件，由于 Hadoop 2.x 使用了 YARN 框架，因此必须在 mapreduce.framework.name 属性下配置 YARN。mapreduce.jobhistory.address 和 mapreduce.jobhistory.webapp.address 是 JobHistoryServer 的相关配置，JobHistoryServer 是运行 MapReduce 任务的日志相关服务，进行该服务相关配置时，同样需要注意修改 master，将 master 修改为实际服务所在虚拟机的主机名。

mapred-site.xml 文件是从 mapred-site.xml.template 文件复制得到的，执行命令“cp mapred-site.xml.template mapred-site.xml”，修改 mapred-site.xml 文件，如代码 2-4 所示。

代码 2-4　修改 mapred-site.xml 文件

```
<configuration>
<property>
  <name>mapreduce.framework.name</name>
  <value>yarn</value>
</property>
<!-- jobhistory properties -->
<property>
  <name>mapreduce.jobhistory.address</name>
  <value>master:10020</value>
</property>
<property>
   <name>mapreduce.jobhistory.webapp.address</name>
   <value>master:19888</value>
</property>
</configuration>
```

⑤ 修改 yarn-site.xml 文件。yarn-site.xml 文件为 YARN 框架的配置文件，在文件中需要命名一个 yarn.resourcemanager.hostname 的变量，在后面 YARN 的相关配置中可以直接引用该变量，其他配置保持不变即可，修改后的 yarn-site.xml 文件如代码 2-5 所示。

代码 2-5　修改后的 yarn-site.xml 文件

```
<configuration>
<!-- Site specific YARN configuration properties -->
<property>
    <name>yarn.resourcemanager.hostname</name>
    <value>master</value>
  </property>
  <property>
    <name>yarn.resourcemanager.address</name>
    <value>${yarn.resourcemanager.hostname}:8032</value>
  </property>
  <property>
    <name>yarn.resourcemanager.scheduler.address</name>
    <value>${yarn.resourcemanager.hostname}:8030</value>
  </property>
```

```
  <property>
    <name>yarn.resourcemanager.webapp.address</name>
    <value>${yarn.resourcemanager.hostname}:8088</value>
  </property>
  <property>
    <name>yarn.resourcemanager.webapp.https.address</name>
    <value>${yarn.resourcemanager.hostname}:8090</value>
  </property>
  <property>
    <name>yarn.resourcemanager.resource-tracker.address</name>
    <value>${yarn.resourcemanager.hostname}:8031</value>
  </property>
  <property>
    <name>yarn.resourcemanager.admin.address</name>
    <value>${yarn.resourcemanager.hostname}:8033</value>
  </property>
  <property>
    <name>yarn.nodemanager.local-dirs</name>
    <value>/data/hadoop/yarn/local</value>
  </property>
  <property>
    <name>yarn.log-aggregation-enable</name>
    <value>true</value>
  </property>
  <property>
    <name>yarn.nodemanager.remote-app-log-dir</name>
    <value>/data/tmp/logs</value>
  </property>
<property>
 <name>yarn.log.server.url</name>
 <value>http://master:19888/jobhistory/logs/</value>
 <description>URL for job history server</description>
</property>
<property>
   <name>yarn.nodemanager.vmem-check-enabled</name>
    <value>false</value>
  </property>
 <property>
    <name>yarn.nodemanager.aux-services</name>
    <value>mapreduce_shuffle</value>
  </property>
  <property>
    <name>yarn.nodemanager.aux-services.mapreduce.shuffle.class</name>
      <value>org.apache.hadoop.mapred.ShuffleHandler</value>
      </property>
<property>
        <name>yarn.nodemanager.resource.memory-mb</name>
        <value>2048</value>
 </property>
 <property>
```

```
        <name>yarn.scheduler.minimum-allocation-mb</name>
        <value>512</value>
  </property>
  <property>
        <name>yarn.scheduler.maximum-allocation-mb</name>
        <value>4096</value>
  </property>
  <property>
     <name>mapreduce.map.memory.mb</name>
     <value>2048</value>
  </property>
  <property>
     <name>mapreduce.reduce.memory.mb</name>
     <value>2048</value>
  </property>
  <property>
     <name>yarn.nodemanager.resource.cpu-vcores</name>
     <value>1</value>
  </property>
</configuration>
```

⑥ 修改 slaves 文件。slaves 文件中保存有 slave 节点的信息，这里需修改 slaves 文件，先删除 slaves 文件中原有的内容，并添加代码 2-6 所示的内容。

代码 2-6　修改 slaves 文件

```
slave1
slave2
slave3
```

⑦ 修改 hdfs-site.xml 文件。hdfs-site.xml 文件是 HDFS 相关的配置文件，dfs.namenode.name.dir 和 dfs.datanode.data.dir 分别指定了 NameNode 元数据和 DataNode 数据的存储位置。dfs.namenode.secondary.http-address 配置的是 Secondary NameNode 所在虚拟机的主机名和端口号，同样需要注意修改 master 为实际的 Secondary NameNode 所在虚拟机的主机名。dfs.replication 配置了文件块的副本数，默认为 3 个，所以可以不进行配置。修改 hdfs-site.xml 文件，如代码 2-7 所示。

代码 2-7　修改 hdfs-site.xml 文件

```
<configuration>
  <property>
    <name>dfs.namenode.name.dir</name>
    <value>file:///data/hadoop/hdfs/name</value>
  </property>
  <property>
    <name>dfs.datanode.data.dir</name>
    <value>file:///data/hadoop/hdfs/data</value>
  </property>
  <property>
    <name>dfs.namenode.secondary.http-address</name>
    <value>master:50090</value>
  </property>
  <property>
    <name>dfs.replication</name>
```

```
    <value>3</value>
  </property>
</configuration>
```

⑧ 除了上述 7 个文件外，还需要修改/etc/hosts 文件。该文件配置的是主机名与 IP 地址的映射。集群共有 4 个节点，其主机名及 IP 地址如表 2-2 所示，因此可在/etc/hosts 文件的末尾添加代码 2-8 所示的内容。

代码 2-8 修改/etc/hosts 文件

```
192.168.128.130 master master.centos.com
192.168.128.131 slave1 slave1.centos.com
192.168.128.132 slave2 slave2.centos.com
192.168.128.133 slave3 slave3.centos.com
```

（2）配置 SSH 协议无密码登录

安全外壳（Secure Shell，SSH）协议是建立在 TCP/IP（Transmission Control Protocol/Internet Protocol，传输控制协议/互联网协议）的应用层和传输层基础上的安全协议。SSH 协议保障了远程登录和网络传输服务的安全性，起到了防止信息泄露等作用。通过 SSH 协议可以对文件进行加密处理，SSH 协议可以运行于多平台。配置 SSH 协议无密码登录的步骤（以下步骤都是在主节点 master 上进行的）如下。

① 使用 ssh-keygen 产生公钥与私钥对。执行命令“ssh-keygen -t rsa”，并按 3 次“Enter”键，如图 2-4 所示，生成私钥 id_rsa 和公钥 id_rsa.pub 两个文件。ssh-keygen 用于生成 RSA 类型的密钥以及管理该密钥，选项“-t”用于指定要创建的 SSH 密钥的类型。

```
[root@master ~]# ssh-keygen -t rsa
Generating public/private rsa key pair.
Enter file in which to save the key (/root/.ssh/id_rsa):
Created directory '/root/.ssh'.
Enter passphrase (empty for no passphrase):
Enter same passphrase again:
Your identification has been saved in /root/.ssh/id_rsa.
Your public key has been saved in /root/.ssh/id_rsa.pub.
The key fingerprint is:
03:a0:69:1d:c8:05:8d:60:ab:41:f6:7f:4f:44:43:44 root@master.centos.com
The key's randomart image is:
+--[ RSA 2048]----+
|.*.*+    +E      |
|+ *+.o   . .     |
|..+.. .   .      |
|.o  .   ..       |
|.     . .S.      |
|       . o.      |
|          .      |
|                 |
|                 |
+-----------------+
```

图 2-4 设置 SSH 协议无密码登录

② 使用 ssh-copy-id 将公钥复制到远程虚拟机中。执行命令“ssh-copy-id -i/root/.ssh/id_rsa.pub master”可以将公钥复制到远程虚拟机 master 中，其中，选项“-i”用于指定公钥文件。各个节点将公钥复制到远程虚拟机中的命令如代码 2-9 所示。

代码 2-9 各个节点将公钥复制到远程虚拟机中

```
ssh-copy-id -i /root/.ssh/id_rsa.pub master//依次输入“yes,123456”(123456 为 root 用户的密码)
ssh-copy-id -i /root/.ssh/id_rsa.pub slave1
ssh-copy-id -i /root/.ssh/id_rsa.pub slave2
ssh-copy-id -i /root/.ssh/id_rsa.pub slave3
```

③ 验证 SSH 协议是否能够无密钥登录。在 master 中分别执行命令“ssh slave1”“ssh slave2”“ssh slave3”，若结果如图 2-5 所示，则说明配置 SSH 协议无密码登录成功。

```
[root@master ~]# ssh slave1
Last login: Fri Apr 28 23:51:32 2017 from 192.168.128.1
[root@slave1 ~]# exit
logout
Connection to slave1 closed.
[root@master ~]# ssh slave2
Last login: Tue Apr 25 18:04:44 2017 from 192.168.128.1
[root@slave2 ~]# exit
logout
Connection to slave2 closed.
[root@master ~]# ssh slave3
Last login: Tue Apr 25 18:04:49 2017 from 192.168.128.1
[root@slave3 ~]# exit
logout
Connection to slave3 closed.
```

图 2-5　验证 SSH 协议无密钥登录

（3）复制 Hadoop 安装文件到集群的其他节点中

在 master 节点中安装完 Hadoop 后，通过执行代码 2-10 所示的命令将 Hadoop 安装文件复制到集群的其他节点中。

代码 2-10　将 Hadoop 安装文件复制到集群的其他节点中

```
scp -r /usr/local/hadoop-2.6.4 slave1:/usr/local
scp -r /usr/local/hadoop-2.6.4 slave2:/usr/local
scp -r /usr/local/hadoop-2.6.4 slave3:/usr/local
```

（4）配置时间同步服务

网络时间协议（Network Time Protocol，NTP）是用于使计算机时间同步化的一种协议，它可以使计算机对其服务器或时钟源进行同步操作，提供高精准度的时间校正。Hadoop 集群对时间精准度要求很高，主节点与各从节点的时间都必须同步。配置时间同步服务主要是为了进行集群间的时间同步。配置 Hadoop 集群时间同步服务的步骤如下。

① 安装 NTP 服务。在前面已经配置了 YUM 源，可以直接使用 YUM 安装 NTP 服务，在各节点上执行命令“yum install -y ntp”即可，若最终出现了“Complete”信息，则说明 NTP 服务安装成功。

② 设置 master 节点为 NTP 服务主节点。执行命令“vim /etc/ntp.conf”，打开/etc/ntp.conf 文件，注释掉以 server 开头的行，并添加代码 2-11 所示的内容。

代码 2-11　修改 master 节点的 ntp.conf 文件

```
restrict 192.168.0.0 mask 255.255.255.0 nomodify notrap
server 127.127.1.0
fudge 127.127.1.0 stratum 10
```

③ 修改从节点的 ntp.conf 文件。分别在 slave1、slave2、slave3 中配置 NTP 服务，同样需要修改/etc/ntp.conf 文件，注释掉以 server 开头的行，并添加代码 2-12 所示的内容。

代码 2-12　修改从节点的 ntp.conf 文件

```
server master
```

④ 关闭防火墙。执行命令“service iptables stop & chkconfig iptables off”，永久性关闭防火墙，且主节点和从节点都要关闭防火墙。

⑤ 启动NTP服务。

a. 在master节点中执行命令“service ntpd start & chkconfig ntpd on”，若结果如图2-6所示，则说明主节点NTP服务启动成功。

```
[root@master ~]# service ntpd start & chkconfig ntpd on
[1] 1450
Starting ntpd:
[1]+  Done                    service ntpd start
```

图2-6 主节点NTP服务启动成功

b. 在slave1、slave2、slave3中分别执行命令“ntpdate master”即可同步时间，如图2-7所示。

```
[root@slave1 ~]# ntpdate master
29 Apr 00:47:46 ntpdate[1554]: adjust time server 192.168.128.130 offset -0.000670 sec
```

图2-7 从节点同步时间

c. 在slave1、slave2、slave3中分别执行“service ntpd start & chkconfig ntpd on”即可启动且永久启动NTP服务，如图2-8所示。

```
[root@slave1 ~]# service ntpd start & chkconfig ntpd on
[1] 1555
[root@slave1 ~]# Starting ntpd:                          [  OK  ]

[1]+  Done                    service ntpd start
```

图2-8 从节点启动且永久启动NTP服务

（5）启动/关闭Hadoop集群

完成Hadoop的所有配置后，即可执行格式化NameNode的操作，该操作会在NameNode所在虚拟机中初始化HDFS的相关配置，并且该操作在集群搭建过程中只需执行一次。执行格式化之前可以先配置环境变量，格式化完成之后即可启动集群。

① 配置环境变量。配置环境变量需在master、slave1、slave2、slave3中修改/etc/profile文件，在文件末尾配置环境变量，如代码2-13所示。文件修改完后保存并退出，执行命令“source /etc/profile”使配置生效。

代码2-13 配置环境变量

```
export HADOOP_HOME=/usr/local/hadoop-2.6.4
export PATH=$HADOOP_HOME/bin:$PATH:/usr/java/jdk1.7.0_80/bin
```

② 格式化NameNode。格式化NameNode只需执行命令“hdfs namenode -format”，若提示“Storage directory /data/hadoop/ hdfs/name has been successfully formatted.”，则格式化成功，如图2-9所示。

```
17/04/29 00:58:45 INFO util.GSet: Computing capacity for map NameNodeRetryCache
17/04/29 00:58:45 INFO util.GSet: VM type       = 64-bit
17/04/29 00:58:45 INFO util.GSet: 0.029999999329447746% max memory 966.7 MB = 297.0 KB
17/04/29 00:58:45 INFO util.GSet: capacity      = 2^15 = 32768 entries
17/04/29 00:58:45 INFO namenode.NNConf: ACLs enabled? false
17/04/29 00:58:45 INFO namenode.NNConf: XAttrs enabled? true
17/04/29 00:58:45 INFO namenode.NNConf: Maximum size of an xattr: 16384
17/04/29 00:58:45 INFO namenode.FSImage: Allocated new BlockPoolId: BP-299710164-192.168.128.130-1493398725649
17/04/29 00:58:45 INFO common.Storage: Storage directory /data/hadoop/hdfs/name has been successfully formatted.
17/04/29 00:58:46 INFO namenode.NNStorageRetentionManager: Going to retain 1 images with txid >= 0
17/04/29 00:58:46 INFO util.ExitUtil: Exiting with status 0
17/04/29 00:58:46 INFO namenode.NameNode: SHUTDOWN_MSG:
/************************************************************
SHUTDOWN_MSG: Shutting down NameNode at master.centos.com/192.168.128.130
************************************************************/
```

图2-9 格式化成功

③ 启动 Hadoop 集群。格式化完成之后即可启动 Hadoop 集群，启动 Hadoop 集群只需要在 master 节点中直接进入 Hadoop 安装目录，分别执行代码 2-14 所示的命令即可。

代码 2-14　启动 Hadoop 集群的命令

```
cd $HADOOP_HOME                 // 进入 Hadoop 安装目录
sbin/start-dfs.sh               // 启动 HDFS 相关服务
sbin/start-yarn.sh              // 启动 YARN 相关服务
sbin/mr-jobhistory-daemon.sh start historyserver          // 启动日志相关服务
```

④ 查看节点进程。集群启动之后，在主节点 master，以及从节点 slave1、slave2、slave3 中分别执行命令 “jps”，若出现图 2-10 所示的信息，则说明集群启动成功。

```
[root@master hadoop-2.6.4]# jps
2082 JobHistoryServer
2340 Jps
1697 NameNode
1989 ResourceManager
1845 SecondaryNameNode
[root@master hadoop-2.6.4]# ssh slave1
Last login: Thu May  4 05:18:23 2017 from 192.168.128.1
[root@slave1 ~]# jps
1563 NodeManager
1502 DataNode
1754 Jps
```

图 2-10　查看节点进程

⑤ 关闭 Hadoop 集群。关闭 Hadoop 集群只需要在 master 节点中直接进入 Hadoop 安装目录，分别执行代码 2-15 所示的命令即可。

代码 2-15　关闭 Hadoop 集群的命令

```
cd $HADOOP_HOME              // 进入 Hadoop 安装目录
sbin/stop-yarn.sh            // 关闭 YARN 相关服务
sbin/stop-dfs.sh             // 关闭 HDFS 相关服务
sbin/mr-jobhistory-daemon.sh stop historyserver       // 关闭日志相关服务
```

2. 监控集群

Hadoop 集群的相关监控服务如表 2-3 所示，表中从上到下分别是 HDFS 监控服务、YARN 监控服务及日志监控服务。

表 2-3　Hadoop 集群的相关监控服务

服务	Web 接口	默认端口
NameNode	http://namenode_host:port/	50070
ResourceManager	http://resourcemanager_host:port/	8088
MapReduce JobHistoryServer	http://jobhistoryserver_host:port/	19888

为了能够顺利地通过浏览器进入 Hadoop 集群相关监控服务的监控界面，需要修改本地计算机 host 文件，本地计算机的 host 文件可在 C:\Windows\System32\drivers\etc 目录下找到，在文件末尾添加代码 2-16 所示的内容。host 文件是一个没有扩展名的系统文件，其作用是对一些常用的网址域名与其对应的 IP 地址建立一个关联“数据库”，当用户在浏览器

的地址栏中输入一个网址时，系统会先自动从 host 文件中寻找网址对应的 IP 地址，一旦找到，系统会立即打开对应网页。

代码 2-16　修改本地计算机 host 文件

```
192.168.128.130 master master.centos.com
192.168.128.131 slave1 slave1.centos.com
192.168.128.132 slave2 slave2.centos.com
192.168.128.133 slave3 slave3.centos.com
```

（1）HDFS 监控

在浏览器的地址栏中输入“http://master:50070”并按“Enter”键，即可进入 HDFS 的监控界面，如图 2-11 所示。

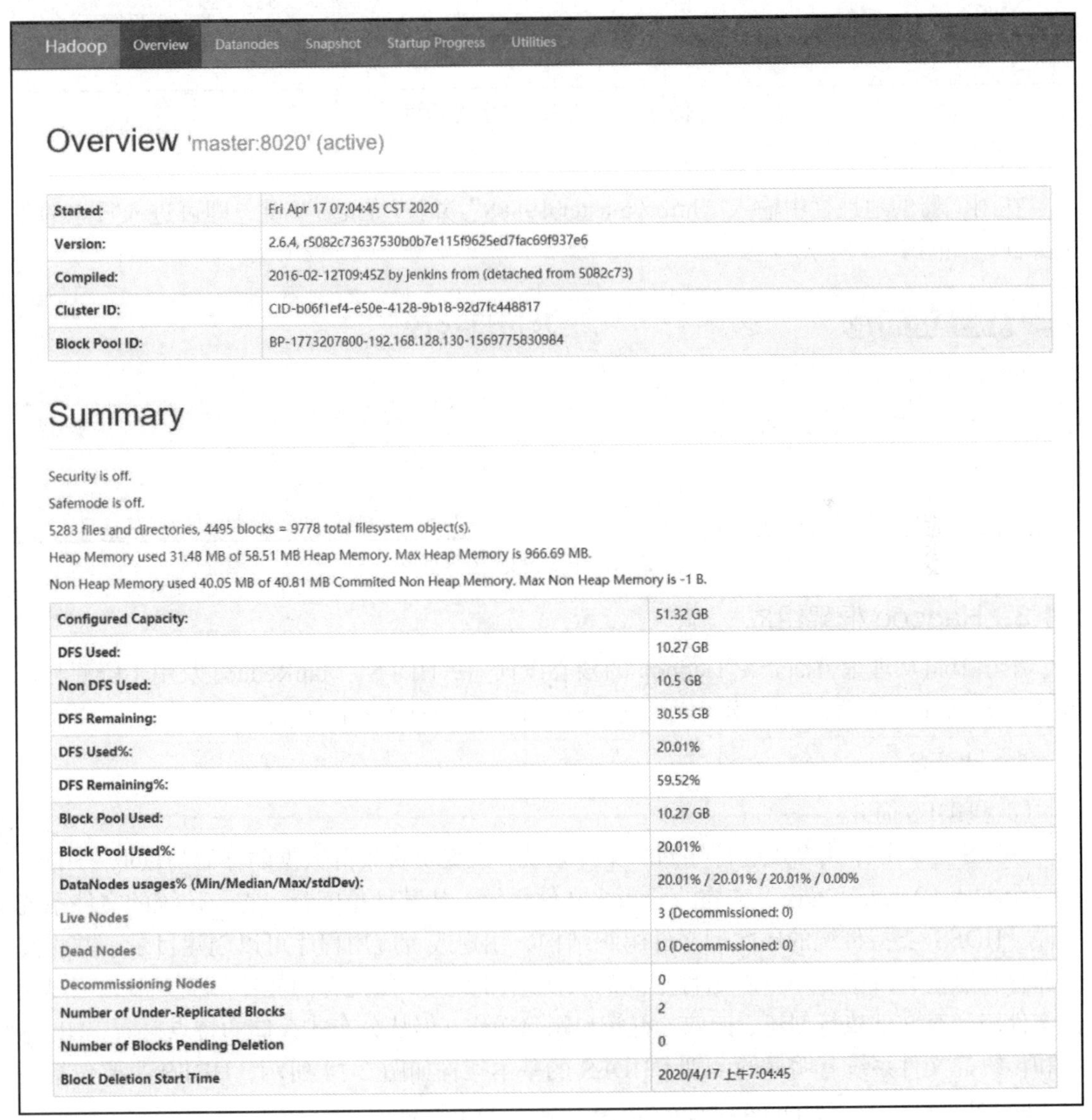

图 2-11　HDFS 的监控界面

（2）YARN 监控

在浏览器的地址栏中输入“http://master:8088”并按“Enter”键，即可进入 YARN 的监控界面，如图 2-12 所示。

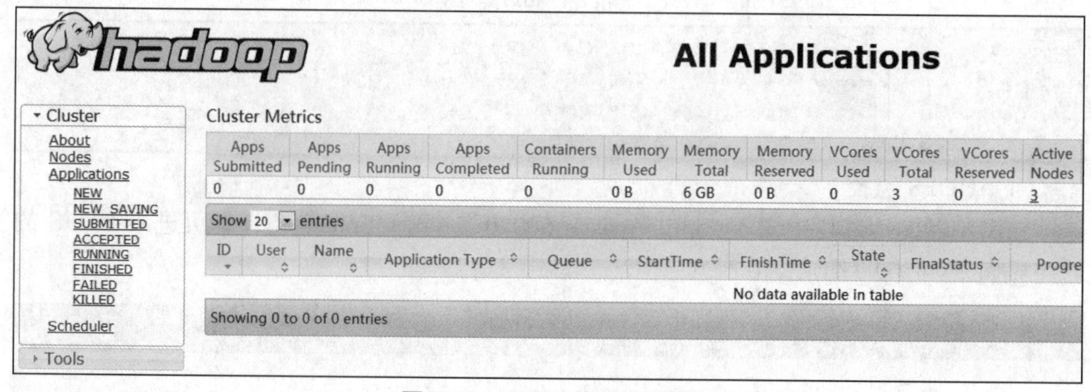

图 2-12　YARN 的监控界面

（3）日志监控

在浏览器的地址栏中输入“http://master:19888”并按“Enter”键，即可进入日志的监控界面，如图 2-13 所示。

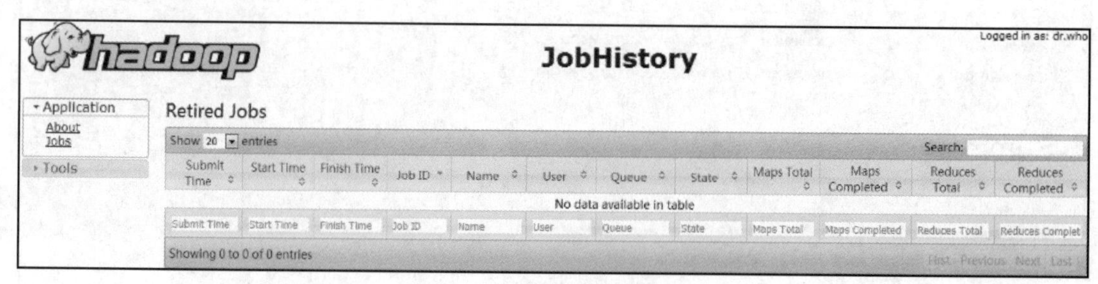

图 2-13　日志的监控界面

2.1.3　Hadoop 框架组成

本小节将从理论方面介绍 Hadoop 的核心设计——HDFS、MapReduce 及用于资源与任务调度的 YARN 框架，以便读者深入地了解 Hadoop 的整体架构组成。

1. HDFS

（1）HDFS 简介

HDFS 是以分布式方式进行存储的文件系统，主要负责集群数据的存储与读取。HDFS 是一个主从（Master-Slave）结构的分布式文件系统，从某种角度看，它与传统的文件系统一样。HDFS 支持传统的层次型文件组织结构，用户或者应用程序可以创建目录，并将文件保存在这些目录中。HDFS 名称空间的层次结构和大多数现有的文件系统类似，可以通过文件路径对文件执行创建、读取、更新和删除操作。但从分布式存储的特点来看，HDFS 又和传统的文件系统有明显的区别。HDFS 的基本架构如图 2-14 所示，HDFS 主要包括一个 NameNode、一个 Secondary NameNode 和多个 DataNode。

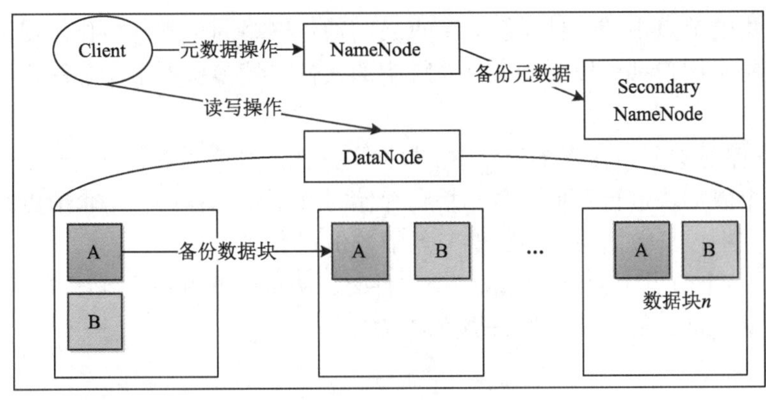

图 2-14　HDFS 的基本架构

① 元数据。元数据不是具体的文件内容，它有 3 类重要信息：第一类是文件和目录自身的属性信息，如文件名、目录名、父目录信息、文件大小、创建时间、修改时间等；第二类是文件记录的信息的存储相关的信息，如文件分块情况、副本个数、每个副本所在的 DataNode 信息等；第三类则是 HDFS 中所有 DataNode 的信息，用于管理 DataNode。

② NameNode。NameNode 用于存储元数据以及处理客户端发出的请求。在 NameNode 中存放元数据的文件是 fsimage 文件。在系统运行期间，所有对元数据的操作都保存在内存中，并被持久化到另一个文件 edits 中。当 NameNode 启动的时候，fsimage 会被加载到内存中，并对内存中的数据执行 edits 所记录的操作，以确保内存所保留的数据是最新的。

③ Secondary NameNode。Secondary NameNode 用于备份 NameNode 的数据，周期性地将 edits 文件合并到 fsimage 文件中并在本地备份，将新的 fsimage 文件存储到 NameNode 中，以取代原来的 fsimage，删除 edits 文件，创建一个新的 edits 以继续存储文件的修改状态。

④ DataNode。DataNode 是真正存储数据的地方。在 DataNode 中，文件以数据块的形式进行存储。当文件上传到 HDFS 中时，系统会以 128MB 为一个数据块对文件进行分割，将每个数据块存储到不同的或相同的 DataNode 中并备份副本，每个数据块一般默认备份 3 个副本，NameNode 负责记录文件的分块信息，确保在读取该文件时可以找到并整合所有数据块。

⑤ 数据块。文件上传到 HDFS 中时，系统会根据默认数据块大小把文件分成一个个数据块，Hadoop 2.x 默认一个数据块大小为 128MB，当存储大小为 129MB 的文件时，该文件将被分为两个数据块进行存储。数据块会被存储到各个从节点中，每个数据块都会备份副本。

（2）HDFS 工作原理

分布式系统会划分多个子系统或模块，各自运行在不同的机器上，子系统或模块之间通过网络通信进行协作，最终实现整体功能。利用多个节点协作完成一个或多个具体业务功能的系统就是分布式系统。其中，分布式文件系统是分布式系统的一个子集，分布式文件系统解决的是数据存储问题，换句话说，它是横跨在多台计算机上的存储系统。存储在分布式文件系统中的数据会自动分布在不同的节点中。

作为分布式文件系统，HDFS 主要体现在以下 3 个方面。

① HDFS 是分布在多个集群节点中的文件系统。HDFS 并不是一个单机文件系统，它

是分布在多个集群节点中的文件系统。节点之间通过网络通信进行协作，提供多个节点的文件信息，使每个用户都可以看到文件系统中的文件，使多机器上的多用户可以分享文件和存储空间。

② 数据存储时被分布在多个节点中。数据不是按一个文件存储的，而是以一个文件分成的一个或多个数据块进行存储。数据块在存储时并不是都存储在一个节点中，而是分别存储在各个节点中，且数据块会在其他节点中备份副本。

③ 数据从多个节点中读取。读取一个文件时，从多个节点中找到该文件的数据块，分别读取所有数据块直到最后一个数据块读取完毕。

（3）HDFS 特点

① 优点。

a. 高容错性：上传至 HDFS 的数据会自动保存多个副本，可通过增加副本的数量来增加系统的容错性。如果某一个副本丢失，HDFS 则会复制其他节点中的副本。

b. 适合大数据的处理：HDFS 能够处理规模达 GB 级、TB 级甚至 PB 级的数据。

c. 流式数据访问：HDFS 以流式数据访问模式存储超大文件，特点是"一次写入，多次读取"，即文件一旦写入，不能修改，只能增加。这样可以保证数据的一致性。

② 缺点。

a. 不适合低延迟数据访问：HDFS 是为了处理大规模数据集分析任务而设计的，目的是达到较大的数据吞吐量，这可能以高延迟作为代价。

b. 无法高效存储大量小文件：因为 NameNode 把 HDFS 的元数据存放在内存中，所以 HDFS 所能容纳的文件数目是由 NameNode 的内存大小来决定的，即每存入一个文件都会在 NameNode 中写入该文件信息，如果写入太多小文件，NameNode 内存则会被占满而无法继续写入文件信息。而与多个小文件的总大小相同的单一文件只会一次性写入文件信息到内存中，因此 HDFS 更适合大文件存储。

c. 不支持多用户写入及在文件任意位置修改文件：HDFS 的一个文件同一时刻只能有一个写入者，且写操作只能在文件末尾进行，即只能执行追加操作。目前，HDFS 还不支持多个用户对同一文件的写操作，以及在文件任意位置进行修改操作。

2. MapReduce

（1）MapReduce 简介

MapReduce 是 Hadoop 的核心计算框架，适用于大规模（大于 1TB）数据集并行运算的编程模型，主要包括 Map（映射）和 Reduce（规约）两部分。当启动一个 MapReduce 任务时，Map 端会读取 HDFS 中的数据，并将数据映射成所需要的键值对类型传到 Reduce 端。Reduce 端接收 Map 端传送过来的键值对类型的数据，根据不同键进行分组，对每一组键相同的数据进行处理，得到新的键值对并输出到 HDFS 中，这就是 MapReduce 的核心思想。

（2）MapReduce 工作原理

一个完整的 MapReduce 过程包含数据的输入与分片、Map 阶段的数据处理、Shuffle/Sort 阶段、Reduce 阶段的数据处理、数据输出等。MapReduce 作业执行流程如图 2-15 所示。

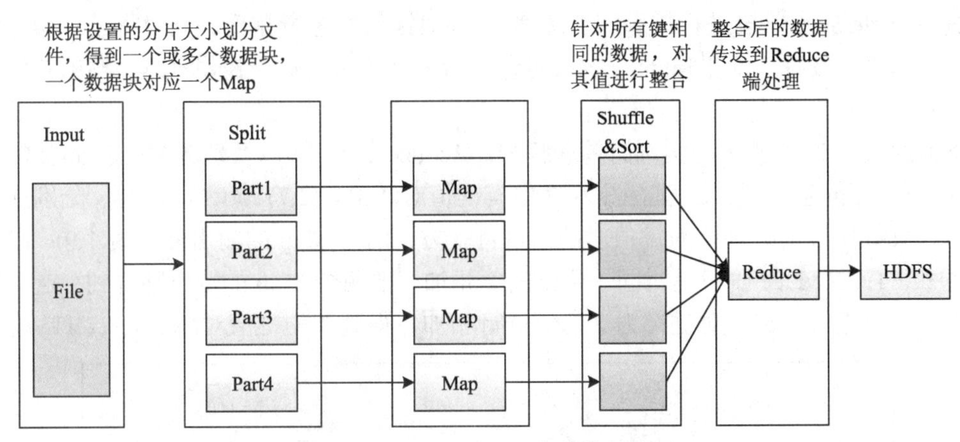

图 2-15　MapReduce 作业执行流程

① 读取输入数据。在 MapReduce 过程中，数据是从 HDFS 中读取的。文件在上传到 HDFS 时，一般按照 128MB 分成了一个或多个数据块，因此在运行 MapReduce 程序时，每个数据块都会生成一个 Map。可以通过重新设置文件分片大小来调整 Map 的个数，在运行 MapReduce 时，系统会根据所设置的分片大小对文件重新进行分割，一个分片大小的数据块会对应一个 Map。

② Map 阶段。程序有一个还是多个 Map，由分片大小或数据块个数决定。Map 阶段，数据以键值对的形式读入，键的值一般为每行首字符与文件最初始位置的偏移量，即中间所隔字符个数，值为这一行的数据记录。根据业务需求对键值对进行处理，将其映射成新的键值对，再将新的键值对传送到 Reduce 端。在此阶段还能设置数据分区 Partition，通过指定分区，同一个分区的数据会被发送到同一个 Reduce 中进行处理。

③ Shuffle/Sort 阶段。Shuffle 阶段是指从 Map 输出开始，传送 Map 输出结果到 Reduce 的过程。该过程会将同一个 Map 中输出的键相同的值进行进一步整合，即“洗牌”，减少传输的数据量，并在整合后将数据按照键进行排序。

④ Reduce 阶段。Reduce 任务可以有多个，由 Map 阶段设置的数据分区确定，一个分区数据被一个 Reduce 任务处理。每一个 Reduce 任务会接收到不同 Map 任务发送的数据，并且每个 Map 传送来的数据都是有序的。一个 Reduce 任务中的每一次处理都是针对所有键相同的键值对组进行的，其会对数据进行规约，以新的键值对输出到 HDFS 中。

根据上述内容，MapReduce 的本质可以用图 2-16 完整地表现出来。

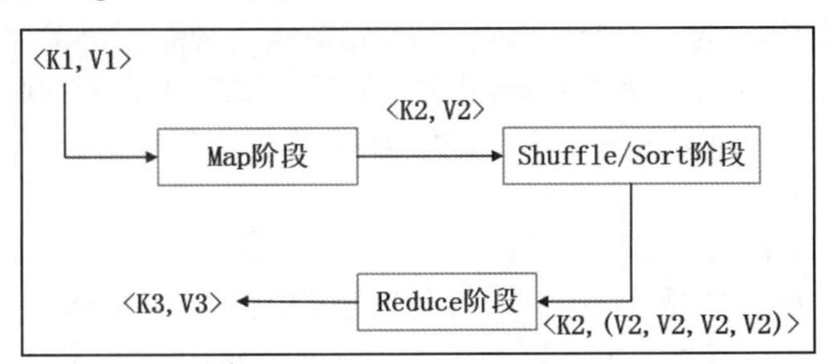

图 2-16　MapReduce 的本质

MapReduce 的本质是把一组键值对<K1,V1>经过 Map 阶段映射成新的键值对<K2,V2>，再经过 Shuffle/Sort 阶段进行排序和“洗牌”，即对键值对进行排序，同时将具有相同的键的值整合起来，最后，经过 Reduce 阶段对整合后的键值对组进行逻辑处理，输出新的键值对<K3,V3>。

下面通过图 2-17 所示的实例简单理解 MapReduce 的本质，以及理解 MapReduce 的 Map 端和 Reduce 端的基本运行原理。图 2-17 中有键值对(1,3)、(2,7)、(1,4)、(2,8)，分别在两个 Map 中，Map 阶段的处理是对键值对的值进行平方，两个 Map 的输出分别为(1,9)、(2,49)、(1,16)、(2,64)。Reduce 阶段是对同一个键的值相加，把两个 Map 的输出整合到一起，即对键都为 1 的值进行相加，对键都为 2 的值进行相加，得到新的键值对(1,25)、(2,113)。

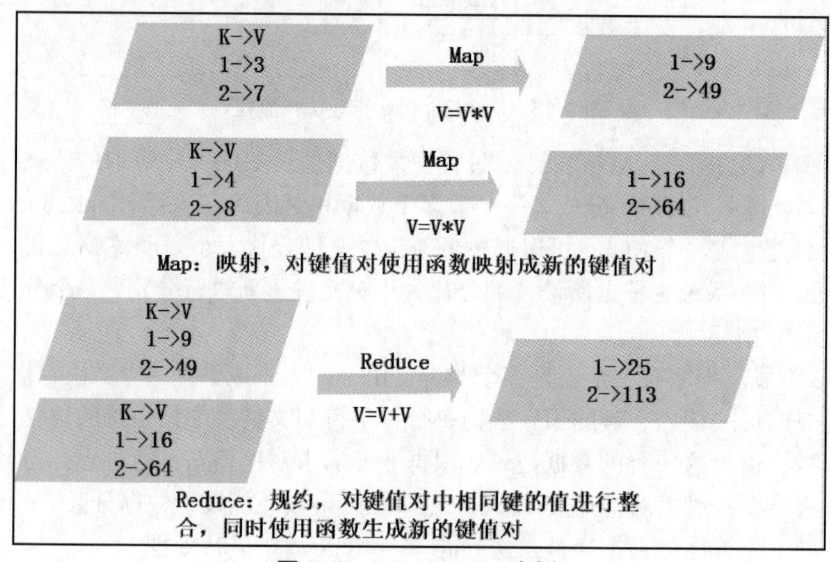

图 2-17　MapReduce 实例

3. YARN

（1）YARN 简介

HDFS 是 Hadoop 的数据存储框架，Hadoop MapReduce 是数据处理框架。然而，MapReduce 已经不能满足现今多种多样的数据处理需求，如实时/准实时计算、图计算等。而 Hadoop YARN 提供了一个更加通用的资源管理和分布式应用框架。在这个框架中，用户可以根据自己的需求实现定制化的数据处理框架。Hadoop MapReduce 也是 YARN 中的一个框架。YARN 的另一个目标是拓展 Hadoop，使得 Hadoop 不仅可以支持 MapReduce 计算，还能很方便地管理诸如 Hive、HBase、Pig、Spark/Shark 等应用。YARN 的这种新的架构设计能够让各种类型的应用运行在 Hadoop 中，并通过 YARN 从系统层面进行统一管理，也就是说，有了 YARN，各种应用可以互不干扰地运行在同一个 Hadoop 系统中，共享整个集群资源。

（2）YARN 基本架构

YARN 总体上仍然是主从结构，在资源管理框架中，ResourceManager（简称 RM）为 Master，NodeManager（简称 NM）为 Slave，ResourceManager 负责对各个 NodeManager 中的资源进行统一管理和调度。当用户提交一个应用程序时，ResourceManager 需要提供一个用以跟踪和管理这个程序的 ApplicationMaster（简称 AM），它负责向 ResourceManager

申请资源，并要求 NodeManager 启动可以占用一定资源的任务。因为不同的 ApplicationMaster 被分布到不同的节点中，所以它们之间不会相互影响。图 2-18 所示为 YARN 基本架构，YARN 主要由 ResourceManager、NodeManager、ApplicationMaster 和 Client Application 等构成。

① ResourceManager。ResourceManager 是一个全局的资源管理器，负责整个系统的资源管理和分配。它主要由两个组件构成：调度器（Scheduler）和应用程序管理器（Applications Manager）。

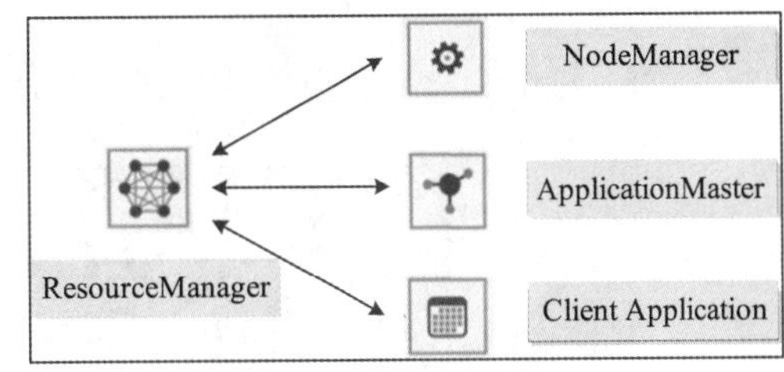

图 2-18 YARN 基本架构

调度器负责将系统中的资源分配给各个正在运行的应用程序，它不负责任何与具体应用程序相关的工作，如不负责监控或者跟踪应用程序的执行状态等，也不负责重新启动因应用程序运行失败或者硬件故障而产生的任务。

应用程序管理器负责处理客户端提交的作业（Job）请求以及协商第一个 Container（封装资源的对象）以供 ApplicationMaster 运行，并在 ApplicationMaster 运行失败的时候重新启动 ApplicationMaster。

② NodeManager。NodeManager 是每个节点上的资源和任务管理器。一方面，它会定时地向 ResourceManager 汇报本节点的资源使用情况和各个 Container 的运行状态；另一方面，它会接收并处理来自 ApplicationMaster 的 Container 启动、停止等各种请求。其中，Container 是 YARN 中的资源抽象对象，它封装了某个节点中的多维度资源，如内存、CPU、磁盘、网络等。当 ApplicationMaster 向 ResourceManager 申请资源时，ResourceManager 为 ApplicationMaster 返回的资源便是使用 Container 表示的。YARN 会为每个任务分配一个 Container，且该任务只能使用该 Container 中描述的资源。

③ ApplicationMaster。用户每提交一个应用程序，系统都会生成一个 ApplicationMaster 程序包含到提交的程序中。ApplicationMaster 的主要功能如下：与 ResourceManager 调度器协商以获取资源（用 Container 表示）；将得到的任务进一步分配给内部的任务；与 NodeManager 通信以启动、停止任务；监控所有任务运行状态，并在任务运行失败时重新为任务申请资源以重启任务。

④ Client Application。Client Application 是客户端应用程序，客户端会先创建一个应用程序上下文对象，并设置 ApplicationMaster 必需的资源请求信息，再将应用程序提交到 ResourceManager 中。

（3）YARN 工作流程

YARN 从提交任务到完成任务的整个工作流程如图 2-19 所示，YARN 的工作流程分为以下几个步骤。

① 用户通过客户端提交一个应用程序到 ResourceManager 中进行处理，应用程序包括 ApplicationMaster 程序、启动 ApplicationMaster 的命令、用户程序等。

② ResourceManager 为该应用程序分配第一个 Container，并与分配的 Container 所在节点的 NodeManager 通信，要求 NodeManager 在这个 Container 中启动应用程序的 ApplicationMaster。该 Container 用于启动 ApplicationMaster 和执行 ApplicationMaster 后续命令。

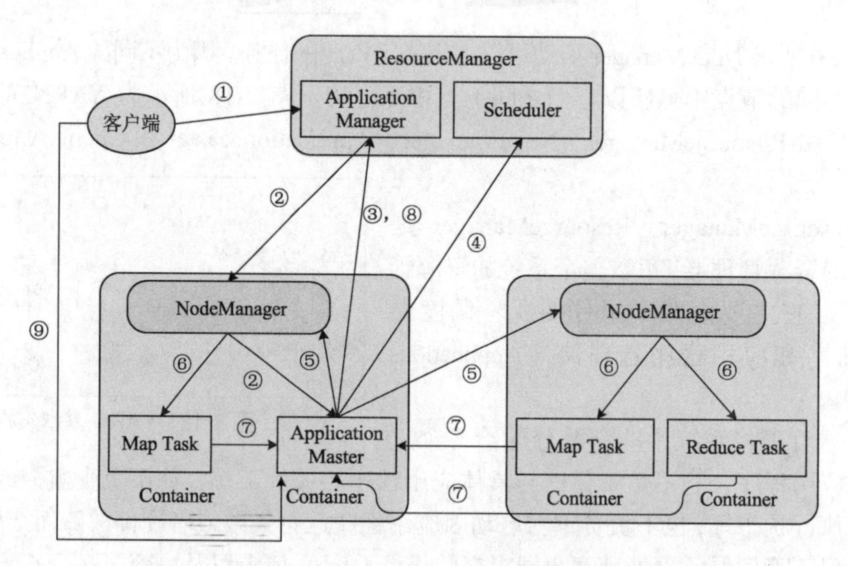

图 2-19　YARN 从提交任务到完成任务的整个工作流程

③ ApplicationMaster 启动后先向 ResourceManager 注册，这样用户可以直接通过 ResourceManager 查看应用程序的运行状态，再为提交的应用程序所需要执行的各个任务申请资源，并监控它的运行状态，直到运行结束，即重复步骤④～步骤⑦。这个示例应用中需要执行两个 Map 任务和一个 Reduce 任务，所以需要执行步骤④～步骤⑦共 3 次，先执行 Map 任务，再执行 Reduce 任务。

④ ApplicationMaster 采用轮询的方式通过远程过程调用（Remote Procedure Call，RPC）向 ResourceManager 申请和领取资源。因此，多个应用程序同时提交时，不一定是第一个提交的先执行。

⑤ 一旦 ApplicationMaster 申请到资源，便与资源所在节点对应的 NodeManager 通信，要求 NodeManager 在分配的资源中启动任务。

⑥ NodeManager 为任务设置好运行环境（包括环境变量、JAR 包、二进制程序等）后，将任务启动命令写到一个脚本中，并通过运行该脚本启动任务。

⑦ 被启动的任务开始运行，各个任务直接或通过某个远程过程调用（Remote Procedure Call，RPC）协议向 ApplicationMaster 汇报自己的状态和进度，以使 ApplicationMaster 随时掌握各个任务的运行状态，从而可以在任务失败时重新启动任务。在应用程序运行过程中，用户可随时通过 RPC 向 ApplicationMaster 查询应用程序的当前运行状态。

⑧ 应用程序运行完成后，ApplicationMaster 向 ResourceManager 注销自己，释放资源。

⑨ 用户通过客户端申请关闭 ApplicationMaster。

2.1.4　Hadoop 应用实践

本小节将使用 Hadoop 示例包的 WordCount 源程序进行实践，进一步帮助读者了解 Hadoop 及其核心思想，学会应用 Hadoop 去解决实际问题。

1．实践环境

（1）CentOS 6.8 的 Linux 操作系统。

（2）JDK 1.7。

（3）Hadoop 2.6.4 的 4 个节点的集群（其中包含 1 个主节点和 3 个从节点）。

2. 实践内容

HDFS 中有一个日志文件/user/root/email_log.txt，如图 2-20 所示，其中，每一行记录是一个用户名，表示用户登录网站的记录。要求统计每个用户的登录次数，提交 MapReduce 任务给集群运行，并将统计结果存储到 HDFS 中。

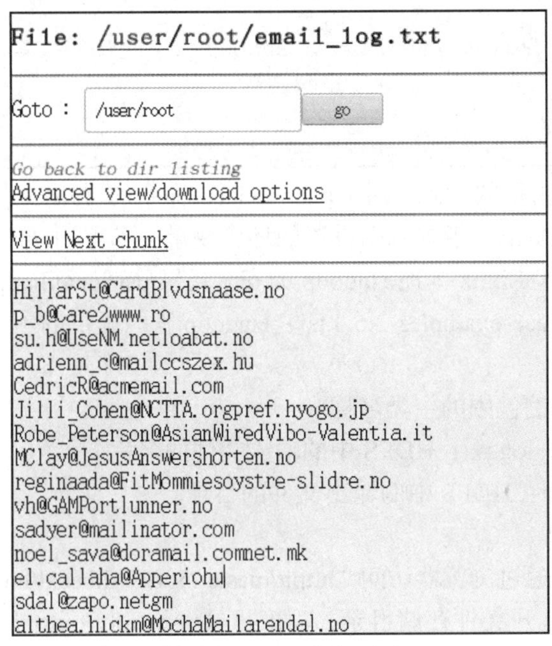

图 2-20　email_log.txt

3. 实践步骤

使用 Hadoop 示例包的 WordCount 源程序实现用户登录次数统计的步骤如下。

（1）启动 Hadoop 集群

提交 MapReduce 任务给集群运行前，需要先在主节点中启动 Hadoop 集群。启动 Hadoop 集群的命令如代码 2-17 所示。

代码 2-17　启动 Hadoop 集群的命令

```
[root@master ~]#cd /usr/local/hadoop-2.6.4/sbin
[root@master sbin]#./start-dfs.sh
[root@master sbin]#./start-yarn.sh
[root@master sbin]#./mr-jobhistory-daemon.sh start historyserver
```

（2）提交 MapReduce 任务命令及参数介绍

① 提交 MapReduce 任务时通常使用 hadoop jar 命令。其语法格式如代码 2-18 所示。

代码 2-18　hadoop jar 的语法格式

```
hadoop jar <jar> [mainClass] args
```

② 因为 hadoop jar 命令的附带参数较多，所以需要结合实际任务对其各项参数进行说明。提交 MapReduce 任务的命令如代码 2-19 所示，通过统计数据中用户名出现的次数可统计用户的登录次数，并将统计结果存储到 HDFS 中。其中，每个参数之间需要用空格分开，反斜杠“\”的作用为换行。执行 MapReduce 任务命令时会有相应的日志输出，其中的关键信息有助于检查执行的过程与状态。

代码 2-19 提交 MapReduce 任务的命令

```
hadoop jar \
$HADOOP_HOME/share/hadoop/mapreduce/hadoop-mapreduce-examples-2.6.4.jar \
wordcount \
/user/root/email_log.txt \
/user/root/output
```

③ 针对以上示例执行的代码，对“hadoop jar”命令的常用参数进行如下说明。

a. $HADOOP_HOME：指主机中设置的环境变量（参考/etc/profile 的内容）。此处的 $HADOOP_HOME 指代 Linux 本地 Hadoop 的安装目录/usr/local/hadoop-2.6.4。

b. hadoop-mapreduce-examples-2.6.4.jar：Hadoop 官方提供的示例包，其中包括词频统计模块（wordcount）。

c. wordcount：示例包中的主类名称。

d. /user/root/email_log.txt：HDFS 中的输入文件的名称。

e. /user/root/output：HDFS 中的输出文件的目录。

（3）查看统计结果

任务执行完成后，通过浏览器访问“http://master:50070/dfshealth.jsp”，单击“Browse the filesystem”链接可以打开文件存储目录。输出目录/user/root/output/中有两个新文件生成：一个是_SUCCESS，这是一个标识文件，表示任务执行完成；另一个是 part-r-00000，即任务执行完成后生成的结果文件。查看 part-r-00000 的内容，如图 2-21 所示，统计结果中有两列数据，第 1 列表示用户名，第 2 列表示该用户的登录次数。

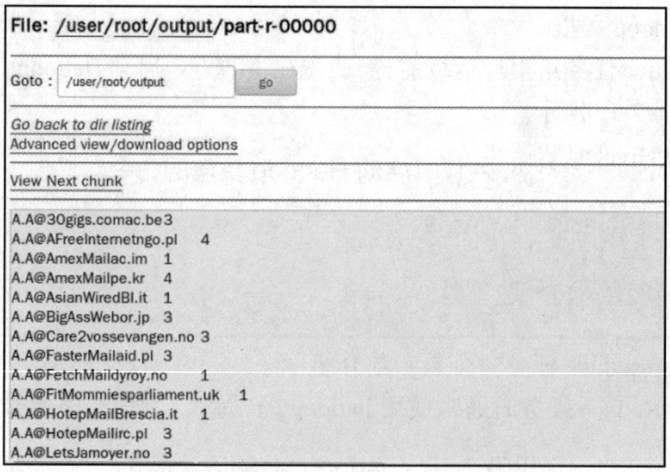

File: /user/root/output/part-r-00000

Goto : /user/root/output go

Go back to dir listing
Advanced view/download options

View Next chunk

```
A.A@30gigs.comac.be	3
A.A@AFreeInternetngo.pl	4
A.A@AmexMailac.im	1
A.A@AmexMailpe.kr	4
A.A@AsianWiredBI.it	1
A.A@BigAssWebor.jp	3
A.A@Care2vossevangen.no	3
A.A@FasterMailaid.pl	3
A.A@FetchMaildyroy.no	1
A.A@FitMommiesparliament.uk	1
A.A@HotepMailBrescia.it	1
A.A@HotepMailirc.pl	3
A.A@LetsJamoyer.no	3
```

图 2-21 查看 part-r-00000 的内容

2.2 Hive 基础

Hive 是构建在分布式计算框架 Hadoop 之上的 SQL 引擎，是 Hadoop 生态系统中的重要组成部分，适用于数据仓库的统计分析。

2.2.1 Hive 概述

本小节将从理论方面介绍 Hive 的基本概念、Hive 的特点及应用场景，并介绍 Hive 与传统数据库的区别等。

1. Hive 的基本概念

Hive 是基于 Hadoop 开发的数据仓库工具，是对大规模数据的抽取、转换、装载方法（Extract Transformation Load Method，ETL Method）。它可以将结构化的数据文件映射为一张数据库表，用来存储、查询和分析 Hadoop 中的大规模数据。Hive 中定义了简单的类 SQL 查询语言，称为 HiveQL，使用 HiveQL 可将数据操作转换为复杂的 MapReduce 运行在 Hadoop 大数据平台上。Hive 允许熟悉 SQL 的用户基于 Hadoop 框架分析数据，Hive 的优点是学习成本低，对于简单的统计分析，不必开发专门的 MapReduce 程序，直接通过 HiveQL 即可实现。

Hive 是一款分析历史数据的利器，但是 Hive 只有在结构化数据情境下才能大显“神威”。Hive 不支持联机事务处理（Online Transaction Processing，OLTP），也不提供实时查询功能，即 Hive 的处理有延迟性。Hive 较为适用的场景是大数据集的批处理作业，如网络日志分析或者海量结构化数据离线分析。

2. Hive 的特点

（1）优点

① 可扩展性。一般情况下不需要重启 Hive 就可以自由扩展集群的规模。

② 延展性。Hive 支持用户自定义函数，用户可以根据自己的需求实现自定义函数。

③ 良好的容错性。Hive 有良好的容错性，当节点出现问题时，仍可执行 HiveQL 命令。

④ 更友好的接口。Hive 的操作接口采用了类 SQL 语法，具备了快速开发的能力。

⑤ 更低的学习成本。Hive 避免了编写 MapReduce 程序，降低了开发人员的学习成本。

（2）缺点

① Hive 的 HiveQL 表达能力有限。HiveQL 无法表达迭代式算法，不擅长数据挖掘。

② Hive 的效率较低。通常情况下，Hive 自动生成的 MapReduce 作业不够智能化。

③ Hive 调试和优化比较困难，粒度较粗。

3. Hive 的应用场景

Hive 目前主要应用在日志分析，以及海量结构化数据离线分析、数据挖掘、数据分析等方面。Hive 不适用于复杂的机器学习算法、复杂的科学计算等场景。同时，Hive 虽然是针对批量、长时间数据分析设计的，但是它并不能做到交互式实时查询。

4. Hive 与传统数据库的区别

Hive 在很多方面和传统数据库类似，如 Hive 支持 SQL 接口。由于其他底层设计，Hive

对 HDFS 和 MapReduce 有很强的依赖，这也意味着 Hive 的体系结构和传统数据库有很大的区别，这些区别又间接地影响了为 Hive 所支持的一些特性。

在传统数据库中，表的模式是在数据加载时强行确定好的，如果在加载时发现数据不符合模式，那么传统数据库会拒绝加载这些数据。Hive 在加载的过程中不对数据进行任何验证操作，只是简单地将数据复制或者移动到表对应的目录下，从这方面来说，传统数据库在数据加载的过程中比 Hive 慢。传统数据库在数据加载过程中可以进行一些处理，如对某一列建立索引等，这样可以提升数据的查询性能，而 Hive 是不支持这种功能的。

数据库的事务、索引及更新是传统数据库的重要特性。Hive 目前还不支持对行级别的数据进行更新，不支持联机事务处理。Hive 虽然支持建立索引，但是 Hive 的索引与传统数据库中的索引并不相同，Hive 的索引只能建立在表的列上，且不支持主键或外键。Hive 与传统数据库的对比如表 2-4 所示。

表 2-4　Hive 与传统数据库的对比

对比项	Hive	传统数据库
查询语言	HiveQL	SQL
数据存储	HDFS	块设备、本地文件系统
数据更新	不支持	支持
处理数据规模	大	小
可扩展性	高	低
执行	MapReduce	执行引擎
执行延时	高	低
模式	读模式	写模式
事务	不支持	支持

2.2.2　Hive 安装与配置

因为 Hive 查询等操作过程严格遵守 Hadoop MapReduce 的作业执行模式，将用户的 HiveQL 语句通过解释器转换为 MapReduce 作业提交到 Hadoop 集群上，再返回作业执行结果给用户，所以 Hive 启动前需要在 4 个节点上搭建好 Hadoop 集群。

Hive 在安装之前需要先配置好元数据库，本书项目中 Hive 使用的元数据库为 MySQL，因此需要先配置好 MySQL 数据库。本小节主要介绍 Hive 的元数据库 MySQL 的安装配置，包括在线安装和离线安装两种方式，并介绍 Hadoop 集群中 Hive 安装配置的具体步骤。

1. MySQL 安装与配置

MySQL 数据库可以安装在 Windows 或 Linux 中，本小节主要介绍基于 Linux 的 MySQL 的安装配置，安装 MySQL 可以分为在线安装和离线安装两种。

（1）在线安装

在线安装需要保证 Linux 操作系统联网，同时保证 YUM 源可用，执行代码 2-20 所示的命令，即可安装 MySQL。

代码 2-20 Linux 在线安装 MySQL 的命令

```
yum install mysql-server.x86_64 -y
```

（2）离线安装

离线安装虽然不需要 Linux 操作系统联网，但是需要预先下载必要的安装包，包括 MySQL-server-5.6.28-1.el6.x86_64.rpm、MySQL-client-5.6.28-1.el6.x86_64.rpm、MySQL-devel-5.6.28-1.el6.x86_64.rpm。执行代码 2-21 所示的命令，即可安装 MySQL。

代码 2-21 Linux 离线安装 MySQL 的命令

```
rpm -ivh MySQL-server-5.6.28-1.el6.x86_64.rpm
rpm -ivh MySQL-client-5.6.28-1.el6.x86_64.rpm
rpm -ivh MySQL-devel-5.6.28-1.el6.x86_64.rpm
```

（3）MySQL 开启远程访问权限

如果 Hive 服务和 MySQL 不在同一台虚拟机上，而是在远程虚拟机上，那么需要开启 MySQL 远程权限。针对 root 用户，设置密码为 123456，开启远程访问权限，保证在其他虚拟机中也能直接访问 MySQL。执行命令“mysql”，进入 MySQL 终端，并在 MySQL 的终端执行代码 2-22 所示的命令即可。

代码 2-22 MySQL 开启远程访问权限的命令

```
use mysql;
delete from user where 1=1;
GRANT ALL PRIVILEGES ON *.* TO 'root'@'%' IDENTIFIED BY '123456' WITH GRANT OPTION;
FLUSH PRIVILEGES;
```

（4）启动 MySQL 服务

修改完 MySQL 的远程访问权限后，执行代码 2-23 所示的命令开启 MySQL 服务并进入 MySQL 终端。

代码 2-23 启动 MySQL 服务的命令

```
service mysqld start
mysql -uroot -p123456
```

2. Hive 安装与配置

本书项目使用的 Hive 的安装包为 apache-hive-1.2.1-bin.tar.gz，源码包为 apache-hive-1.2.1-src.tar.gz、mysql-connector-java-5.1.42-bin.jar，Hive 的安装与配置包括以下几个步骤。

（1）将 Hive 的安装包 apache-hive-1.2.1-bin.tar.gz 上传到虚拟机 master 的/opt 目录下，并将其解压到/usr/local 目录下，解压 Hive 安装包命令如代码 2-24 所示。

代码 2-24 解压 Hive 安装包命令

```
tar -zxvf apache-hive-1.2.1-bin.tar.gz -C /usr/local/
```

（2）将 Hive 的源码包 apache-hive-1.2.1-src.tar.gz 上传到虚拟机 master 的/opt 目录下，并将其解压到/usr/local 目录下。进入解压后的文件目录/apache-hive-1.2.1-src/hwi/web，生成 WAR 包，并将生成的 WAR 包复制到 Hive 安装包的 lib 目录下，如代码 2-25 所示。

代码 2-25 解压 Hive 源码包并复制生成的 WAR 包

```
tar -zxvf apache-hive-1.2.1-src.tar.gz -C /usr/local/
cd /apache-hive-1.2.1-src/hwi/web
jar -cvf hive-hwi-1.2.1.war *
cp hive-hwi-1.2.1.war /usr/local/apache-hive-1.2.1-bin/lib
```

（3）执行命令“cp hive-env.sh.template hive-env.sh”复制$HIVE_HOME/conf 目录下的 hive-env.sh.template 为 hive-env.sh，并修改 hive-env.sh 文件，添加代码 2-26 所示的内容。

代码 2-26 修改 hive-env.sh 文件

```
export HADOOP_HOME=/usr/local/hadoop-2.6.4/
export PATH=$HADOOP_HOME/bin:$PATH
```

（4）编辑/etc/profile 文件，配置代码 2-27 所示的 Hive 环境变量，文件修改完成后保存并退出，执行命令“source /etc/profile”，使配置生效。

代码 2-27 配置 Hive 环境变量

```
export HIVE_HOME=/usr/local/apache-hive-1.2.1-bin
export PATH=$HIVE_HOME/bin:$PATH
```

（5）执行命令“cp hive-default.xml.template hive-site.xml”复制$HIVE_HOME/conf 目录下的 hive-default.xml.template 为 hive-site.xml，并在文件 hive-site.xml 中添加代码 2-28 所示的内容。

代码 2-28 修改 hive-site.xml 文件

```
<property>
<name>javax.jdo.option.ConnectionURL</name>
<value>jdbc:mysql://master:3306/hive?createDatabaseIfNotExist=true</value>
</property>
<property>
<name>javax.jdo.option.ConnectionDriverName</name>
<value>com.mysql.jdbc.Driver</value>
</property>
<property>
<name>javax.jdo.PersistenceManagerFactoryClass</name>
<value>org.datanucleus.api.jdo.JDOPersistenceManagerFactory</value>
</property>
<property>
<name>javax.jdo.option.DetachAllOnCommit</name>
<value>true</value>
</property>
<property>
<name>javax.jdo.option.NonTransactionalRead</name>
<value>true</value>
</property>
<property>
<name>javax.jdo.option.ConnectionUserName</name>
<value>root</value>
</property>
```

```
<property>
<name>javax.jdo.option.ConnectionPassword</name>
<value>123456</value>
</property>
<property>
<name>javax.jdo.option.Multithreaded</name>
<value>true</value>
</property>
<property>
<name>datanucleus.connectionPoolingType</name>
<value>BoneCP</value>
</property>
<property>
<name>hive.metastore.warehouse.dir</name>
<value>/user/hive/warehouse</value>
</property>
<property>
<name>hive.server2.thrift.port</name>
<value>10000</value>
</property>
<property>
<name>hive.server2.thrift.bind.host</name>
<value>master</value>
</property>
<property>
<name>hive.metastore.uris</name>
<value>thrift://master:9083</value>
</property>
<property>
<name>hive.hwi.listen.host</name>
<value>0.0.0.0</value>
</property>
<property>
<name>hive.hwi.listen.port</name>
<value>9999</value>
</property>
<property>
<name>hive.hwi.war.file</name>
<value>lib/hive-hwi-1.2.1.war</value>
</property>
```

（6）配置 JAR 包。将 MySQL 驱动 mysql-connector-java-5.1.25-bin.jar 上传到虚拟机 master 的$HIVE_HOME/lib 目录下，如代码 2-29 所示，用$HIVE_HOME/lib 目录下的 jline-2.12.jar 替换$HADOOP_HOME/share/hadoop/yarn/lib/目录下的 jline-0.9.94.jar，并将$JAVA_HOME/lib 目录下的 tools.jar 复制到$HIVE_HOME/lib 目录下。

代码 2-29　配置 JAR 包

```
cd /usr/local/hadoop-2.6.4/share/hadoop/yarn/lib
mv jline-0.9.94.jar jline-0.9.94.jar.bak
cp $HIVE_HOME/lib/jline-2.12.jar .
```

```
cd $HIVE_HOME/lib
cp $JAVA_HOME/lib/tools.jar .
```

（7）启动 Hive。启动 Hive 前要先启动 Hadoop 集群和开启 MySQL 服务，再启动 Hive 服务并进入 Hive，如代码 2-30 所示。

代码 2-30　启动 Hive

```
[root@master ~]# cd /usr/local/hadoop-2.6.4
[root@master hadoop-2.6.4]# sbin/start-all.sh
[root@master hadoop-2.6.4]# service mysqld start
[root@master hadoop-2.6.4]# hive --service metastore &
[root@master hadoop-2.6.4]# hive
```

2.2.3　Hive 体系架构

Hive 的体系架构如图 2-22 所示。Hive 的体系架构可以分为用户接口（①）、元数据存储（②）、Driver 组件（③）和数据存储（④）4 部分。其中，用户接口包括命令行接口（Command Line Interface，CLI）、客户端（Client）和 Web 用户接口（Web User Interface，WUI），元数据存储是指 MetaStore，Driver 组件包括编译器（Compiler）、优化器（Optimizer）和执行器（Executor），数据存储是指将数据存储至 Hadoop 集群中。用户通过 CLI、Client 或 WUI 提交 HiveQL 到 Hive 服务，其中 Client 是通过应用程序编程接口（JDBC/ODBC）远程连接 HiveServer2 服务，再提交 HiveQL 到 Hive 服务，WUI 是通过浏览器（Browser）提交 HiveQL 到 Hive 服务，通过编译器、优化器、执行器完成 HiveQL 查询语句从词法分析、语法分析、编译、优化到查询计划的生成，将元数据存储到数据库中，执行完成查询计划的处理，由 Hadoop 集群调用执行。

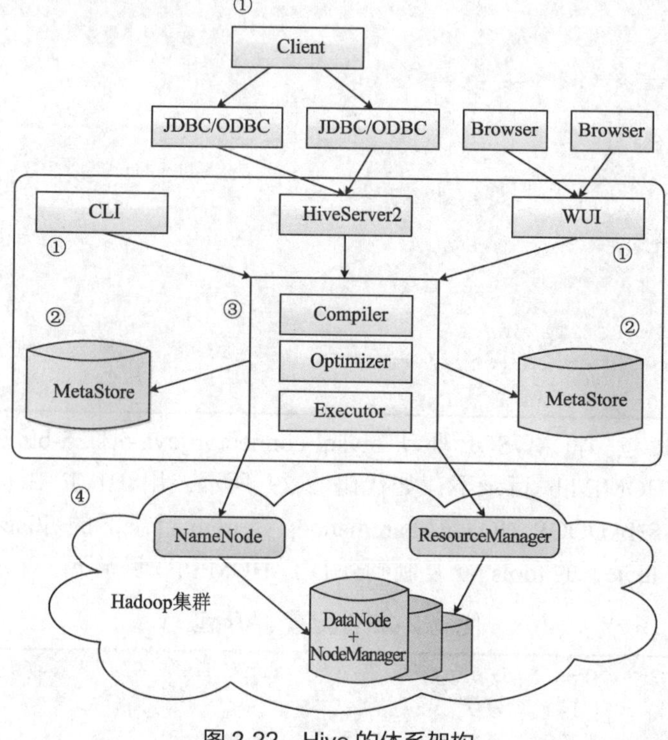

图 2-22　Hive 的体系架构

1. 用户接口

Hive对外提供了3种用户接口，即CLI、WUI和Client。CLI是Hive的命令行模式，WUI是Hive的Web模式，Client是Hive的客户端，可实现Hive的远程服务。

（1）Hive的命令行模式

配置好Hive的环境变量后，通过直接执行命令“hive”可进入Hive。Hive的命令行模式用于Linux平台命令行查询，其查询语句与MySQL查询语句类似，启动之后可通过执行“jps”命令查看RunJar进程，如图2-23所示。

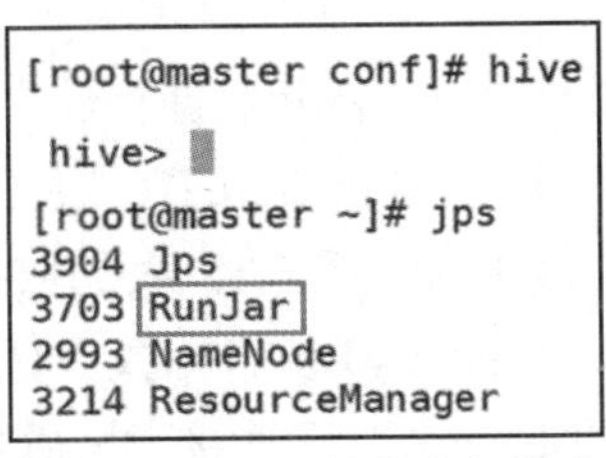

图2-23 Hive的命令行模式

（2）Hive的Web模式

执行Hive Web界面的启动命令“hive --service hwi”之后，通过浏览器访问“http://master:9999/hwi/”（其中，master是Hive服务所在虚拟机的主机名，9999是其默认端口号），可进入Hive Web界面，浏览器运行效果如图2-24所示。

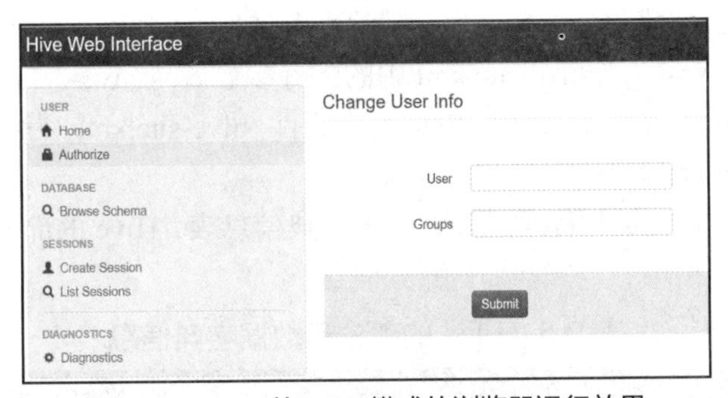

图2-24 Hive的Web模式的浏览器运行效果

（3）Hive的远程服务模式

启动远程服务（端口号默认为10000）时要执行命令“hive --service hiveserver2 &”，然后执行命令“netstat -ntpl |grep 10000”，若可以看到10000端口已绑定IP地址，则表示正常启动。命令“hive --service hiveserver2 &”表示在后台启动Hive远程服务，通过Java数据库互连（Java Database Connectivity，JDBC）访问Hive时就使用了这种启动方式，Hive的JDBC连接和MySQL类似，如代码2-31所示。

代码2-31 通过JDBC连接Hive的代码

```
// 创建emp表
import java.sql.Connection;
import java.sql.DriverManager;
import java.sql.PreparedStatement;
import java.sql.ResultSet;
public class Test {
    public static void main(String[] args) {
            Class.forName("org.apache.hive.jdbc.HiveDriver");
            Connection conn=DriverManager.getConnection
("jdbc:hive2://192.168.0.130:10000/testhive","root","");
            String sql = "create table emp(empno int,ename string,job string)
```

```
row format delimited fields terminated by ','";
            PreparedStatement ps =conn.prepareStatement(sql);
            ps.execute();
    }
}
```

2. Hive 的元数据存储

Hive 将元数据存储在数据库中，常用的数据库是 MySQL 和 Derby，其中元数据存储依赖于 MetaStore DB 服务。Hive 中的元数据包括表名、表的列和分区及其属性、表的属性（是否为外部表）、表的数据所在目录等。

Hive 默认将元数据存储在内嵌的 Derby 中。如果使用默认的数据库 Derby，那么只能允许一个会话连接，只适合进行简单的测试。在实际的生产环境中，为了支持多用户会话，需要一个独立的元数据库，使用 MySQL 作为元数据库可以满足此要求。同时，Hive 内部对 MySQL 提供了很好的支持，因此生成环境一般情况下会配置 MySQL 数据库作为 Hive 的元数据存储库。

Hive 在启动时会加载两个配置文件：默认配置文件 hive-default.xml.template 以及用户自定义文件 hive-site.xml。当 hive-site.xml 中的配置参数值与 hive-default.xml.template 文件中的配置参数值不一致时，以用户自定义文件 hive-site.xml 为准，其文件格式与 hive-default.xml.template 类似。

Hive 启动之后会在元数据库中创建元数据字典信息表，Hive 的部分元数据字典信息如表 2-5 所示。

表 2-5　Hive 的部分元数据字典信息

表名	说明
BUCKETING_COLS	Hive 表 CLUSTERED BY 字段信息（字段名、字段序号）
COLUMNS	Hive 表字段信息（字段注释、字段名、字段类型、字段序号）
DBS	Hive 中所有数据库的基本信息
NUCLEUS_TABLES	元数据表和 Hive 中类（class）的对应关系
PARTITIONS	Hive 表分区信息（创建时间、具体的分区）
PARTITION_KEYS	Hive 分区表分区键（名称、类型、字段注释、序号）
PARTITION_KEY_VALS	Hive 表分区名（键值、序号）
SDS	所有 Hive 表、表分区所对应的 HDFS 数据目录和数据格式
SEQUENCE_TABLE	Hive 对象的下一个可用 ID
SERDES	Hive 表序列化、反序列化使用的类库信息
SERDE_PARAMS	序列化、反序列化信息，如行分隔符、列分隔符、NULL 的表示字符等
SORT_COLS	Hive 表 SORTED BY 字段信息（字段名、排序类型、字段序号）
TABLE_PARAMS	表级属性，如是否是外部表、表注释等
TBLS	所有 Hive 表的基本信息

已知 HiveQL 语句的执行过程，结合元数据字典信息，Hive 创建表的整个过程如下：用户提交 HiveQL 语句→对其进行解析→分解为表、字段、分区等 Hive 对象。根据解析到的信息构建对应的表、字段、分区等对象，从 SEQUENCE_TABLE 中获取对象的最新 ID，与构建对象信息（如名称、类型等）一同通过数据访问对象（Data Access Object，DAO）方法写入元数据表，成功后将 SEQUENCE_TABLE 中对应的最新 ID 加 5。

实际上，常见的 RDBMS 都是通过相同的方法组织的，一般用于存放数据字典和内部结构、数据统计等信息的系统表和 Hive 元数据一样显示了这些 ID 信息，而 Oracle 等数据库隐藏了这些具体的 ID 信息。通过这些元数据可以很容易读到数据，如创建一张表的数据字典信息、导出建表语句等。

3. Driver 组件

Driver 组件包括编译器、优化器和执行器。Driver 的作用是完成 HiveQL 查询语句从词法分析、语法分析、编译、优化到逻辑计划的生成，随后调用底层的 MapReduce 计算框架执行。

（1）编译器。Driver 调用解释器处理 HiveQL 字符串，这些字符串可能是一条数据定义语言（Data Definition Language，DDL）语句、数据操纵语言（Data Manipulation Language，DML）语句或查询语句。编译器将字符串转化为策略。策略仅由元数据操作和 HDFS 操作组成，元数据操作只包含 DDL 语句，HDFS 操作只包含 LOAD 语句。对插入和查询操作而言，策略由 MapReduce 任务中的有向无环图（Directed Acyclic Graph，DAG）组成。编译器主要构成部分及其具体功能如下。

① 解释器：将查询字符串转换为解析树表达式。

② 语义分析器：将解析树表达式转换为基于块的内部查询表达式，将输入表的模式信息从 MetaStore 中恢复出来，使用这些信息验证列名，展开 SELECT *及类型检查。

③ 逻辑策略生成器：将内部查询表达式转换为逻辑策略，这些策略由逻辑操作树组成。

（2）优化器。通过逻辑策略构造多途径并以不同方式重写逻辑计划。优化器的功能如下：将 multiple join 合并为一个 multi-way join；对 join、group-by 和自定义的 map-reduce 操作重新进行划分；削减不必要的列；在表扫描操作中使用断言；对于已分区的表，削减不必要的分区；在抽样查询中，削减不必要的桶（Bucket）；增加局部聚合操作，用于处理大分组聚合，并增加再分区操作，用于处理不对称的分组聚合。

（3）执行器。调用底层的 MapReduce 计算框架执行逻辑计划。

4. 数据存储

首先，Hive 没有专门的数据存储格式，也没有为数据建立索引，用户可以自由地组织 Hive 中的表，只需要在创建表时告诉 Hive 数据中的列分隔符和行分隔符，Hive 就可以解析数据。其次，Hive 中所有的数据都存储在 HDFS 中，Hive 的数据模型包含数据库（DataBase）、表（Table）、分区（Partition）、桶等，其中数据库、分区、表都对应着 HDFS 中的某个目录，Hive 表中的数据存储在表目录下。

Hive 的数据库类似传统数据库的 DataBase，其在第三方数据库中实际是一张表。

Hive 中的表和数据库中的表在概念上类似，每一个表在 HDFS 中都有一个相应的目录来存储数据。例如，存在一个表 people，它在 HDFS 中的路径为/user/hive/warehouse/people，其中/user/hive/warehouse 是在 hive-site.xml 中由参数 hive.metastore.warehouse.dir 指定的数据库的目录，所有的表数据（不包括外部表数据）都保存在这个目录下。

外部表和表的主要区别是对数据的管理。外部表数据存储在建表时由关键字 location 指定的目录下，且当外部表删除时，只删除外部表的结构，而不删除数据。

分区对应数据中分区列的密集索引，但是 Hive 中分区的组织方式和数据库中分区的组织方式不同。在 Hive 中，表中的一个分区对应表中的一个目录，所有分区的数据都存在对应的目录下。

桶对指定列计算 Hash（哈希）值，根据 Hash 值切分数据，其目的是并行处理，每个桶在 HDFS 中对应一个文件。

2.2.4 Hive 应用实践

本小节将使用 Hive 对航空公司客户价值数据进行预处理及分析，进一步帮助读者了解 Hive 及其核心思想，学会应用 Hive 去解决实际问题。

1. 实践环境

（1）CentOS 6.8 的 Linux 操作系统。

（2）JDK 1.8。

（3）Hadoop 2.6.4 的 4 个节点的集群（其中包含 1 个主节点和 3 个从节点）。

（4）apache-hive-1.2.1-bin 版本的 Hive。

2. 实践内容

（1）现有一个航空公司客户价值数据文件 air_data_base.txt，其内容如图 2-25 所示，一个客户的信息为一行数据。

```
1  54993,2006-11-2,2008-12-24,男,6,.,北京,CN,31,2014-3-31,210,505308,0,74460,
2  28065,2007-2-19,2007-8-3,男,6,,北京,CN,42,2014-3-31,140,362480,0,41288,171
3  55106,2007-2-1,2007-8-30,男,6,.,北京,CN,40,2014-3-31,135,351159,0,39711,16
4  21189,2008-8-22,2008-8-23,男,5,Los Angeles,CA,US,64,2014-3-31,23,337314,0,
5  39546,2009-4-10,2009-4-15,男,6,贵阳,贵州,CN,48,2014-3-31,152,273844,0,4226
6  56972,2008-2-10,2009-9-29,男,6,广州,广东,CN,64,2014-3-31,92,313338,0,27323
7  44924,2006-3-22,2006-3-29,男,6,乌鲁木齐市,新疆,CN,46,2014-3-31,101,248864,
8  22631,2010-4-9,2010-4-9,女,6,温州市,浙江,CN,50,2014-3-31,73,301864,0,39834
9  32197,2011-6-7,2011-7-1,男,5,DRANCY,,FR,50,2014-3-31,56,262958,0,31700,725
10 31645,2010-7-5,2010-7-5,女,6,温州,浙江,CN,43,2014-3-31,64,204855,0,47052,8
11 58877,2010-11-18,2010-11-20,女,6,PARIS,PARIS,FR,34,2014-3-31,43,298321,0,3
12 37994,2004-11-13,2004-12-2,男,6,北京,.,CN,47,2014-3-31,145,256093,0,44539,
13 28012,2006-11-23,2007-11-18,男,5,SAN MARINO,CA,US,58,2014-3-31,29,210269,0
14 54943,2006-10-25,2007-10-27,男,6,深圳,广东,CN,47,2014-3-31,118,241614,0,26
```

图 2-25 air_data_base.txt 文件的内容

（2）要求对该数据进行数据探索分析，统计 SUM_YR_1（观测窗口的票价收入）、SEG_KM_SUM（观测窗口的总飞行千米数）、AVG_DISCOUNT（平均折扣率）3 个字段的空值记录数及最小值。

（3）对数据进行数据清洗，对数据中存在的缺失值、票价为 0 或平均折扣率为 0 的数据等进行丢弃处理。

（4）对数据进行属性规约，根据航空公司客户价值 LRFMC 模型，选择与 LRFMC 指标相关的 6 个字段：FFP_DATE、LOAD_TIME、FLIGHT_COUNT、AVG_DISCOUNT、SEG_KM_SUM、LAST_TO_END。

（5）对数据进行数据转换，构造 LRFMC 的 5 个指标，并统计 5 个指标的取值范围。其中，5 个指标分别如下。

① 会员入会时间距离观测窗口结束的月数=观测窗口的结束时间-入会时间 [单位：月]，即 L = LOAD_TIME－FFP_DATE。

② 客户最近一次乘坐公司飞机距观测窗口结束的月数 = 最后一次乘机时间至观测窗口末端时长 [单位:月]，即 R = LAST_TO_END。

③ 客户在观测窗口内乘坐公司飞机的次数 = 观测窗口的飞行次数 [单位:次]，即 F = FLIGHT_COUNT。

④ 客户在观测时间内在公司累计的飞行里程 = 观测窗口总飞行千米数 [单位:千米]，即 M = SEG_KM_SUM。

⑤ 客户在观测时间内乘坐舱位所对应的折扣率的平均值 = 平均折扣率，即 C = AVG_DISCOUNT。

3. 实践步骤

使用 Hive 对航空公司客户价值数据进行预处理及分析的步骤如下。

（1）将文件 air_data_base.txt 通过 Xshell 的文件资源管理器上传到 master 节点的 /opt/data 目录下，若目录不存在，则执行命令“mkdir -p /opt/data”创建目录。

（2）启动 Hive 前，先启动 Hadoop 集群和开启 MySQL 服务，然后执行命令“hive --service metastore &”启动 Hive 服务，再执行命令“hive”以进入 Hive，如代码 2-32 所示。

代码 2-32 启动 Hive

```
[root@master ~]# cd /usr/local/hadoop-2.6.4
[root@master hadoop-2.6.4]# sbin/start-all.sh
[root@master hadoop-2.6.4]# service mysqld start
[root@master hadoop-2.6.4]# hive --service metastore &
[root@master hadoop-2.6.4]# hive
```

（3）创建 Hive 数据表前应通过执行“create database”命令创建数据库，并在数据库中进行表的创建等操作，创建数据库及 Hive 数据表的命令如代码 2-33 所示。

代码 2-33 创建数据库及数据表的命令

```
hive> create database air;
hive> use air;
hive> create table air_data_base(
member_no string,
ffp_date string,
first_flight_date string,
gender string,
ffp_tier int,
work_city string,
work_province string,
work_country string,
```

```
age int,
load_time string,
flight_count int,
bp_sum bigint,
ep_sum_yr_1 int,
ep_sum_yr_2 bigint,
sum_yr_1 bigint,
sum_yr_2 bigint,
seg_km_sum bigint,
weighted_seg_km double,
last_flight_date string,
avg_flight_count double,
avg_bp_sum double,
begin_to_first int,
last_to_end int,
avg_interval float,
max_interval int,
add_points_sum_yr_1 bigint,
add_points_sum_yr_2 bigint,
exchange_count int,
avg_discount float,
p1y_flight_count int,
l1y_flight_count int,
p1y_bp_sum bigint,
l1y_bp_sum bigint,
ep_sum bigint,
add_point_sum bigint,
eli_add_point_sum bigint,
l1y_eli_add_points bigint,
points_sum bigint,
l1y_points_sum float,
ration_l1y_flight_count float,
ration_p1y_flight_count float,
ration_p1y_bps float,
ration_l1y_bps float,
point_notflight int
)row format delimited fields terminated by ',';
```

（4）创建表后，需将数据导入 air_data_base 表，如代码 2-34 所示，数据导入完成后可通过执行“select”命令查询 air_data_base 表的内容。

代码 2-34　数据导入

```
hive> load data local inpath '/opt/data/air_data_base.txt' overwrite into table
air_data_base;
hive> select * from air_data_base limit 5;
```

（5）数据导入完成后，统计 SUM_YR_1、SEG_KM_SUM、AVG_DISCOUNT 这 3 个字段的空值记录数，并将统计结果保存到 null_count 表中，如代码 2-35 所示。

代码 2-35　统计字段的空值记录数

```
hive> create table null_count as
select * from
(select count(*) as sum_yr_1_null_count from air_data_base where sum_yr_1 is null)
sum_yr_1_null,
(select count(*) as seg_km_sum_null_count from air_data_base where seg_km_sum
is null)
seg_km_sum_null,
(select count(*) as avg_discount_null_count from air_data_base where avg_discount
is null)
avg_discount_null ;
```

（6）统计 SUM_YR_1、SEG_KM_SUM、AVG_DISCOUNT 这 3 个字段的最小值并将统计结果保存到 min_count 表中，如代码 2-36 所示。

代码 2-36　统计字段的最小值

```
hive>create table min_count as \
select min(sum_yr_1) as sum_yr_1_min,min(seg_km_sum) as \
seg_km_sum_min,min(avg_discount) as avg_discount_min from air_data_base ;
```

（7）通过“select”操作查看统计结果，如图 2-26 和图 2-27 所示。

```
hive> select * from null_count;
OK
591     0       0
Time taken: 0.031 seconds, Fetched: 1 row(s)
```

图 2-26　字段空值记录数统计结果

```
hive> select * from min_count;
OK
0       368     0.0
Time taken: 0.054 seconds, Fetched: 1 row(s)
```

图 2-27　字段最小值统计结果

（8）通过数据探索分析，发现数据中存在缺失值、票价为 0 或平均折扣率为 0 的数据等。因为原始数据量大，这类数据所占比例较小，对问题影响不大，所以在进行数据清洗时，对这部分数据进行丢弃处理。

（9）对数据进行数据清洗，丢弃票价为空的记录，将结果存储到 sum_yr_1_not_null 表中，如代码 2-37 所示。

代码 2-37　存储结果到 sum_yr_1_not_null 表中

```
hive> create table sum_yr_1_not_null as select * from air_data_base where sum_
yr_1 is not null ;
```

（10）对数据进行数据清洗，丢弃平均折扣率为 0 的记录，将结果存储到 avg_discount_not_0 表中，如代码 2-38 所示。

代码 2-38　存储结果到 avg_discount_not_0 表中

```
hive> create table avg_discount_not_0 as select * from sum_yr_1_not_null where
avg_discount <> 0.0 ;
```

（11）对数据进行数据清洗，丢弃票价为 0、平均折扣率不为 0、总飞行千米数大于 0 的记录，将结果存储到 sum_0_seg_avg_not_0 表中，如代码 2-39 所示。

代码 2-39　存储结果到 sum_0_seg_avg_not_0 表中

```
hive> create table sum_0_seg_avg_not_0 as select * from avg_discount_not_0
where !(sum_yr_1 = 0 and avg_discount <> 0.0 and seg_km_sum <> 0);
```

（12）对数据进行属性规约，从数据清洗结果中选择 6 个属性——FFP_DATE、LOAD_TIME、FLIGHT_COUNT、AVG_DISCOUNT、SEG_KM_SUM、LAST_TO_END，形成数据集，存储到 flfasl 表中，如代码 2-40 所示。

代码 2-40　存储数据集到 flfasl 表中

```
hive>create table flfasl as \
select FFP_DATE,LOAD_TIME,FLIGHT_COUNT,AVG_DISCOUNT,SEG_KM_SUM,LAST_TO_END from
sum_0_seg_avg_not_0 ;
```

（13）根据属性构造，基于属性规约数据结果，对数据进行数据转换，构造 LRFMC 的 5 个指标，并将结果存储到 lrfmc 表中，如代码 2-41 所示。

代码 2-41　数据转换

```
hive> create table lrfmc as select
round((unix_timestamp(LOAD_TIME,'yyyyMMdd')-unix_timestamp(FFP_DATE,'yyyyMMdd'))/
(30*24*60*60),2) as l,
round(last_to_end/30,2) as r,
FLIGHT_COUNT as f,
SEG_KM_SUM as m,
round(AVG_DISCOUNT,2) as c
from flfasl ;
```

（14）根据 lrfmc 表，统计 LRFMC 的 5 个指标的取值范围，并将结果存储到 max_min 表中，如代码 2-42 所示。

代码 2-42　统计指标的取值范围并存储到 max_min 表中

```
hive> create table max_min as
select max(l),min(l),max(r),min(r),max(f),min(f),max(m),min(m),max(c),min(c) from
lrfmc;
```

（15）通过执行“select”命令查看统计结果，如图 2-28 所示。

```
hive> select * from max_min;
OK
118.67  9.13    24.37   0.03    213     2       580717  368     1.5     0.11
Time taken: 0.04 seconds, Fetched: 1 row(s)
```

图 2-28　查看统计结果

2.3 Spark 基础

MapReduce 计算框架在海量离线数据的分析处理方面应用比较广泛，但是它更适用于线下批量数据处理，它对于线上海量数据分析处理的效率不能满足日益增长的业务需求。

因此，Spark 计算框架逐渐受到了各行业的关注。

2.3.1　Spark 概述

大数据技术的蓬勃发展，使得基于开源技术的 Hadoop 在相关行业中的应用越来越广泛。但是 Hadoop 本身存在着诸多缺陷，其最主要的缺陷是 MapReduce 计算模型延迟过高，无法满足实时、快速计算的需求。Spark 借鉴了 MapReduce 分布式计算的优点的同时，弥补了 MapReduce 的明显缺陷。Spark 的中间数据计算结果可以保存在内存中，大大减少了对 HDFS 进行读写操作的次数，因此 Spark 可以更适用于数据挖掘与机器学习中需要迭代的算法。

1. Spark 的基本概念

基于内存计算的大数据并行计算框架——Spark，于 2009 年在美国加利福尼亚大学伯克利分校的 AMP 实验室诞生，它可用于构建大型的、低延迟的数据分析应用程序。2013 年，Spark 加入 Apache 孵化器项目后，发展十分迅速。目前，Spark 是 Apache 软件基金会旗下的顶级开源项目之一。Spark 最初的设计目标是提高运行速度，使数据分析更快，以及快速简便地编写程序。Spark 提供了内存计算和基于 DAG 的任务调度执行机制，使程序运行更快，同时减少了迭代计算时的 I/O 开销。Spark 使用简练、优雅的 Scala 语言编写而成，使编写程序更为容易，且 Spark 基于 Scala 提供了交互式的编程体验。Spark 支持 Scala、Java、Python、R 等多种编程语言。

2. Spark 的发展历程

Spark 的发展历程如图 2-29 所示。Spark 是一个具有一定技术门槛与复杂度的平台，它从诞生到正式版本的发布，经历的时间很短，让人感到惊诧。

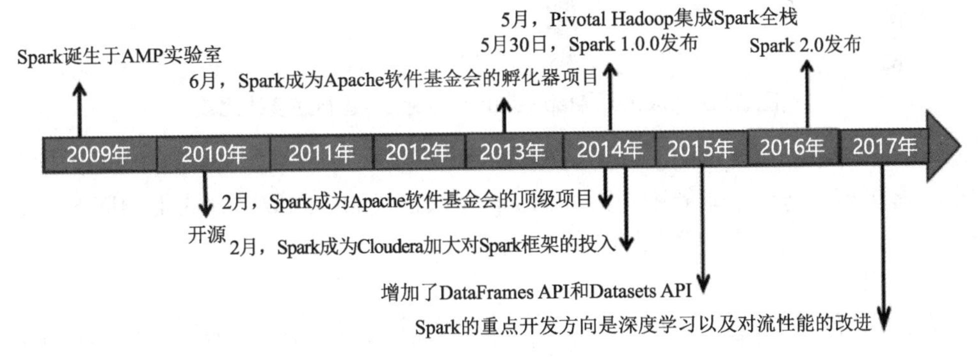

图 2-29　Spark 的发展历程

Spark 于 2009 年诞生，最初属于美国加利福尼亚大学伯克利分校的研究性项目，AMP 实验室的研究人员基于之前 Hadoop MapReduce 的工作，发现了 MapReduce 对于迭代和交互式计算任务执行效率不高，因此他们对 Spark 的研究主要是交互式查询和迭代算法设计，支持内存存储和高效的容错恢复等。

2010 年，Spark 正式开源。

2013 年 6 月，Spark 成为 Apache 软件基金会的孵化器项目。

2014 年 2 月，Spark 成为 Apache 软件基金会的顶级项目。同时，大数据公司 Cloudera

宣称加大对 Spark 框架的投入，使之取代 MapReduce。

2014 年 5 月，Pivotal Hadoop 集成 Spark 全栈。同月 30 日，Spark 1.0.0 发布。

2015 年，Spark 增加了新的 DataFrames API 和 Datasets API。

2016 年，Spark 2.0 发布。Spark 2.0 与 Spark 1.0 的主要区别是，Spark 2.0 修订了 API 的兼容性问题。

2017 年，在美国旧金山举行的 Spark Summit 2017 会议介绍了 2017 年 Spark 的重点更新内容是深度学习以及对流性能的改进。

2017 年以后，Spark 的发展主要是针对 Spark 的可用性、稳定性进行改进，并持续润色代码。随着 Spark 的逐渐成熟，并在社区的推动下，Spark 所提供的强大功能受到了越来越多技术团队和企业的青睐。

3. Spark 的特点

作为新一代的轻量级大数据处理平台，Spark 具有以下特点。

（1）快速

分别使用 Hadoop MapReduce 和 Spark 运行逻辑回归算法，两者运行速度的比较如图 2-30 所示，Hadoop 运行逻辑回归算法所使用的运行时间是 110s，Spark 运行逻辑回归算法所使用的运行时间是 0.9s。逻辑回归算法一般需要多次迭代，可以看出，Spark 运行逻辑回归算法的速度是 Hadoop MapReduce 运行速度的 100 多倍。由此可知，在一般情况下，对于迭代次数较多的应用程序，Spark 程序在内存中的运行速度是 Hadoop MapReduce 运行速度的 100 多倍，在磁盘中的运行速度是 Hadoop MapReduce 运行速度的 10 多倍。

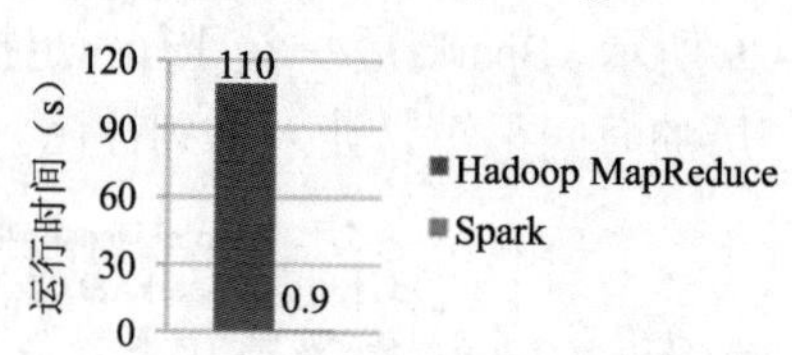

图 2-30　Hadoop MapReduce 和 Spark 运行速度的比较

Spark 与 Hadoop MapReduce 的运行速度差异较大的原因是，Spark 的中间数据存放于内存中，有更高的迭代运算效率；而 Hadoop 每次迭代的中间数据存放于 HDFS 中，涉及硬盘的读写，运算效率明显降低了。

（2）易用

Spark 支持使用 Scala、Python、Java 和 R 等简便的编程语言快速编写应用程序，同时提供超过 80 个高级运算符，使编写并行应用程序变得更加容易，并且可以在 Scala、Python 或 R 的交互模式下使用。

（3）通用

Spark 可以与 SQL、Streaming 及其他复杂的分析技术很好地结合在一起。Spark 还有一系列的高级组件，包括 Spark SQL（分布式查询）、MLlib（机器学习库）、GraphX（图计算）和 Spark Streaming（实时计算），并且支持在一个应用中同时使用这些组件。Spark 高级工具的架构如图 2-31 所示。

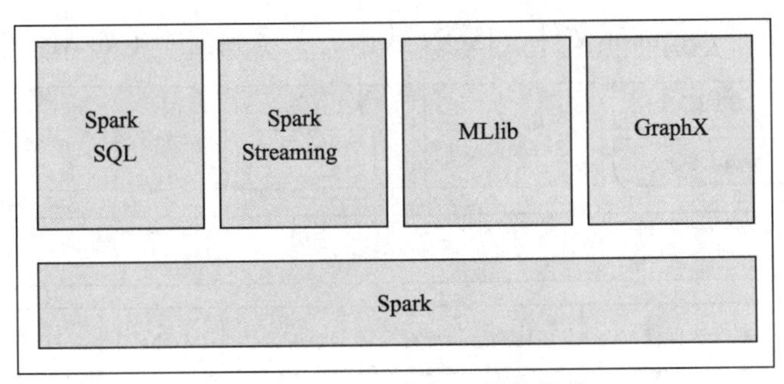

图 2-31 Spark 高级工具的架构

（4）随处运行

用户可以使用 Spark 的独立集群模式运行 Spark，并从 HDFS、Cassandra、HBase、Hive、Tachyon（基于内存的分布式文件系统）和任何分布式文件系统中读取数据，也可以在亚马逊弹性计算云（Amazon Elastic Compute Cloud，Amazon EC2）、Hadoop YARN 或者 Apache Mesos 上运行 Spark。

（5）代码简洁

Spark 支持使用 Scala、Python 等语言编写应用程序。Scala 或 Python 的代码相对 Java 更为简洁，因此，Spark 使用 Scala 或者 Python 编写应用程序要比使用 MapReduce 编写应用程序更加简洁。例如，MapReduce 实现单词计数可能需要 60 多行代码，而 Spark 使用 Scala 语言实现单词计数只需要一行，如代码 2-43 所示。

代码 2-43 Spark 使用 Scala 语言实现单词计数

```
sc.textFile("/user/root/test.txt").flatMap(_.split(" ")).map((_,1)).reduceByKey(_+_).
saveAsTextFile("/user/root/output")
```

4. Spark 生态系统

发展到现在，Spark 已经形成了一个丰富的生态系统，包括官方和第三方开发的组件或工具。Spark 生态系统如图 2-32 所示。Spark 生态系统也称为伯克利数据分析栈（Berkeley Data Analytics Stack，BDAS），它是一个可以在算法（Algorithm）、机器（Machine）、人（People）之间通过大规模集成展现大数据应用的平台。Spark 生态系统以 Spark Core 为核心，可以从 HDFS、Amazon S3 和 HBase 等持久层获取数据，利用本地运行模式 Apache Mesos、Hadoop YARN 或自身携带的 Standalone（独立运行模式）为资源管理器调度作业完成 Spark 应用程序的计算；这些应用程序可以来自不同的组件，如 Spark Shell/Spark Submit 的批处理、Spark Streaming 的实时处理应用、Spark SQL 的即席查询、BlinkDB 的权衡查询、MLlib/MLBase 的机器学习、GraphX 的图处理和 SparkR 的数学计算等。

Spark 生态系统中的重要组件简单描述如下。

（1）Spark Core：Spark 核心，对底层的框架及核心提供支持。

（2）BlinkDB：一个用于在海量数据上运行交互式 SQL 查询的大规模并行查询引擎。BlinkDB 允许用户通过权衡数据精度提升查询响应时间，其数据的精度被控制在允许的误差范围内。

（3）Spark SQL：可以执行 SQL 查询，包括基本的 SQL 语法和 HiveQL 语法。读取的数据源包括 Hive 表、Parquent 文件、JSON 数据、关系数据库（如 MySQL）等。

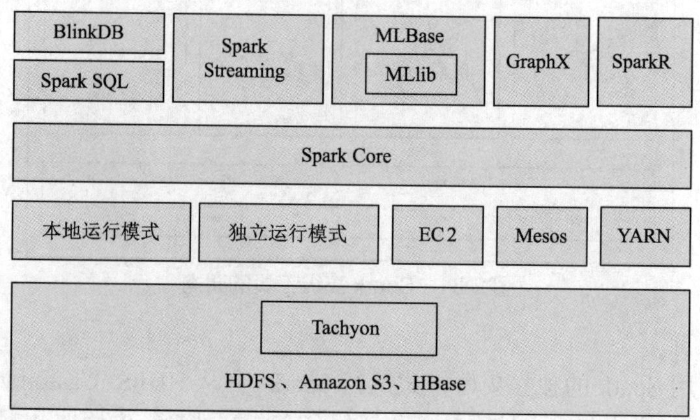

图 2-32　Spark 生态系统

（4）Spark Streaming：流式计算。例如，一个网站的流量是每时每刻都在发生的，如果需要知道过去 15min 或 1h 的流量，可以使用 Spark Streaming 解决这个问题。

（5）MLBase：Spark 生态系统中的 MLBase 专注于机器学习，MLBase 十分适用于机器学习的初学者，让一些可能并不了解机器学习的用户也能方便地使用。MLBase 分为 4 部分：MLlib、MLI、ML Optimizer 和 MLRuntime。

（6）MLlib：MLlib 是 MLBase 的一部分，是 Spark 的数据挖掘算法包，它并行化了一些常见的机器学习算法，包括分类、聚类、回归、协同过滤、降维以及底层优化等。

（7）GraphX：图计算的应用在很多情况下处理的都是大规模数据，如移动社交的关系等都可以使用图相关算法进行处理和挖掘。如果用户要自行编写相关的图计算算法，并且要在集群中应用，难度是非常大的，而使用 GraphX 就可以解决这个问题，因为其中内置了很多图相关算法。

（8）SparkR：AMP 实验室发布的一个 R 语言开发包，使得 R 语言可以“摆脱”单机运行，并作为 Spark 的作业运行在集群上，这极大地扩展了 R 语言的数据处理能力。

5. Spark 的应用场景

目前，大数据的应用非常广泛。大数据应用场景的普遍特点是数据计算量大，要求效率高，而 Spark 计算框架刚好可以满足这些计算要求。Spark 项目一经推出便受到开源社区的广泛关注和好评。目前，Spark 已发展成为大数据处理领域最炙手可热的开源项目之一。

国内外应用 Spark 成功解决企业实际业务的公司主要有以下几个。

（1）腾讯

广点通是最早使用 Spark 的应用之一。腾讯大数据精准推荐借助 Spark 快速迭代的优势，围绕“数据+算法+系统”技术方案，实现了“数据实时采集、算法实时训练、系统实时预测”的全流程实时并行高维算法，最终成功应用于广点通广告投放系统上，支持每天上百亿的请求量。

（2）淘宝

淘宝技术团队使用 Spark 解决多次迭代的机器学习算法、高计算复杂度的算法等，将 Spark 运用于淘宝的推荐相关算法上，同时利用 GraphX 解决了许多生产问题，包括以下计算场景：基于度分布的中枢节点发现、基于最大连通图的社区发现、基于三角形计数的关系衡量和基于随机游走的用户属性传播等。

（3）优酷土豆

目前，Spark 已经广泛应用在优酷土豆的视频推荐、广告业务等方面。Spark 交互查询响应快，性能是 Hadoop 的若干倍。使用 Spark 模拟广告投放不但计算效率高，而且延迟低（同 Hadoop 相比，延迟至少降低一个数量级）。此外，优酷土豆的视频推荐往往涉及机器学习及图计算，而使用 Spark 进行机器学习、图计算等迭代计算能够大大减少网络传输、数据落地等的次数，极大地提高计算性能。

（4）Yahoo

Yahoo 将 Spark 用在 Audience Expansion 中。Audience Expansion 是广告者寻找目标用户的一种方法，广告者先提供一些观看了广告并且购买产品的样本客户，据此进行学习，寻找更多可能转化的用户，对这些用户定向投放广告。Yahoo 采用的算法是逻辑回归。同时，由于某些 SQL 负载需要更高的服务质量，Yahoo 又增加了专门运行 Shark 的大内存集群，用于取代商务智能（Business Intelligence，BI）/联机分析处理（Online Analytical Processing，OLAP）工具，承担报表/仪表盘和交互式/即席查询，并与桌面 BI 工具对接。

2.3.2 Spark 集群安装与配置

Spark 环境可分为单机版环境、单机伪分布式环境和完全分布式环境，本小节介绍的是通过 VMware 模拟的完全分布式环境下的基于 Hadoop 集群的 Spark 安装与配置。读者可从官网下载 Spark 的安装包，本书使用的 Spark 安装包是 spark-1.6.3-bin-hadoop2.6.tgz。

完全分布式环境使用 Master-Slave 模式，即其中一台虚拟机作为 master 节点，其他几台虚拟机作为 slave 节点。本书项目搭建的完全分布式环境有 1 个主节点和 3 个从节点，Spark 集群的拓扑如图 2-33 所示。

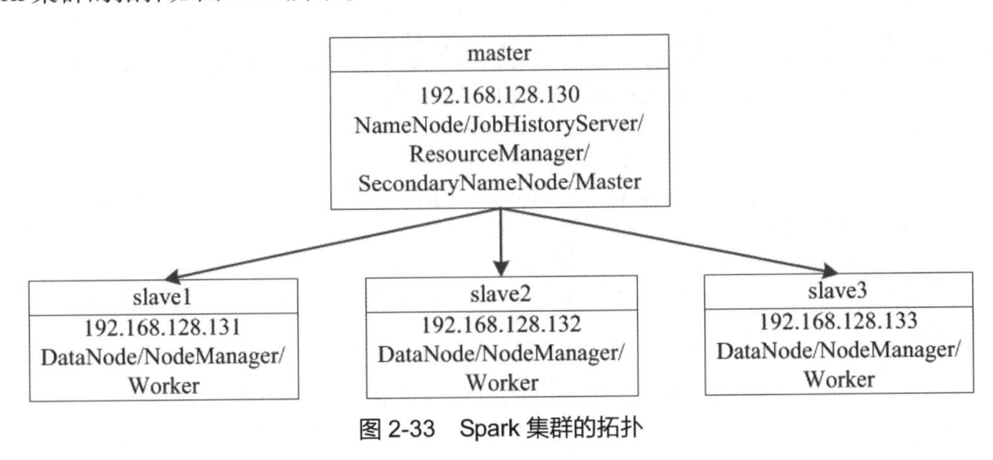

图 2-33 Spark 集群的拓扑

搭建 Spark 完全分布式环境的步骤如下。

（1）将 Spark 安装包解压到/usr/local 目录下。

（2）切换到 Spark 安装包的 conf 目录，复制 spark-env.sh.template 并将其重命名，得到 spark-env.sh 文件，如代码 2-44 所示。

代码 2-44　复制 spark-env.sh.template 并将其重命名为 spark-env.sh

```
cd /usr/local/spark-1.6.3-bin-hadoop2.6/conf/
cp spark-env.sh.template spark-env.sh
```

（3）通过执行命令“vi spark-env.sh”打开 spark-env.sh 文件，并配置代码 2-45 所示的内容。

代码 2-45　spark-env.sh 文件配置内容

```
export JAVA_HOME=/usr/java/jdk1.7.0_80          #Java 的安装路径
export HADOOP_CONF_DIR=/usr/local/hadoop-2.6.4/etc/hadoop/   #Hadoop 配置文件的路径
export SPARK_MASTER_IP=master                   #Spark 主节点的 IP 地址或主机名
export SPARK_MASTER_PORT=7077                   #Spark 主节点的端口号
export SPARK_WORKER_MEMORY=512m                 #Worker 节点能给予 Executors 的内存数
export SPARK_WORKER_CORES=1                     #每台虚拟机使用的 CPU 核数
export SPARK_EXECUTOR_MEMORY=512m               #每个 Executor 的内存
export SPARK_EXECUTOR_CORES=1                   #Executors 的核数
export SPARK_WORKER_INSTANCES=1                 #每个节点的 Worker 进程数
```

（4）复制 slaves.template 并将其重命名，得到 slaves 文件，如代码 2-46 所示。

代码 2-46　复制 slaves.template 并将其重命名为 slaves

```
cp slaves.template slaves
```

（5）通过执行命令“vi slaves”打开 slaves 文件，配置代码 2-47 所示的内容，每行代表一个从节点主机名。

代码 2-47　slaves 文件配置内容

```
slave1
slave2
slave3
```

（6）复制 spark-defaults.conf.template 并将其重命名，得到 spark-default.conf 文件，如代码 2-48 所示。

代码 2-48　复制 spark-defaults.conf.template 并将其重命名为 spark-default.conf

```
cp spark-defaults.conf.template spark-default.conf
```

（7）通过执行命令“vi spark-default.conf”打开 spark-default.conf 文件，配置代码 2-49 所示的内容。

代码 2-49　spark-default.conf 文件配置内容

```
spark.master            spark://master:7077
                        # Spark 主节点所在虚拟机的主机名及端口号，默认写法是 spark://
spark.eventLog.enabled     true     #是否打开任务日志功能，默认为 false，即不打开
spark.eventLog.dir         hdfs://master:8020/spark-logs
                           #任务日志默认存放目录，配置为一个 HDFS 路径即可
spark.history.fs.logDirectory    hdfs://master:8020/spark-logs
                                 #存放历史应用日志文件的目录
```

（8）在 master 节点上将配置好的 Spark 目录复制到 slave1、slave2、slave3 中，如代码 2-50 所示。

代码 2-50　将 Spark 目录复制到从节点中

```
scp -r /usr/local/spark-1.6.3-bin-hadoop2.6/ slave1:/usr/local/
scp -r /usr/local/spark-1.6.3-bin-hadoop2.6/ slave2:/usr/local/
scp -r /usr/local/spark-1.6.3-bin-hadoop2.6/ slave3:/usr/local/
```

（9）启动 Hadoop 集群，创建 spark-logs 目录，在 master 节点上执行代码 2-51 所示的命令。

代码 2-51　创建 spark-logs 目录的命令

```
hdfs dfs -mkdir /spark-logs
```

（10）启动 Spark 集群之前（已启动 Hadoop 集群），先通过执行命令“jps”查看进程，如图 2-34 所示（slave2 和 slave3 的进程类似于 slave1）。

```
[root@master ~]# jps          [root@slave1 ~]# jps
1939 JobHistoryServer         1550 DataNode
1676 SecondaryNameNode        1611 NodeManager
3499 Jps                      2737 Jps
1846 ResourceManager
1528 NameNode
```

图 2-34　启动 Spark 集群之前的进程

（11）进入 Spark 安装目录的 sbin 目录，启动 Spark 独立集群的命令，如代码 2-52 所示。

代码 2-52　启动 Spark 独立集群的命令

```
cd /usr/local/spark-1.6.3-bin-hadoop2.6/sbin
./start-all.sh
./start-history-server.sh
```

（12）通过执行命令“jps”查看进程，如图 2-35 所示。对比图 2-35，在 master 节点上运行着 Master 进程，而在从节点（如 slave1）上运行着 Worker 进程。

```
[root@master sbin]# jps       [root@slave1 ~]# jps
2337 JobHistoryServer         1643 DataNode
1922 NameNode                 1739 NodeManager
2070 SecondaryNameNode        1892 Worker
2244 ResourceManager          1960 Jps
2451 Master
2636 HistoryServer
2670 Jps
```

图 2-35　启动 Spark 集群之后的进程

Spark 独立集群启动后，通过“http://master:7077”访问主节点，即可进入图 2-36 所示的监控界面。其中，master 指代主节点所在虚拟机的 IP 地址。

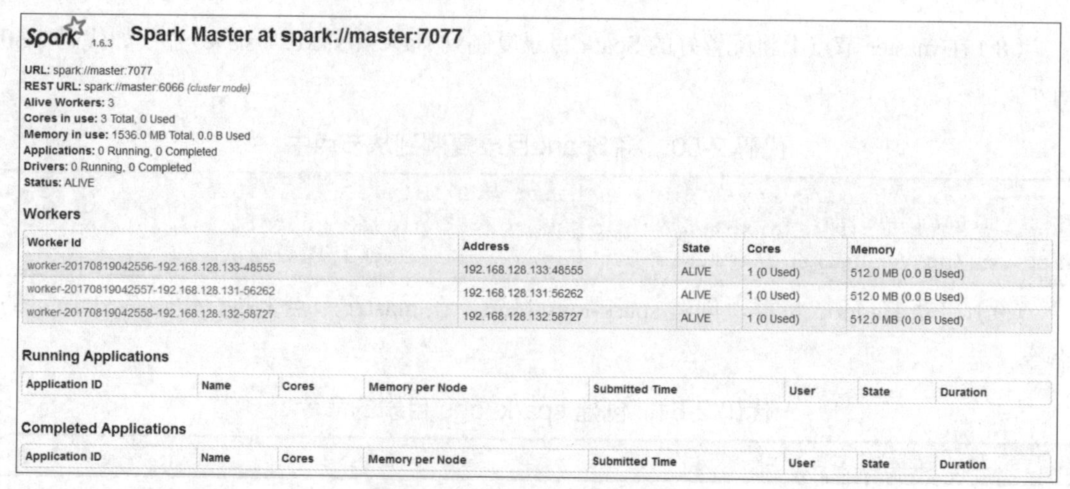

Spark 1.6.3 Spark Master at spark://master:7077

URL: spark://master:7077
REST URL: spark://master:6066 (cluster mode)
Alive Workers: 3
Cores in use: 3 Total, 0 Used
Memory in use: 1536.0 MB Total, 0.0 B Used
Applications: 0 Running, 0 Completed
Drivers: 0 Running, 0 Completed
Status: ALIVE

Workers

Worker Id	Address	State	Cores	Memory
worker-20170819042556-192.168.128.133-48555	192.168.128.133:48555	ALIVE	1 (0 Used)	512.0 MB (0.0 B Used)
worker-20170819042557-192.168.128.131-56262	192.168.128.131:56262	ALIVE	1 (0 Used)	512.0 MB (0.0 B Used)
worker-20170819042558-192.168.128.132-58727	192.168.128.132:58727	ALIVE	1 (0 Used)	512.0 MB (0.0 B Used)

Running Applications

Application ID	Name	Cores	Memory per Node	Submitted Time	User	State	Duration

Completed Applications

Application ID	Name	Cores	Memory per Node	Submitted Time	User	State	Duration

图 2-36 Spark 主节点监控界面

通过浏览器访问“http://master:8020”，即可进入图 2-37 所示的界面，从中可以看到有两个历史任务已经完成。

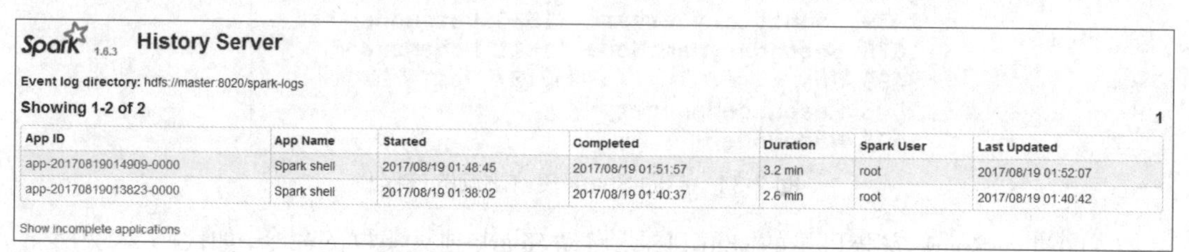

Spark 1.6.3 History Server

Event log directory: hdfs://master:8020/spark-logs

Showing 1-2 of 2

1

App ID	App Name	Started	Completed	Duration	Spark User	Last Updated
app-20170819014909-0000	Spark shell	2017/08/19 01:48:45	2017/08/19 01:51:57	3.2 min	root	2017/08/19 01:52:07
app-20170819013823-0000	Spark shell	2017/08/19 01:38:02	2017/08/19 01:40:37	2.6 min	root	2017/08/19 01:40:42

Show incomplete applications

图 2-37 History Server 监控界面

2.3.3 Spark 集群架构

Spark 集群架构如图 2-38 所示。

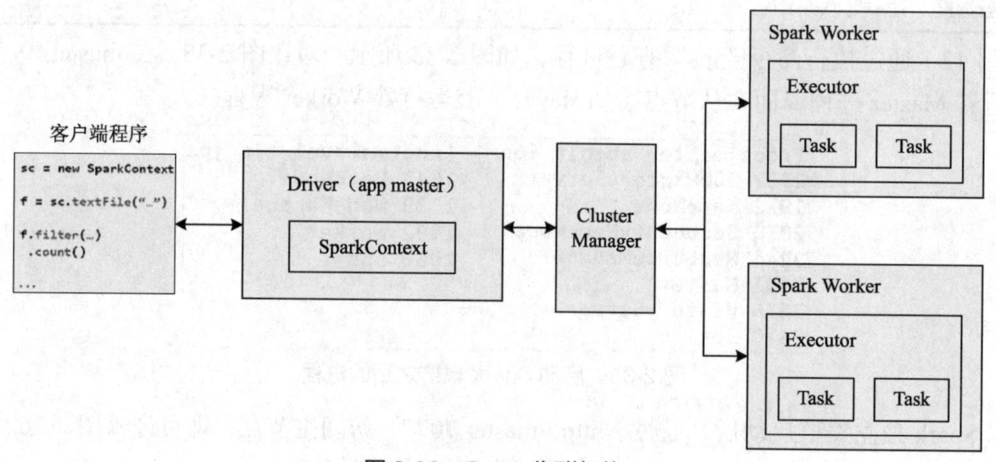

图 2-38 Spark 集群架构

对于 Spark 集群架构中的核心概念的简要说明如下。

（1）客户端程序：用户提交作业的客户端。

（2）Driver：运行 Application 的 main 函数并创建 SparkContext。Application 是用户编

写的 Spark 应用程序，包含 Driver 功能的代码和分布在集群中多个节点上的 Executor 代码。

（3）SparkContext：应用上下文，负责控制整个 Spark 生命周期。

（4）Cluster Manager：资源管理器，指在集群上获取资源的外部服务。目前，Spark 原生的资源管理器主要有 Standalone、Hadoop YARN 和 Apache Mesos。由 master 负责资源的分配，可以理解为使用 Standalone 时 Cluster Manager 是主节点。若使用 YARN 模式，则由 ResourceManager 负责资源的分配。

（5）Spark Worker：集群中任何可以运行 Application 的节点，可运行一个或多个 Executor 进程。若在 Standalone 模式中指的是通过 slaves 文件配置的 Spark Worker 节点，若在 YARN 模式下指的是 NodeManager 节点。

（6）Executor：运行在 Spark Worker 上的 Task 执行器。Executor 启动线程池运行 Task，并且负责将数据存在内存或磁盘中，每个 Application 都会申请各自的 Executor 处理任务。

（7）Task：运行 Application 的基本单位，指的是被送到某个 Executor 的具体工作任务。

2.3.4 Spark 应用实践

本小节将介绍在完全分布式的 Spark 环境中通过 spark-shell 交互式终端进行单词词频计数统计。

1. 实践环境

（1）CentOS 6.8 的 Linux 操作系统。

（2）Hadoop 2.6.4 的 4 个节点的集群（其中包含 1 个主节点和 3 个从节点）。

（3）Spark 1.6.3 的 Spark 集群。

2. 实践内容

数据文件 words.txt 的内容如图 2-39 所示，文件中包含了多行句子，现要求对文件中的单词进行词频统计，并把单词词频超过 3 的结果存储到 HDFS 中。

```
What is going on there?
I talked to John on email. We talked about some computer stuff that's it.
I went bike riding in the rain, it was not that cold. We went to the museum in SF
yesterday it was $3 to get in and they had free food. At the same time was a SF
Giants game, when we got done we had to take the train with all the Giants
fans, they are 1/2 slept.
```

图 2-39 words.txt 的内容

3. 实训步骤

使用 Spark 集群中的 spark-shell 交互式终端实现单词词频计数统计的步骤如下。

（1）进行单词词频计数统计前，需要在 master 节点中先启动 Hadoop 集群，再启动 Spark 集群。

a. 进入 Hadoop 安装目录的 sbin 目录，启动 Hadoop 集群，如代码 2-53 所示。

代码 2-53 启动 Hadoop 集群

```
[root@master ~]#cd /usr/local/hadoop-2.6.4/sbin
[root@master sbin]#./start-dfs.sh
[root@master sbin]#./start-yarn.sh
```

```
[root@master sbin]#./mr-jobhistory-daemon.sh start historyserver
```

b. 启动 Hadoop 集群后，进入 Spark 安装目录的 sbin 目录，启动 Spark 集群，如代码 2-54 所示。

代码 2-54 启动 Spark 集群

```
[root@master sbin]# cd /usr/local/spark-1.6.3-bin-hadoop2.6/sbin
[root@master sbin]# ./start-all.sh
[root@master sbin]#./start-history-server.sh
```

（2）按“Ctrl+Alt+F”快捷键弹出文件传输框，将文件 words.txt 上传到 Linux 本地/opt/data 目录下，如图 2-40 所示。

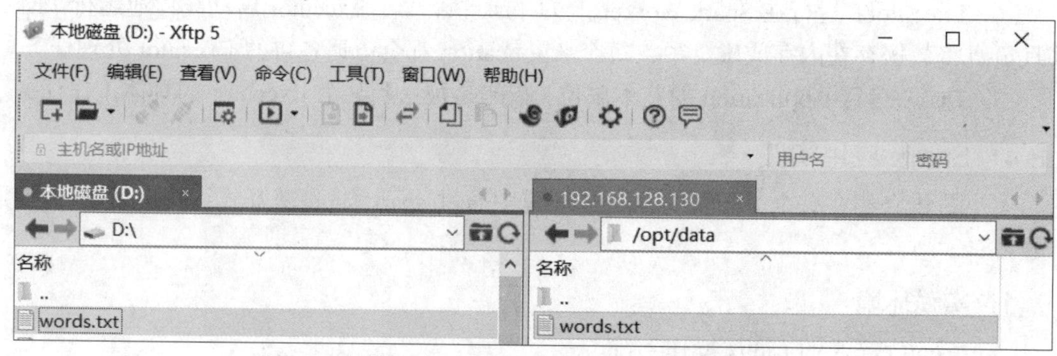

图 2-40 将文件 words.txt 上传到 Linux 本地目录下

（3）创建 HDFS 的目录/user/root/，将/opt/data 目录下的 words.txt 文件上传到 HDFS 的/user/root 目录下，如代码 2-55 所示。

代码 2-55 上传 words.txt 到 HDFS 的目录下

```
[root@ master ~]# hdfs dfs -mkdir -p /user/root/
[root@ master ~]# hdfs dfs -put /opt/data/words.txt /user/root/
```

（4）执行命令“hdfs dfs -mkdir /spark-logs”创建 HDFS 中的日志输出目录 spark-logs。

（5）进入 Spark 的安装目录，通过执行代码 2-56 所示的命令进入 spark-shell 交互式终端，执行命令“sc.setLogLevel("WARN")”，设置日志输出级别为“WARN”。

代码 2-56 进入 spark-shell 交互式终端

```
[root@ master ~]# cd /usr/local/spark-1.6.3-bin-hadoop2.6/sbin
[root@ master ~]#bin/spark-shell
scala> sc.setLogLevel("WARN")
```

（6）通过 textFile()方法读取数据创建弹性分布式数据集（Resilient Distributed Dataset，RDD），再通过 flatMap()方法根据空格分割数据，并使用 map()方法将数据转化为（单词,1）的键值对，通过 reduceByKey()方法汇总操作统计词频，如代码 2-57 所示。

代码 2-57 词频统计

```
scala>val wordcount=sc.textFile("words.txt").flatMap(x=>x.split(" ")).map(x=>
(x,1)).reduceByKey((a,b)=>a+b)
```

（7）通过 filter()方法过滤出单词词频大于 3 的记录，通过 repartition()方法重设分区数为 1，将结果存储到一个分区中，并将结果保存在 HDFS 的/user/root/wordcount 目录中，执行的命令如代码 2-58 所示。

代码 2-58 过滤出单词词频大于 3 的记录并保存到 HDFS 中的命令

```
scala>  wordcount.filter(x=>x._2>3).map(x=>x._1+","+x._2).repartition(1).
saveAsTextFile("/user/root/wordcount")
```

（8）在浏览器地址栏中输入网址“http://master:50070/dfshealth.jsp”，单击页面中的“Browse the filesystem”链接，进入/user/root/wordcount 目录，可查看单词词频统计结果，如图 2-41 所示。

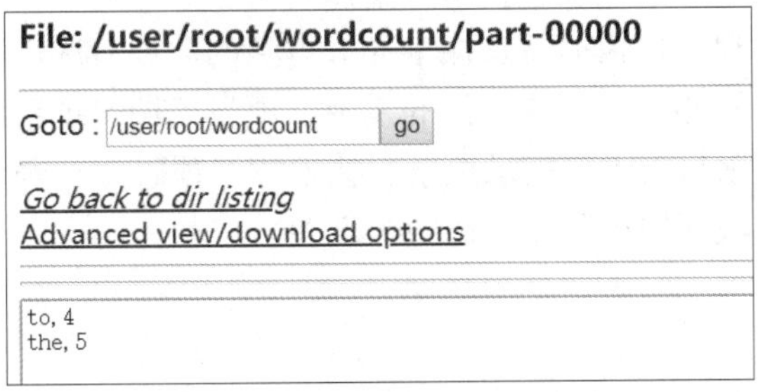

File: /user/root/wordcount/part-00000

Goto : /user/root/wordcount go

Go back to dir listing
Advanced view/download options

to, 4
the, 5

图 2-41 单词词频统计结果

小结

本章主要介绍了 Hadoop 框架、Hadoop 生态系统中 Hive 组件和 Spark 组件的基础知识，并通过应用实践帮助读者使用 Hadoop、Hive、Spark 进行简单的数据分析和存储。本章先介绍了 Hadoop 的发展历程、优点、生态系统常用组件和 Hadoop 集群的安装与配置，然后介绍了 Hadoop 的 HDFS、MapReduce 和 YARN 三大主要的核心框架，最后详细介绍了 Hadoop 生态系统中的 Hive 组件和 Spark 组件，内容包括组件的基本概念、特点、应用场景、安装配置和架构等。

第3章 广电大数据用户画像——需求分析

广电大数据用户画像项目实战主要是通过大数据技术对广电公司客户进行画像分析。项目前期涉及项目需求、需求探索、技术方案预研与确立等环节。项目中期需要先考虑数据传输、数据存储等过程，再根据需求探索环节得到的数据清洗规则对数据进行预处理；预处理之后就是项目的核心，即用户画像的实现。项目后期实现系统上线以及后期运维。本章将对项目前期中的项目需求、需求探索、技术方案预研 3 个环节进行详细分析，确定整个项目的任务指标、数据情况和技术选型等事项。

学习目标

（1）了解广电大数据用户画像项目的前期工作内容。

（2）了解广电大数据用户画像项目的项目背景和项目目标。

（3）熟悉项目需求探索的探索分析思路和过程。

（4）掌握对数据的基础探索和业务需求探索，总结数据的清洗处理规则并对数据进行预处理。

（5）熟悉项目前期的项目整体结构和项目所需的技术方案设计的过程。

3.1 项目需求

在广电大数据用户画像项目的项目需求环节，读者首先需要了解该项目的背景，了解广电行业目前的发展现状和潜在的危机，总结广电行业的需求，并根据项目背景提出项目目标。

3.1.1 项目背景

随着互联网技术的快速发展和应用扩展，国家积极推进并完成“三网融合”，三网融合是指电信网、广播电视网、互联网在向宽带通信网、数字电视网、下一代互联网演进过程中，通过技术改造，使技术趋于一致，实现三大网络互联互通、资源共享，为用户提供语音、数据和广播电视等多种服务。随着三网融合的深入推进和互联网电视（Internet Protocol Television，IPTV）的加速布局，OTT（Over The Top，是指越过运营商，通过互联网向用户提供各种应用服务）风起云涌，新媒体业务的飞速发展对传统媒体造成了巨大冲击。

复杂激烈的竞争环境，使广电的客户流失问题变得异常突出。如何减少客户流失、挽留客户并挖掘客户的潜在需求，是广电公司目前急需解决的问题！

在传统媒体播送时代，广电公司“不知道用户在哪里，不知道用户是谁，也不知道用户想看什么”，因此难以精准把握用户需求，而随着数字电视机顶盒等设备的普及，广电公司具备了获取用户基本信息数据、实时收视数据的能力。

现如今，广电公司已经积累了海量的用户数据，包括用户基本信息数据、用户收视数据、用户订单数据、用户账单数据等。因此，广电公司可以根据用户的特点，从人群、时间、地点、产品和付费方式 5 个维度来挖掘分析用户数据，对用户进行全面画像；从人群维度分析明确用户的年龄段，如少儿、青少年、青年、中年或老年等，以及分析收视语言是外语还是汉语等；从时间维度分析用户每天观看电视的时长或者用户观看某一电视节目的时长；从地点维度分析明确用户的收视常在地；从产品维度分析用户喜欢观看的电视频道，如点播频道、回看频道或直播频道等，以及用户喜欢观看的节目类型，如体育、电视剧、购物、少儿等；从付费方式维度分析用户是收费用户还是免费用户。

广电公司通过用户画像可以把握用户群体的特征和收视行为习惯，了解用户的实际特征和实际需求，并提供个性化、精准化和智能化的推荐服务，使用户获得更直接、更方便、更个性化的用户体验，以此来挽留用户，减少用户的流失。

3.1.2 项目目标

基于项目背景中涉及的利用海量数据进行用户画像，提出的项目目标如下。

（1）实时统计广电公司的订单信息，以此观察公司产品的销售趋势。

（2）挖掘分析用户相关数据，标签化用户数据，建立用户画像模型，可提供标签的增加和删除功能。

（3）用户画像可视化，将挖掘出来的用户标签在页面中展现出来。

（4）利用支持向量机（Support Vector Machine，SVM）算法建立分类模型，预测用户是否值得挽留，并将预测结果作为用户画像的一个标签。

（5）提出的系统架构能适应用户数据的大量增长，而无须调整系统架构，即支持动态横向扩展。

（6）任务实现自动化，任务可以自定义定时、编辑、监控任务状态等，并且要求设置每个月更新一次标签。

3.2 需求探索

在需求探索环节中，乙方需要先对甲方提供的业务数据有一定程度的了解，再对业务数据进行基础探索和业务需求探索，最后总结得到数据的清洗规则。

3.2.1 数据说明

在此项目中，甲方的业务数据表有用户基本信息表、用户状态信息变更表、账单信息表、订单信息表和用户收视行为信息表。这些表的数据由甲方业务人员从 Elasticsearch 集群中以 CSV 格式导出并提供给乙方人员，供乙方人员进行前期预研。首先，对各表进行简要的说明。

1. 用户基本信息表

用户基本信息表记录的是用户最新状态信息。用户基本信息表对应的 CSV 文件名为 mediamatch_usermsg.csv，文件大小约为 800MB，数据时间范围为 1991 年 1 月至 2018 年 6 月。用户基本信息表字段说明如表 3-1 所示。

表 3-1 用户基本信息表字段说明

字段	说明	字段	说明
terminal_no	用户地址编号	owner_code	用户等级编号
phone_no	用户编号	run_time	状态变更时间
sm_name	品牌名称	addressoj	完整地址
run_name	状态名称	estate_name	街道或小区地址
sm_code	品牌编号	open_time	开户时间
owner_name	用户等级名称	force	宽带是否生效

用户基本信息表的部分记录如图 3-1 所示。

phone_no	run_time	sm_name	run_name	terminal_no	owner_name	open_time
5132880	2013-05-31 11:59:22	模拟有线电视	正常	2000417991	HC级	2013-05-31 11:59:22
5162217	2013-07-29 17:45:38	模拟有线电视	正常	2000552769	HC级	NULL
5134817	2016-06-18 10:02:11	数字电视	正常	2000156870	HC级	2012-10-29 12:21:12
5163904	2018-01-18 08:51:58	珠江宽频	欠费暂停	2000157200	HC级	2013-08-10 15:08:12
5138439	2015-11-30 23:33:04	互动电视	主动暂停	2000404015	HC级	2012-11-28 21:12:44
5143217	2013-01-19 16:35:13	互动电视	正常	2000067688	HC级	2013-01-19 16:35:13
5143998	2015-06-19 10:06:57	互动电视	欠费暂停	2000543426	HC级	NULL
5145435	2014-07-21 16:06:21	数字电视	主动销户	2000542932	HC级	NULL
5146901	2013-02-20 11:52:12	互动电视	正常	2000276061	HC级	2013-02-20 11:52:12
5164344	2013-08-19 11:38:24	互动电视	正常	2000362522	HC级	2013-08-19 11:38:24

图 3-1 用户基本信息表的部分记录

2. 用户状态信息变更表

用户状态信息变更表记录的是用户所有时段的状态信息。用户状态信息变更表对应的 CSV 文件名为 mediamatch_userevent.csv，文件大小约为 2GB，数据时间范围为 1991 年 1 月至 2018 年 6 月。用户状态信息变更表字段说明如表 3-2 所示。

表 3-2 用户状态信息变更表字段说明

字段	说明	字段	说明
run_name	状态名称	sm_name	品牌名称
run_time	状态变更时间	open_time	开户时间
owner_code	用户等级编号	phone_no	用户编号
owner_name	用户等级名称	—	—

用户状态信息变更表的部分记录如图 3-2 所示。

run_name	run_time	owner_code	sm_name	open_time	phone_no	owner_name
正常	2014-02-20 16:45:47	00	模拟有线电视	2014-02-20 16:45:47	3514607	HC级
欠费暂停	2016-10-11 15:07:22	00	珠江宽频	2014-02-25 11:17:11	3514693	HC级
正常	2014-04-06 11:35:21	00	数字电视	2014-04-06 11:35:21	3515712	EE级
正常	2014-02-22 12:30:24	00	模拟有线电视	2014-02-22 12:30:24	3514827	HC级
欠费暂停	2015-02-28 10:35:47	00	数字电视	2014-03-08 10:49:58	3515835	HC级
正常	2014-02-23 14:39:44	00	模拟有线电视	2014-02-23 14:39:44	3514996	HC级
创建	2014-03-05 11:30:55	00	模拟有线电视	2014-03-05 18:45:23	3516482	HC级
正常	2016-07-21 10:43:13	00	互动电视	2014-03-09 14:21:22	3516941	HC级
欠费暂停	2015-07-21 11:38:29	00	数字电视	2014-03-08 10:49:58	3515835	HC级
欠费暂停	2015-09-16 10:20:29	00	数字电视	2014-03-10 12:37:01	3517045	HC级

图 3-2 用户状态信息变更表的部分记录

3. 账单信息表

账单信息表记录的是用户每月的账单信息，这些账单信息会在每月 1 日生成。账单信息表对应的 CSV 文件名为 mmconsume_billevents.csv，文件大小约为 2GB，数据时间范围为 2018 年 1 月至 2018 年 7 月。账单信息表字段说明如表 3-3 所示。

表 3-3 账单信息表字段说明

字段	说明
fee_code	费用类型
phone_no	用户编号
owner_code	用户等级编号
owner_name	用户等级名称
sm_name	品牌名称
year_month	账单时间
terminal_no	用户地址编号
favour_fee	优惠金额（+代表优惠，-代表额外费用），单位为元
should_pay	应付金额，单位为元

账单信息表的部分记录如图 3-3 所示。

year_month	terminal_no	sm_name	favour_fee	owner_code	should_pay	fee_code	phone_no	owner_name
2018-03-01 00:00:00	2000304671	互动电视	0.0	00	26.5	0B	1603021	HE级
2018-06-01 00:00:00	2000304671	互动电视	0.0	00	26.5	0B	1603021	HC级
2018-04-01 00:00:00	2000016684	数字电视	5.0	00	27.0	0Y	1603120	HC级
2018-07-01 00:00:00	2000355663	数字电视	0.0	00	5.0	0Y	1603318	HC级
2018-03-01 00:00:00	2000355663	数字电视	0.0	00	5.0	0Y	1603318	HE级
2018-04-01 00:00:00	2000355663	数字电视	0.0	00	5.0	0Y	1603318	HC级
2018-01-01 00:00:00	2000355663	数字电视	0.0	NULL	5.0	0Y	1603318	HE级
2017-12-01 00:00:00	2000355663	数字电视	0.0	NULL	5.0	0Y	1603318	HE级
2018-05-01 00:00:00	2000355137	数字电视	0.0	00	5.0	0Y	1603354	HC级
2017-12-01 00:00:00	2000355137	数字电视	0.0	NULL	5.0	0Y	1603354	HE级

图 3-3 账单信息表的部分记录

4. 订单信息表

订单信息表记录的是用户订购产品的信息。用户每订购一个产品，就会有相应的记录。

订单信息表对应的 CSV 文件名为 order_index_v3.csv，文件大小约为 7GB，数据时间范围为 2010 年 1 月至 2018 年 5 月。订单信息表字段说明如表 3-4 所示。

表 3-4　订单信息表字段说明

字段	说明
phone_no	用户编号
owner_name	用户等级名称
optdate	产品订购状态更新时间
prodname	订购产品名称
sm_name	品牌名称
offerid	订购套餐编号
offername	订购套餐名称
business_name	订购业务状态
owner_code	用户等级编号
prodprcid	订购产品名称（带价格）的编号
prodprcname	订购产品名称（带价格）
effdate	产品生效时间
expdate	产品失效时间
orderdate	产品订购时间
cost	订购产品价格
mode_time	产品标识，辅助标识电视主、附销售品
prodstatus	订购产品状态
run_name	状态名称
orderno	订单编号
offertype	订购套餐类别

订单信息表的部分记录如图 3-4 所示。

offerid	offername	owner_name	orderdate	prodname	expdate	business_name
GZ122216	互动标准包(副卡)	HC级	2015-01-30 09:44:21	标清直播基本包_广州	2050-01-01 00:00:00	欠费暂停状态
GZ122216	互动标准包(副卡)	HE级	2015-01-30 09:44:21	所有基本节目_时移	2050-01-01 00:00:00	欠费暂停状态
GZ122216	互动标准包(副卡)	HE级	2015-01-30 09:44:21	个人用户免费专区_点播	2050-01-01 00:00:00	正常状态
GZ122216	互动标准包(副卡)	HE级	2015-01-30 09:44:21	基本组_点播	2015-03-31 00:00:00	到期暂停状态
00118041	[互动]优惠购机(388元)(26.5元/月)	HE级	2013-12-02 15:10:03	个人用户免费专区_点播	2014-12-31 23:59:59	到期暂停状态
GZ101369	支持单片点播权限(按片付费)	HE级	2013-12-02 15:10:03	广州基本点播组	2050-01-01 00:00:00	正常状态
GZ122216	互动标准包(副卡)	HE级	2015-01-30 09:44:21	标清直播基本包_广州	2050-01-01 00:00:00	正常状态
GZ122560	互动+联合宽带-59元包	HE级	2016-01-19 11:07:02	精彩点_点播	2050-01-01 00:00:00	正常状态
GZ122560	互动+联合宽带-59元包	HE级	2016-01-19 11:07:02	宝贝家	2050-01-01 00:00:00	正常状态
GZ122560	互动+联合宽带-59元包	HE级	2016-01-19 11:07:02	应用类点播组	2050-01-01 00:00:00	正常状态

图 3-4　订单信息表的部分记录

5．用户收视行为信息表

用户收视行为信息表记录了用户观看电视的收视信息，其中，用户观看的媒体节目类型可分为直播、点播和回看，用户每切换一个频道就会生成一条新的记录。用户收视行为信息表对应的 CSV 文件名是 media_index_3m.csv，文件大小约为 30GB，数据时间范围为 2018 年 5 月至 2018 年 7 月。用户收视行为信息表字段说明如表 3-5 所示。

表 3-5 用户收视行为信息表字段说明

字段	说明
terminal_no	用户地址编号
phone_no	用户编号
duration	观看时长，单位为 ms
station_name	直播频道名称
origin_time	观看行为开始时间
end_time	观看行为结束时间
owner_code	用户等级
owner_name	用户等级名称
vod_cat_tags	视频点播（Video On Demand，VOD）节目包相关信息（该字段的数据类型为嵌套对象类型，即 nested object），按不同的节目包目录组织
resolution	点播节目的清晰度
audio_lang	点播节目的语言类别
region	节目地区信息
res_name	设备名称
res_type	媒体节目类型：0 表示直播，1 表示点播或回看
vod_title	VOD 节目名称
category_name	节目所属分类
program_title	直播节目名称
sm_name	品牌名称
first_show_time	第一次收视的时间

用户收视行为信息表的部分记录如图 3-5 所示。

terminal_no	phone_no	duration	station_name	origin_time	end_time	owner_code	owner_name
2000148366	5179844	434000	中央1台-高清	2018-10-13 22:18:21	2018-10-13 22:25:35	00	HC级
2000256434	1645503	660000	广东体育-高清	2018-10-13 21:19:00	2018-10-13 21:30:00	NULL	HE级
2000228389	1658031	3360000	广东体育-高清	2018-10-13 22:31:00	2018-10-13 23:27:00	00	HC级
2400214315	1709427	57000	湖北卫视-高清	2018-10-13 22:32:31	2018-10-13 22:33:28	00	HC级
2000212569	1405629	265000	上海纪实-高清	2018-10-13 22:22:57	2018-10-13 22:27:22	NULL	HE级
1200256367	1028894	1473000	广东体育-高清	2018-10-13 20:10:57	2018-10-13 20:35:30	00	HC级
12000499	1571196	1800000	CGTN	2018-10-13 23:30:00	2018-10-13 00:00:00	00	EE级
1200049788	1565737	76000	北京卫视-高清	2018-10-13 21:13:11	2018-10-13 21:14:27	00	HC级
1300173753	1993722	154000	北京纪实-高清	2018-10-13 23:44:38	2018-10-13 23:47:12	00	HC级
1100047425	2122375	3060000	湖北卫视-高清	2018-10-13 20:33:00	2018-10-13 21:24:00	00	HC级

图 3-5 用户收视行为信息表的部分记录

为了还原实际项目的环境，需要先把所有 CSV 文件导入 Elasticsearch 集群，再从 Elasticsearch 集群中将数据导入 Hive 的 user_profile 库。如何把 CSV 文件导入 Elasticsearch 集群和将 Elasticsearch 集群中的数据导入 Hive 可以参考 4.2 节，此处不论述。在 Hive 的 user_profile 库中，这 5 个表分别为 mediamatch_usermsg（用户基本信息表）、mediamatch_userevent（用户状态信息变更表）、mmconsume_billevents（账单信息表）、order_index_v3（订单信息表）、media_index_3m（用户收视行为信息表）。

3.2.2 基础探索

对甲方提供的业务数据有了一定程度的了解后，需要对业务数据进行基础探索，包括对总用户数、数据记录数和用户观看时长的统计，从而对数据基本情况有一个大致的了解，并对数据中可能出现的异常值进行探索分析。

1. 数据总体概述

（1）统计各表的记录数和用户基本信息表的用户数

为了初步了解各表的数据量及用户基本信息表的用户数量，现统计各表的数据记录数及用户基本信息表的用户数，如代码 3-1 所示。

代码 3-1 统计各表的记录数及用户基本信息表的用户数

```
// 从 Hive 中读取数据
scala> val usermsgData = sqlContext.sql("select * from user_profile.mediamatch_usermsg")
scala> val usereventData = sqlContext.sql("select * from user_profile.mediamatch_userevent")
scala> val billeventsData = sqlContext.sql("select * from user_profile.mmconsume_billevents")
scala> val orderData = sqlContext.sql("select * from user_profile.order_index_v3")
scala> val mediaData = sqlContext.sql("select * from user_profile.media_index_3m")
// 统计各表的记录数
scala> usermsgData.count
res0: Long = 5401493
scala> usereventData.count
res1: Long = 29454901
scala> billeventsData.count
res2: Long = 21097373
scala> orderData.count
res3: Long = 30973419
scala> mediaData.count
res4: Long = 254125817
// 统计用户基本信息表的用户数
scala> usermsgData.select("phone_no").distinct().count
res5: Long = 4401488
```

代码 3-1 的 res 统计结果显示，用户基本信息表的记录数约为 540 万，用户状态信息变更表的记录数约为 3000 万，账单信息表的记录数约为 2000 万，订单信息表的记录数约为 3000 万，用户收视行为信息表的记录数约为 2.5 亿，总用户数约为 440 万。从统计结果来看，用户基本信息表的记录数和用户数相差约 100 万，因此数据是有异常的，需要进行异常数据探索。

（2）统计用户收视记录中观看时长的均值及最值

为了获取用户收视记录中的观看时长的取值范围，以便为后续3.2.3小节业务需求探索中的用户收视行为无效数据的分析探索提供帮助，加之用户收视行为信息表的记录数比较大，因此有必要对用户收视记录中的观看时长进行基本的探索分析，如代码3-2所示。

代码3-2 统计用户收视记录中观看时长的均值及最值

```
scala> val mediaData = sqlContext.sql("select * from user_profile.media_
index_3m")
// 统计用户收视记录中观看时长的均值及最值
scala>mediaData.select(avg(col("duration")/1000).alias("avg_duration"),min
(col("duration")/1000).alias("min_duration"),max(col("duration")/1000).alias
("max_duration"),stddev(col("duration")/1000).alias("std_duration")).show
+-----------------+------------+------------+------------------+
|     avg_duration|min_duration|max_duration|      std_duration|
+-----------------+------------+------------+------------------+
|1076.755915126089|         0.0|     18230.0|1416.9481815514657|
+-----------------+------------+------------+------------------+
```

根据代码3-2的统计结果可以发现，用户收视行为信息表中平均每条记录的观看时长约为1076s（约18min），记录中观看时长最小值约为0s，最大值超过5h，标准差约为1417s（约24min）。统计结果说明了用户收视行为中观看时长的时间范围是比较大的（因为观看时长为0s～5h），观看时长的离散程度较小（因为观看时长的标准差约为24min）。

（3）统计用户月均收视时长

用户收视行为信息表记录了用户每次观看节目的时长，根据用户收视行为信息表的数据，可以统计每个用户平均每月的收视时长，以掌握每个用户对电视的依赖度。统计每个用户的月均收视时长，如代码3-3所示。

代码3-3 统计每个用户的月均收视时长

```
scala> val mediaData = sqlContext.sql("select distinct * from user_profile.
media_index_3m")
mediaData: org.apache.spark.sql.DataFrame = [terminal_no: string, phone_no:
string, duration: string, station_name: string, origin_time: string, end_time:
string, owner_code: string, owner_name: string, vod_cat_tags:array<struct<
level1_name:string,level2_name:string,level3_name:string,level4_name:string,
level5_name:string>>, resolution: string, audio_lang: string, region: string,
res_name: string, res_type: string, vod_title: string, category_name: string,
program_title: string, sm_name: string, first_show_time: string]
// 统计每个用户的月均收视时长
scala>  mediaData.groupBy("phone_no").agg((sum("duration")/(3*1000*60*60)).
alias("duration_avg"))
res7: org.apache.spark.sql.DataFrame = [phone_no: string, duration_avg:
double]
scala> val perUserDuration = \
mediaData.groupBy("phone_no").agg((sum("duration")/(3*1000*60*60)).alias
("duration_avg"))
perUserDuration: org.apache.spark.sql.DataFrame = [phone_no: string, duration_ avg:
double]
```

```
// 按统计出来的字段 duration_avg 降序排序并显示第 5 条数据
scala> perUserDuration.orderBy(col("duration_avg").desc).show(5)
+--------+------------------+
|phone_no|      duration_avg|
+--------+------------------+
| 5401487|51158.131944444445|
| 1557544|373.56305555555554|
| 2448946| 328.1268518518518|
| 1480023| 324.0813888888889|
| 1368105|308.89824074074073|
+--------+------------------+
```

根据代码 3-3 的统计结果可以发现，用户编号（phone_no）为 5401487 的用户月均收视时长约为 51158h，而一个月的最多时长为 24×31=744h，因此可知此用户的数据是异常数据。除此之外，用户月均收视时长最大约为 374h，这个数据是在合理范围内的。

2. 异常数据探索

根据甲方业务人员提供的信息和业务的实际情况，广电大数据用户画像项目中需要清洗的记录主要分为以下几类。

（1）重复记录的用户

经过数据总体概述中的数据统计后，发现用户基本信息表的记录数和用户数不相等，这说明用户基本信息表中的部分用户不止一条记录（本小节中的重复数据指的就是此类数据）。在正常的情况下，用户基本信息表中每个用户应只有一条记录，因此需要探索用户基本信息表中是否存在重复记录的用户，如代码 3-4 所示。

代码 3-4　探索用户基本信息表中重复记录的用户

```
scala> val usermsgData = sqlContext.sql("select * from user_profile.mediamatch_usermsg")
// 统计用户基本信息表中用户记录数大于 1 的个数
scala> usermsgData.groupBy("phone_no").count().filter("count>1").count
res5: Long = 999999
// 分组统计每个 phone_no 的记录数
scala>  usermsgData.groupBy("phone_no").count().orderBy(col("count").desc).
show(3)
+--------+-----+
|phone_no|count|
+--------+-----+
| 5401487|    8|
| 1048142|    2|
| 1134687|    2|
+--------+-----+
only showing top 3 rows
scala> usermsgData.filter("phone_no=1048142").show()
+-----------+--------+-------+--------+-------+----------+----------+-----
--------------+------------------+-----------+-------------------+-----+
|terminal_no|phone_no|sm_name|run_name|sm_code|owner_name|owner_code|
run_time|        addressoj|estate_name|          open_time|force|
+-----------+--------+-------+--------+-------+----------+----------+-----
```

```
-------------+-----------------+----------+-----------------+-----+
| 2000051647| 1048142| 模拟有线电视|      主动暂停|        a0|        HC 级|
00|2002-07-06 00:00:00|海珠区昌岗中路***号之*-***房|          NULL|1995-11-17
00:00:00| NULL|
| 1200239978| 1048142| 模拟有线电视|          正常|       a0|        HC 级|
00|2008-08-20 11:07:26|  天河区沙和路**号**栋***|       NULL|2008-08-20 11:08:02|
NULL|
+----------+-------+-------+-------+-------+---------+---------+-----
-------------+-----------------+----------+-----------------+-----+
```

根据代码 3-4 的统计结果可以发现，用户基本信息表中记录数大于 1 的用户有 999999 个，查看用户编号（phone_no）为 1048142 的用户数据，发现该用户的多条记录的状态变更时间（run_time）字段值是不同的。从统计的结果可以确定，用户基本信息表中的部分用户不止一条记录，而用户基本信息表的设计规定一个用户只能有一条记录，通过与甲方业务人员沟通后，了解到导致部分用户的记录重复的原因是数据管理员操作不当。甲方的业务人员提出的解决方案如下：根据用户编号分组保留状态变更时间最大值的记录来实现用户去重。根据甲方的业务人员提出的解决方案编写代码，如代码 3-5 所示。

代码 3-5 用户基本信息表中的用户去重实现

```
scala> val usermsgData = sqlContext.sql("select * from user_profile.mediamatch_ usermsg")
scala> val usermsgMaxTime = usermsgData.groupBy("phone_no").agg(max("run_time").
alias("run_time"))
scala> val usermsgFiltered = usermsgData.join(usermsgMaxTime,Seq("phone_no",
"run_time"))
scala> usermsgFiltered.groupBy("phone_no").count.filter("count>1").count
res33: Long = 246
scala>  usermsgFiltered.groupBy("phone_no").count.filter("count>1").show(3)
+--------+-----+
|phone_no|count|
+--------+-----+
| 1797345|    2|
| 1090952|    2|
| 1125935|    2|
+--------+-----+
only showing top 3 rows
scala> usermsgFiltered.filter("phone_no=1797345").show()
+--------+-------------------+----------+-------+--------+-------+-------
---+----------+--------------+-----------+--------------------+-----+
|phone_no| run_time|terminal_no|sm_name|run_name|sm_code|owner_name|owner_code|
addressoj|estate_name|open_time|force|
+--------+-------------------+----------+-------+--------+-------+-------
---+----------+--------------+-----------+--------------------+-----+
| 1797345|2001-09-01 00:00:00| 2000267786| 模拟有线电视|      正常|     a0|
HC 级|       00|海珠区江南大道北**号***房|       NULL|2001-09-06 00:00:00| NULL|
| 1797345|2001-09-01 00:00:00| 1300014043| 模拟有线电视|    主动销户|     a0|
HC 级|       00|            ****|       NULL|2001-09-01 00:00:00| NULL|
+--------+-------------------+----------+-------+--------+-------+-------
---+----------+--------------+-----------+--------------------+-----+
```

```
scala> usermsgFiltered.filter("sm_name!='模拟有线电视'")\
.groupBy("phone_no").count.filter("count>1").count
res34: Long = 0
```

根据代码 3-5 的统计结果可以发现，虽然用户基本信息表中只保留了每个用户状态变更时间最大值的记录，但是仍然还有 200 多个用户有重复的记录，查看其中某个用户的记录，发现其 run_time 字段的值是相同的。发现这个问题后，继续与甲方业务人员沟通，甲方的业务人员指出这 200 多个用户的品牌名称（sm_name）都是模拟有线电视，品牌名称为模拟有线电视的用户都是需要删除的，至于需要保留哪些品牌名称的用户数据，在后面将进行说明。将品牌名称是模拟有线电视的用户删除后，再根据用户编号分组统计记录数大于 1 的记录，得到的统计结果为 0。因此，通过保留每个用户的状态变更时间最大值的记录的方式来实现用户去重是可行的。

除了用户基本信息表有重复数据，在进行其他 4 张表探索时也出现了同样的问题，因此，在数据预处理时也应该删除其他 4 张表中重复的数据。

（2）特殊线路的用户

根据甲方的业务人员提供的信息，用户等级编号（owner_code）的值为 02、09 或 10 的记录是特殊路线的用户，他们负责测试、检验产品，对这些用户进行分析探索是没有意义的，因此需要删除这些用户的记录。因为甲方提供的 5 张表中都存在 owner_code 字段，所以需要对这 5 张表是否存在特殊线路的用户及数量进行探索分析，如代码 3-6 所示。

代码 3-6　统计各表 owner_code 的类型数

```
scala> val usermsgData = sqlContext.sql("select * from user_profile.mediamatch_usermsg")
scala> val usereventData = sqlContext.sql("select * from user_profile.mediamatch_userevent")
scala> val billeventsData = sqlContext.sql("select * from user_profile.mmconsume_billevents")
scala> val orderData = sqlContext.sql("select * from user_profile.order_index_v3")
scala> val mediaData = sqlContext.sql("select * from user_profile.media_index_3m")
// 统计各表的 owner_code 的类型数
scala> val usermsgCode = usermsgData.groupBy("owner_code").count()
scala> val usereventCode = usereventData.groupBy("owner_code").count()
scala> val billeventsCode = billeventsData.groupBy("owner_code").count()
scala> val orderCode = orderData.groupBy("owner_code").count()
scala> val mediaCode = mediaData.groupBy("owner_code").count()
// 用户基本信息表的 owner_code 类型结果
scala> usermsgCode.show()
+----------+-------+
|owner_code|  count|
+----------+-------+
|        00|5205530|
|        01|  58527|
|        02|   2352|
|        04|    663|
|        05|  26039|
|        06|   1788|
|        07|    307|
|        08|    134|
|        09|    701|
```

```
|         10|    328|
|         15|   4139|
|       NULL|  96779|
|         30|    806|
|         31|   3400|
+----------+-------+
```

根据代码 3-6 的统计结果可以发现，owner_code 字段确实存在 02、09 和 10 的值，它们所占的比例较小。此外，owner_code 还存在空值（NULL），经过与甲方业务人员沟通确认，owner_code 存在空值是正常的。因此，在数据预处理时，只需要清洗 owner_code 的值为 02、09 和 10 的记录。

（3）政企用户

由于甲方的用户主要是家庭用户，因此政企用户不纳入分析范围。根据甲方业务人员提供的信息，政企用户的标识是用户等级名称（owner_name）的值为“EA 级”“EB 级”“EC 级”“ED 级”或“EE 级”。根据 3.2.1 小节中 5 张表的信息可知，这 5 张表中都存在 owner_name 字段，因此需要探索这些表是否存在政企用户及其存在的数量，如代码 3-7 所示。

代码 3-7　统计各表 owner_name 的类型数

```
scala> val usermsgData = sqlContext.sql("select * from user_profile.mediamatch_usermsg")
scala> val usereventData = sqlContext.sql("select * from user_profile.mediamatch_userevent")
scala> val billeventsData = sqlContext.sql("select * from user_profile.mmconsume_billevents")
scala> val orderData = sqlContext.sql("select * from user_profile.order_index_v3")
scala> val mediaData = sqlContext.sql("select * from user_profile.media_index_3m")
// 统计各表的 owner_name 的类型数
scala>val usermsgOwnerName = usermsgData.groupBy("owner_name").count()
scala>val usereventOwnerName = usereventData.groupBy("owner_name").count()
scala>val billeventsOwnerName = billeventsData.groupBy("owner_name").count()
scala>val orderOwnerName = orderData.groupBy("owner_name").count()
scala>val mediaOwnerName = mediaData.groupBy("owner_name").count()
// 用户基本信息表的结果
scala> usermsgOwnerName.show(20)
+----------+-------+
|owner_name|  count|
+----------+-------+
|      EA级|   2225|
|      NULL|     50|
|      HE级|     24|
|      HC级|5140332|
|      HB级|   1689|
|      HA级|   1580|
|      EE级| 250864|
|      EB级|   4729|
+----------+-------+
// 用户状态信息变更表的结果
scala> usereventOwnerName.show(20)
```

```
+----------+-------+
|owner_name|  count|
+----------+-------+
|      EA级|   2225|
|      NULL|     50|
|      HE级|     24|
|      HC级|5140332|
|      HB级|   1689|
|      HA级|   1580|
|      EE级| 250864|
|      EB级|   4729|
+----------+-------+
// 账单信息表的结果
scala> billeventsOwnerName.show(20)
+----------+--------+
|owner_name|   count|
+----------+--------+
|      EA级|    7863|
|      HE级| 4185291|
|      HD级|       5|
|      HC级|16133563|
|      HB级|    5010|
|      HA级|    1040|
|      EE级|  747042|
|      EB级|   17559|
+----------+--------+
// 订单信息表的结果
scala> orderOwnerName.show(20)
+----------+--------+
|owner_name|   count|
+----------+--------+
|      HD级|       1|
|      EA级|   10996|
|      HC级| 6701443|
|      HA级|   25188|
|      EE级|  986492|
|      HE级|12113455|
|      HB级|    9535|
|      EB级|   22832|
+----------+--------+
// 用户收视行为信息表的结果
scala> mediaOwnerName.show(20)
+----------+---------+
|owner_name|    count|
+----------+---------+
|      EA级|    12614|
|      HE级| 21337359|
```

```
|       HC级|229947441|
|       HB级|    73178|
|       HA级|    73037|
|       EE级|  2682066|
|       EB级|      122|
+----------+---------+
```

代码 3-7 的统计结果显示，各表中 owner_name 字段的值为“EA 级”“EB 级”“EC 级”“ED 级”“EE 级”的总记录数相对较少，即政企用户在各表中的数量较少，这也说明了甲方的用户主要是家庭用户。虽然目前只出现了 owner_name 字段的值为“EA 级”“EB 级”和“EE 级”政企用户的记录，但是后续的数据中可能会出现 owner_name 为“EC 级”“ED 级”的记录，因此，在数据预处理时，需要清洗 owner_name 字段的值为“EA 级”“EB 级”“EC 级”“ED 级”和“EE 级”的政企用户。

3. 筛选主要业务类型的数据

甲方目前的业务类型主要是数字电视、互动电视、甜果电视和珠江宽频这 4 种，品牌名称可以通过 sm_name 字段标识。统计用户基本信息表中 sm_name 的所有业务类型及其每种业务类型的用户数，如代码 3-8 所示。

代码 3-8　统计用户基本信息表中 sm_name 的所有业务类型及其每种业务类型的用户数

```
// 统计总记录数
scala>val nums = usermsgData.count
// 统计用户基本信息表中 sm_name 各业务类型的数量和占比
scala> usermsgData.groupBy("sm_name").count().withColumn("percent",col("count")/
nums).show
+----------+-------+---------------------+
|   sm_name|  count|              percent|
+----------+-------+---------------------+
|模拟有线电视|1823856|   0.3376577549947764 6|
|    互动电视| 547834|   0.1014226992425983|
|    数字电视|1697136|   0.3141975746335319|
|      NULL|     50| 9.256699953142585E-6|
|      番通|      3| 5.554019971885551E-7|
|    甜果电视| 337068|   0.0624027467961173|
|    珠江宽频| 994190|  0.18405837052829652|
|  移动互动用户|   1356| 2.510417027292269E-4|
+----------+-------+---------------------+
```

代码 3-8 的统计结果显示，品牌名称（包含空值）共有 8 种，模拟有线电视的用户最多，约占总数的 34%。因为甲方以前的主要业务类型是模拟有线电视，而现在的主要业务类型是互动电视、数字电视、甜果电视、珠江宽频这 4 种，约占总数的 65%，所以在数据预处理时需要保留这 4 种品牌名称的用户数据，删除其他品牌名称的用户数据。

根据甲方的业务要求，除了筛选指定品牌名称的用户数据，还要对用户状态名称进行过滤，只保留状态名称为正常、欠费暂停、主动暂停和主动销户的用户数据，其余的不需要进行分析处理。状态名称的字段标识为 run_name，对用户基本信息表中的用户状态进行探索，如代码 3-9 所示。

代码 3-9　统计用户基本信息表中用户状态各类型的记录数

```
// 统计用户基本信息表中 run_name 的各类型的记录数
scala> usermsgData.groupBy("run_name").count().show()
+--------+--------+
|run_name|   count|
+--------+--------+
|  被动销户|   34796|
|  主动销户|  431299|
|      创建|    8962|
|  主动暂停|  598748|
|      冲正|  249468|
|      正常| 3645073|
|    NULL|      50|
|      销号|      68|
|    未激活|     457|
|  欠费暂停|  432572|
+--------+--------+
```

代码 3-9 的统计结果显示，状态名称（包含空值）共有 10 种类型，只保留其中正常、欠费暂停、主动暂停和主动销户的用户数据，其余的状态名称对应的数据不需要进行分析处理。

3.2.3　业务需求探索

基于甲方提供的数据，对数据进行基础探索后，还需要结合项目目标对数据进行业务需求探索。用户群体的特征、收视行为习惯、消费水平特征等，都能作为用户画像的标签使用。因此，下面将从用户收视行为无效数据、消费水平、入网程度 3 个方面对数据进行业务需求探索分析。

1. 用户收视行为无效数据

用户收视行为无效数据是指用户观看时长过短或过长的记录，导致这种现象的原因可能是用户频繁切换频道或者只关闭电视机而忘记关闭机顶盒。在用户收视行为信息表中，duration 字段记录了用户每次的观看时长。因为每个用户的观看时长值的大小差异较大，所以需要把观看时长以小时为区间来划分，统计各区间的记录数，如代码 3-10 所示。

代码 3-10　以小时为区间统计观看时长的记录数

```
scala> val mediaData = sqlContext.sql("select * from user_profile.media_index_3m")
// 统计总记录数
scala>val total= mediaData.count.toDouble
// 增加一个字段 hours：将观看时长转换为以小时为区间并向下取整
scala> val mediaHours= mediaData.withColumn("hours",floor(col("duration")/
(1000*60*60)))
// 统计各 hours 值的记录数和各类所占的比例
```

```
scala> val hoursNum = mediaHours.groupBy("hours").count().withColumn("percent",
col("count")/total)
// 结果显示
scala> hoursNum.show
+-----+---------+--------------------+
|hours|    count|             percent|
+-----+---------+--------------------+
|    0|239480472|  0.9423697081512974|
|    1| 14163099|0.055732625544298794|
|    2|   270098|0.001062851477227...|
|    3|   130100|5.119511332451515E-4|
|    4|    82034|3.228086031101673E-4|
|    5|       14|5.509082140993176E-8|
+-----+---------+--------------------+
```

代码 3-10 的统计结果显示，绝大部分记录的观看时长小于 1h，约占总记录数的 94%；观看时长大于或等于 1h 且小于 2h 的记录数约占总数的 5.6%；观看时长大于或等于 5h 且小于 6h 的记录数最少，只有 14 条记录。观看时长小于 1h 的记录数占了绝大部分，需要将这部分记录以分钟为一个区间来划分，分析每个区间的记录数的分布情况，如代码 3-11 所示。

代码 3-11 对观看时长小于 1h 的数据以分钟为区间统计记录数

```
// 筛选观看时长小于 1h 的记录
scala>val oneHour = mediaHours.where("hours=0")
// 增加一个字段 minutes：将观看时长转换为以分钟为区间并向下取整
scala> val mediaMinutes = oneHour.withColumn("minutes",floor(col("duration")/
(1000*60)))
// 统计各 minutes 值的记录数
scala>val minutesNum = mediaMinutes.groupBy("minutes").count().withColumn
("percent",col("count")/total)
// 按字段 minutes 升序排列并显示前 10 条结果
scala>minutesNum.orderBy(col("minutes").asc).show(10)
+-------+--------+--------------------+
|minutes|   count|             percent|
+-------+--------+--------------------+
|      0|46318317|  0.18226529498968616|
|      1|24647648|   0.0969899410102044|
|      2|15301880|0.060213795594014755|
|      3|11467423|  0.04512498232322456|
|      4| 8909929|0.035061093379583706|
|      5|10313389|0.040583790823582475|
|      6| 6256932|0.024621394527577653|
|      7| 5792865|0.022795263654774595|
|      8| 5334246|0.020990570981617346|
|      9| 4682389|0.018425475440773498|
+-------+--------+--------------------+
only showing top 10 rows
```

代码 3-11 的统计结果显示，用户观看时长小于 1h 的各区间分布的记录数较多，为了方便观察各区间的数据分布情况，使用图表的形式展示所有的结果，如图 3-6 所示。

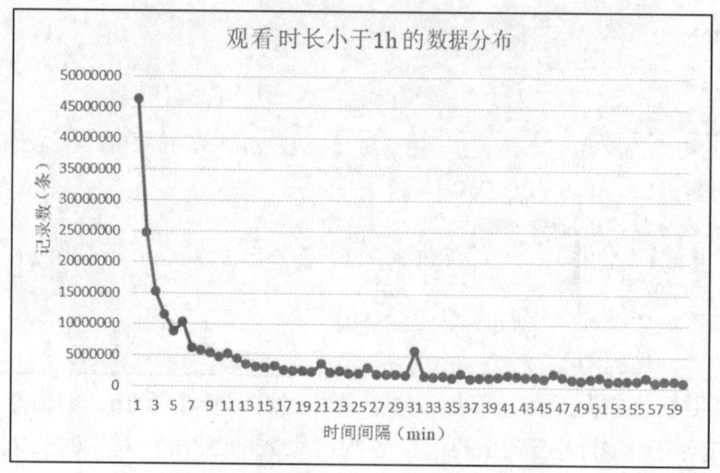

图 3-6 观看时长小于 1h 的数据分布

从图 3-6 中可以看到，观看记录数随着时间间隔的增大而整体呈现出指数递减的趋势，其中观看时长小于 1min 的数据最多。从代码 3-11 的统计结果可以看到观看时长小于 1min 的记录数约占总记录数的 18%。为了进一步了解观看时长小于 1min 的秒级分布情况，需把观看时长小于 1min 的这部分数据以秒为区间来划分，如代码 3-12 所示。

代码 3-12 对观看时长小于 1min 的数据以秒为区间统计记录数

```
// 筛选观看时长小于 1min 的记录
scala>val oneMinute = mediaMinutes.where("minutes=0")
// 新增字段 seconds：将观看时长转换为以秒为区间并向下取整
scala> val mediaSeconds = oneMinute.withColumn("seconds",floor(col("duration")/
1000))
// 统计各 seconds 值的记录数
scala>val  secondsNum  =  mediaSeconds.groupBy("seconds").count().withColumn
("percent",col("count")/total)
// 按字段 seconds 升序排列并显示前 10 条结果
scala>secondsNum.orderBy(col("seconds").asc).show(10)
+-------+-------+--------------------+
|seconds|  count|             percent|
+-------+-------+--------------------+
|      0|1053832|0.004146890750576...|
|      1| 149695|5.890586079256953E-4|
|      2| 139496|5.489249445285601E-4|
|      3| 140227|5.518014724178929E-4|
|      4| 138633| 5.45528988894505E-4|
|      5| 138962|5.468236231976383E-4|
|      6| 138762|5.460366114632108E-4|
|      7| 139463|5.487950875923795E-4|
|      8| 139296|5.481379327941325E-4|
|      9| 139841|5.502825397704477E-4|
+-------+-------+--------------------+
only showing top 10 rows
```

代码 3-12 的统计结果较多，展现的统计结果只显示前 10 条。为了显示所有结果的分布情况，这里使用图表的形式展示，如图 3-7 所示。

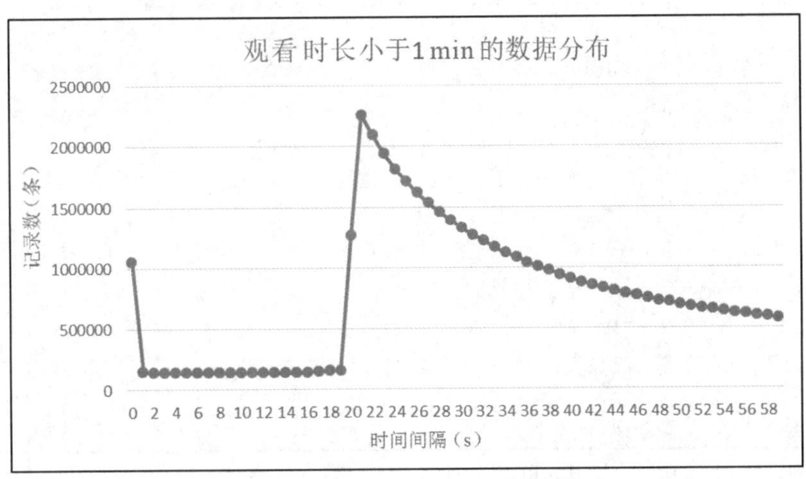

图 3-7 观看时长小于 1min 的数据分布

从图 3-7 中可以看出，观看时长小于 1s 的记录数远高于相邻区间的记录数；观看时长为 1～19s 的每个区间内的观看记录数相差不大，从观看时长为 20s 开始，每个区间的记录数远高于 1～19s 中每个区间的记录数。

综合以上用户观看时长分析统计结果并结合业务的实际情况，把观看时长小于 20s 和观看时长大于 5h 的数据视为无效数据，将这些无效数据排除能够更好地分析用户的收视行为。

在用户收视行为信息表中，还有一部分数据是 res_type=0 时，origin_time 和 end_time 的值结尾为 00 的记录，这些记录是机顶盒自动返回的数据，并不是用户真实的观看记录，属于无效数据，因此，这一部分数据是需要删除的。查询用户收视行为信息表中的无效数据，如代码 3-13 所示。

代码 3-13 查询用户收视行为信息表中的无效数据

```
// 查询 res_type=0，origin_time 和 end_time 结尾为 00 的记录
scala> val invalidData = mediaData.filter("res_type=0").filter(col("origin_ time").
endsWith("00") \
and col("end_time").endsWith("00"))
scala> invalidData.count
res3: Long = 44890060
// 结果显示
scala> invalidData .show(2,false)
+-----------+--------+--------+-----------------+-----------------------+----------
---------+----------+----------+--------------+----------+----------+-------+
--------+--------+-----------+----------------+--------------+-------+----------
-----+
|terminal_no|phone_no|duration|station_name|origin_time                |end_time
|owner_code|owner_name|vod_cat_tags|resolution|audio_lang|region|res_name|
res_type|vod_title|category_name|program_title|sm_name|first_show_time|
```

```
+----------+--------+--------+------------+-------------------+----------
---------+---------+---------+-----------+----------+---------+------+
--------+-------+---------+------------+-----------+------+----------
-----+
|1200298620 |1448938 |2400000 |东方卫视-高清    |2018-05-25 20:23:00|2018-05-25
21:03:00|00        |HC 级      |null        |NULL      |NULL      |NULL  |NULL
|0       |NULL      |NULL         |归去来(23)     |互动电视   |NULL          |
|1200298620 |1448938 |3900000 |东方卫视-高清    |2018-05-27 02:39:00|2018-05-27
03:44:00|00        |HC 级      |null        |NULL      |NULL      |NULL  |NULL
|0       |NULL      |NULL         |梦想改造家Ⅲ(5)    |互动电视   |NULL          |
+----------+--------+--------+------------+-------------------+----------
---------+---------+---------+-----------+----------+---------+------+
--------+-------+---------+------------+-----------+------+----------
-----+
only showing top 2 rows
```

根据代码 3-13 的统计结果可以发现，当用户收视行为信息表中的 res_type=0 时，origin_time 和 end_time 的值结尾为 00 的记录的确存在，并且记录数约为 4489 万，因此在数据预处理时要清洗这部分无效的数据。

2. 消费水平标签阈值探索

账单信息表反映了用户每个月的消费情况，对该表的数据进行分析探索能够更好地了解用户的消费行为，制订消费水平标签的子标签及各子标签的判断阈值，从而给每个用户标注合适的消费水平标签。因为电视用户和宽带用户的业务及费用不同，所以需要单独分析探索这两种用户的消费情况。在统计分析中，为了减小误差，增强数据的稳定性，乙方选择 2018 年 1 月至 2018 年 7 月的数据，计算用户的月均消费金额，以此制订消费水平标签的子标签及其标签阈值。

先对电视用户的账单数据进行探索分析，分组统计每个用户的月均消费金额，再对所有用户的月均消费金额进行基本的统计分析，如代码 3-14 所示。

代码 3-14　统计电视用户月均消费金额

```
// 筛选电视用户
scala> val tvBilleventsData = billeventsData.filter("sm_name like '%电视% and
sm_name !='模拟有线电视''")
// 统计 2018 年 1 月—2018 年 7 月平均每月的每个用户的消费金额
scala>val avgTVBillData = \
tvBilleventsData.groupBy("phone_no").agg((sum(col("should_pay")-col("favour_
fee"))/7).alias("avg_fee"))
// 统计消费金额的最大值、最小值、平均值和标准差
scala>avgTVBillData.select(max("avg_fee"),min("avg_fee"),avg("avg_fee"),stddev
("avg_fee")).show
+-----------------+------------+------------------+----------------------+
|     max(avg_fee)|min(avg_fee)|      avg(avg_fee)|stddev_samp(avg_fee,0,0)|
+-----------------+------------+------------------+----------------------+
|27921.93999999999|     -7909.8|27.580595429910094|      43.24026183195025|
```

```
+----------------+-----------+----------------+--------------------+
// 每月消费金额按 10 的间隔划分
scala> val rangeTVBillData =  avgTVBillData.withColumn("range_fee",floor(col
("avg_fee")/10)*10)
// 按 range_fee 分组统计记录数
scala> val rangeTVCount = rangeTVBillData.groupBy("range_fee").count()
// 按 range_fee 升序显示前 30 条结果
scala> rangeTVCount.orderBy(col("range_fee").asc).show(30)
+---------+------+
|range_fee| count|
+---------+------+
|    -7910|     1|
|    -4880|     1|
|     -270|     2|
......
|       80|  7342|
|       90|  2840|
|      100|  1897|
+---------+------+
only showing top 30 rows
```

根据代码 3-14 的统计结果可以看到，消费金额的标准差约为 43，说明不同电视用户的平均消费金额相差较大。为了探索电视用户月均消费金额的区间分布情况，选择以 10 作为区间的间隔，统计消费金额的区间分布情况，如图 3-8 所示。

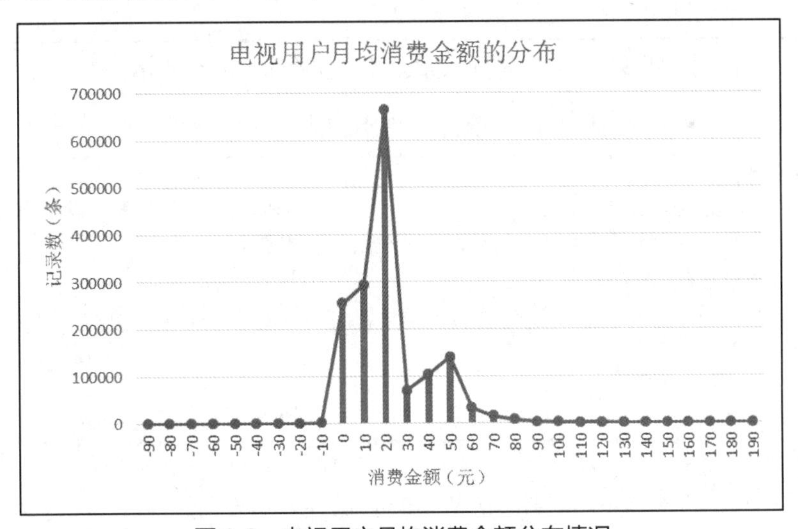

图 3-8　电视用户月均消费金额分布情况

从图 3-8 中可以看出，有些消费金额是负数，就此问题咨询了甲方业务人员，甲方业务人员告知一些优惠方式会导致消费金额出现负数，此种情况属于正常情况。根据图 3-8 可以发现，电视用户月均消费金额大致呈正态分布，主要集中在 0～50（不含）元，其中占比最大的是大于或等于 20 元且小于 30 元的消费区间，这是因为大部分电视用户每月都只支付基本电视费 26.5 元。依据图 3-8 所示电视用户月均消费金额的分布情况，筛选电视用户月均消费金额大于或等于-10 元且小于或等于 90 元的用户数据，统计这部分数据的用户月均消费金额的平均值和标准差，如代码 3-15 所示。

代码 3-15　筛选后的电视用户月均消费金额的平均值和标准差

```
scala> avgTVBillData.filter("avg_fee>=-10 and avg_fee<=90").select(avg("avg_fee"),
stddev("avg_fee")).show
+------------------+------------------------+
|      avg(avg_fee)|stddev_samp(avg_fee,0,0)|
+------------------+------------------------+
|26.914777281258274|       16.70728501623327|
+------------------+------------------------+
```

根据代码 3-15 的统计结果可以发现，月均消费金额为–10~90 元的电视用户月均消费金额的平均值约为 26.9，标准差约为 16.7。根据业务特点，结合月均消费金额的平均值和标准差，以月均消费 26.5 元为基础消费，以标准差的向上取整十数 20 作为浮动阈值，制订 4 个电视消费水平子标签及规则，如表 3-6 所示。

表 3-6　电视消费水平子标签及规则

父标签	子标签	标签规则	备注
电视消费水平	电视超低消费	$-26.5<X<26.5$	X 表示电视用户月均消费金额，单位为元
	电视低消费	$26.5\leqslant X<26.5+20$	
	电视中等消费	$26.5+20\leqslant X<26.5+40$	
	电视高消费	$26.5+40\leqslant X$	

同理，对宽带用户的账单数据进行探索分析，如代码 3-16 所示。

代码 3-16　统计宽带用户月均消费金额

```
// 筛选宽带用户
scala>val netBilleventsData = billeventsData.filter("sm_name like '%珠江宽频%'")
// 统计 2018 年 1 月—2018 年 7 月平均每月的每个用户的消费金额
scala>val avgNetBillData = \
netBilleventsData.groupBy("phone_no").agg((sum(col("should_pay")-col("favour_
fee"))/7).alias("avg_fee"))
// 自定义（50%）分位数函数
scala> import org.apache.commons.math3.stat.descriptive.rank.Percentile
scala> val _50_P = new Percentile(50.0)
scala>  val _50_udf = \
udf{(arr: scala.collection.mutable.WrappedArray[Double]) => _50_P.evaluate
(arr.sorted.toArray)}
// 统计消费金额的最大值、最小值、平均值、标准差和中位数
scala>avgNetBillData.select(max("avg_fee"),min("avg_fee"),avg("avg_fee"),
stddev("avg_fee"),_50_udf(collect_list("avg_fee")).alias("median")).show
+------------+------------+-----------------+------------------------+------------------+
|max(avg_fee)|min(avg_fee)|     avg(avg_fee)|stddev_samp(avg_fee,0,0)|            median|
+------------+------------+-----------------+------------------------+------------------+
|      3800.0|      -566.0|32.49198003325578|       24.91242407676457|26.625714285714285|
+------------+------------+-----------------+------------------------+------------------+
```

```
// 每月消费金额按 10 的间隔划分
scala> val rangeNetBillData =  avgNetBillData.withColumn("range_fee",floor
(col("avg_fee")/10)*10)
// 按 range_fee 分组统计记录数
scala> val rangeNetCount = rangeNetBillData.groupBy("range_fee").count()
// 按 range_fee 升序显示前 30 条结果
scala> rangeNetCount.orderBy(col("range_fee").asc).show(30)
+---------+-----+
|range_fee|count|
+---------+-----+
|     -570|    1|
|     -140|    1|
|      -30|    5|
|      -20|   16|
|      -10|   84|
|        0|35792|
|       10|81299|
|       20|86505|
|       30|56461|
|       40|48672|
|       50|27760|
|       60|13166|
|       70|16399|
|       80| 4804|
|       90| 1732|
|      100|  318|
|      110|  548|
|      120| 1654|
|      130|  932|
|      140|  420|
|      150|  320|
|      160|  123|
|      170|   45|
|      180|    8|
|      190|    5|
|      200|    8|
|      210|    2|
|      230|    2|
|      270|    2|
|      280|    1|
+---------+-----+
only showing top 30 rows
```

根据代码 3-16 的统计结果可以发现，宽带用户的月均消费金额的最大值和最小值相差较大，在不过滤相关记录的情况下，宽带用户的月均消费金额的平均值约为 32，标准差约为 25，中位数约为 26。

代码 3-16 中宽带用户有不同的月均消费金额，统计结果较多，为了便于观察各区间的数据分布情况，使用图表的形式展示部分结果，如图 3-9 所示。

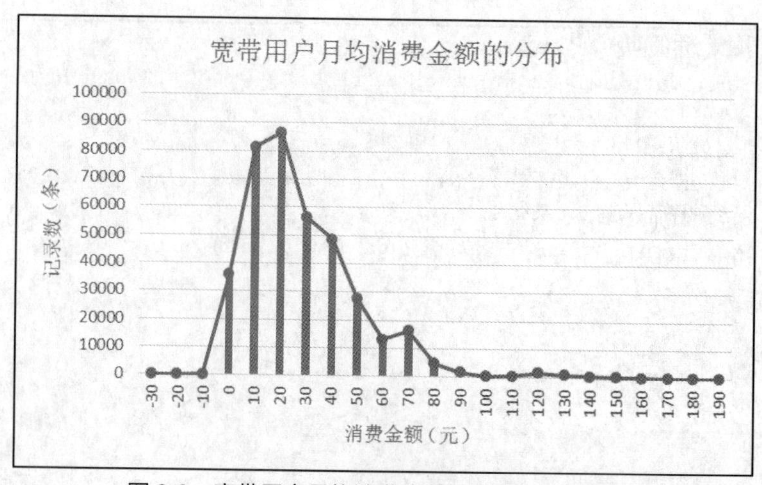

图 3-9　宽带用户月均消费金额分布情况（部分）

图 3-9 展示的是以 10 作为区间间隔的宽带用户部分月均消费金额分布情况，可以看到数据大致呈正态分布，大部分的宽带用户的消费金额集中在 0～90 元，其中消费金额为 10～20 元的用户最多。为了使宽带用户月均消费金额的均值和标准差更加稳定，过滤消费金额小于 0 元且大于或等于 90 元的记录后，再求宽带用户月均消费金额的均值、标准差和中位数，如代码 3-17 所示。

代码 3-17　过滤后宽带用户月均消费金额的统计

```
// 过滤宽带用户月均消费金额大于或等于 0 元且小于 90 元的记录，再统计均值、标准差和中位数
scala> avgNetBillData.filter("avg_fee>=0 and avg_fee<90") \
.select(avg("avg_fee"),stddev("avg_fee"),_50_udf(collect_list("avg_fee")).
alias("median")).show
+-----------------+----------------------+------------------+
|     avg(avg_fee)|stddev_samp(avg_fee,0,0)|            median|
+-----------------+----------------------+------------------+
|30.996529615079027|     19.14213341709401|25.524285714285718|
+-----------------+----------------------+------------------+
```

根据代码 3-17 的统计结果可以发现，过滤部分记录后宽带用户月均消费金额的均值约为 30，标准差约为 20，中位数约为 25。因此，选择 25 元作为宽带用户的基础消费，以标准差 20 作为浮动阈值，制订宽带消费水平的子标签及规则，如表 3-7 所示。

表 3-7　宽带消费水平的子标签及规则

父标签	子标签	标签规则	备注
宽带消费水平	宽带低消费	$Y \leq 25$	Y 表示宽带用户月均消费金额，单位为元
	宽带中消费	$25 < Y \leq 45$	
	宽带高消费	$Y > 45$	

3．入网程度标签阈值探索

在用户基本信息表中，用户的开户时间（open_time）字段记录了用户的开户时间，利用此字段信息可以给用户标注用户入网程度标签（子标签包含老用户、中等用户和新用户）。

具体的入网程度子标签阈值需要通过对 open_time 字段的数据进行分析统计才能确定。用户分为电视用户和宽带用户两种，这两种用户的业务属性不同，需要单独分析这两种用户的入网时长的特征，从而确定子标签的阈值。

下面先对用户基本信息表中的电视用户进行探索分析。因为模拟有线电视的用户不在分析范围内，所以在筛选电视用户的过程中需要过滤模拟有线电视的用户。在探索分析中，需要把 open_time 字段的值与当前时间相减，把差值转化为以年为单位的值，并统计所有用户入网时长的最值、均值、30%分位数和中位数，最后分组统计各入网时长值的用户数，如代码 3-18 所示。

代码 3-18　统计分析电视用户入网时长相关值

```
// 筛选电视用户
scala>val tvUsermsgData = usermsgData.filter("sm_name like '%电视%' and sm_name !=
'模拟有线电视'")
// 过滤 open_time 为空值的记录
scala>val tvFilteredUsermsgData = tvUsermsgData.filter("open_time != 'NULL'")
// 把用户当前时间与开户时间相减并向下取整（单位为年）
scala>val yearTVUserData = \
tvFilteredUsermsgData.groupBy("phone_no").agg(max(floor(datediff(current_d
ate(),col("open_time"))/365)).alias("years"))
// 自定义 30%分位数、中位数
scala> import org.apache.commons.math3.stat.descriptive.rank.Percentile
// 30%分位数
scala> val _30_P = new Percentile(30.0)
scala> val _30_udf = \
udf{(arr: scala.collection.mutable.WrappedArray[Double]) => _30_P.evaluate
(arr.sorted.toArray)}
// 中位数
scala>  val _50_P = new Percentile(50.0)
scala> val _50_udf = \
udf{(arr: scala.collection.mutable.WrappedArray[Double]) => _50_P.evaluate
(arr.sorted.toArray)}
// 统计入网时长的最值、均值、30%分位数和中位数
scala>yearTVUserData.select(max("years"),min("years"),avg("years"),stddev(
"years"),_30_udf(collect_list(col("years").cast("double"))).alias("30%"),_
50_udf(collect_list(col("years").cast("double"))).alias("median")).show
+----------+----------+-----------------+----------------------+---+------+
|max(years)|min(years)|       avg(years)|stddev_samp(years,0,0)|30%|median|
+----------+----------+-----------------+----------------------+---+------+
|        26|         0|7.002325792128553|     4.088445968221869|4.0|   8.0|
+----------+----------+-----------------+----------------------+---+------+
// 分组统计每个入网时长值的用户数
scala> val yearTVUserCount=yearTVUserData.groupBy("years").count
// 显示所有结果
scala> yearTVUserCount.show(30)
+-----+------+
|years| count|
```

```
+-----+------+
|    0|165106|
|    1|236043|
|    2|124335|
|    3|117747|
......
|   24|    60|
|   25|    22|
|   26|     1|
+-----+------+
```

根据代码 3-18 的统计结果可以发现，电视用户入网时长的平均值约为 7，标准差约为 4，30%分位数约为 4，中位数约为 8。

代码 3-18 中不同入网时长的用户数统计结果较多，因此使用图表的形式展示结果，如图 3-10 所示。

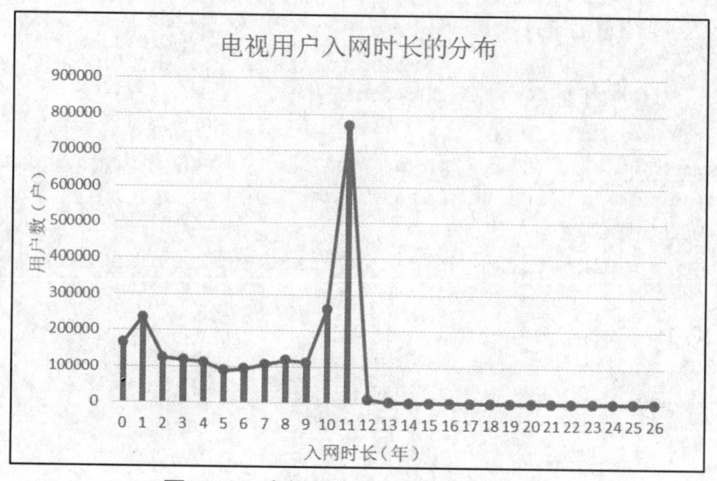

图 3-10　电视用户入网时长分布情况

从图 3-10 中可以发现电视用户的入网时长大部分都分布在 0～11 年，其中入网时长为 11 年的用户最多。根据统计结果并结合实际的业务场景，选择以 30%分位数 4 年为电视用户的入网时长的最低临界值，以标准差 4 为浮动阈值，制订 3 个电视用户入网程度子标签及规则，如表 3-8 所示。

表 3-8　电视用户入网程度子标签及规则

父标签	子标签	标签规则	备注
电视用户入网程度	新用户	$T \leqslant 4$	T表示电视用户入网时长，单位为年
	中等用户	$4 < T \leqslant 8$	
	老用户	$T > 8$	

在对宽带用户入网时长的统计分析中，主要统计分析宽带用户入网时长的最值、均值、标准差、中位数及其分布情况，如代码 3-19 所示。

代码 3-19 统计分析宽带用户入网时长相关值

```
// 筛选宽带用户
scala> val netUsermsgData = usermsgData.filter("sm_name='珠江宽频'")
// 过滤 open_time 为空值的记录
scala> val netFilteredUsermsgData = netUsermsgData.filter("open_time != 'NULL'")
// 把用户当前时间与开户时间相减并向下取整（单位为年）
scala> val yearNetUserData = \
netFilteredUsermsgData.groupBy("phone_no").agg(max(floor(datediff(current_
date(),col("open_time"))/365)).alias("years"))
// 统计入网时长的最值、均值、标准差和中位数
scala>yearNetUserData.select(max("years"),min("years"),avg("years"),stddev
("years"),_50_udf(collect_list(col("years").cast("double"))).alias("median
")).show
+----------+----------+-----------------+---------------------+------+
|max(years)|min(years)|       avg(years)|stddev_samp(years,0,0)|median|
+----------+----------+-----------------+---------------------+------+
|        18|         0|3.887968483816014|     3.713132149873835|   2.0|
+----------+----------+-----------------+---------------------+------+
// 分组统计每个入网时长值的用户数
scala> val yearNetUserCount=yearNetUserData.groupBy("years").count
// 显示所有结果
scala> yearNetUserCount.show(20)
+-----+------+
|years| count|
+-----+------+
|    0|121069|
|    1|255305|
|    2|123113|
......
|   16|   140|
|   17|     9|
|   18|     4|
+-----+------+
```

代码 3-19 中不同入网时长的用户数的统计结果较多，为了使统计结果易于观察，使用图表的形式展示结果，如图 3-11 所示。

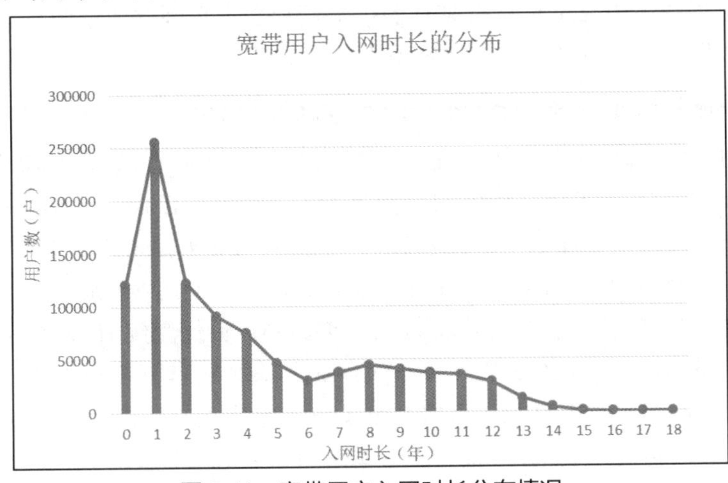

图 3-11 宽带用户入网时长分布情况

根据代码 3-19 的统计结果可以发现，宽带用户中入网时长的均值约为 4，标准差约为 4，中位数为 2。从图 3-11 中可以发现宽带用户的入网时长主要集中在 0～5 年，其中入网时长为 1 年的用户最多。根据统计结果并结合实际的业务场景，选择以中位数 2 年作为入网时长的最低临界值，以标准差 4 作为浮动阈值，制订 3 个宽带用户入网程度子标签及规则，如表 3-9 所示。

表 3-9　宽带用户入网程度子标签及规则

父标签	子标签	标签规则	备注
宽带用户入网程度	新用户	$T \leqslant 2$	T 表示宽带用户入网时长，单位为年
	中等用户	$2 < T \leqslant 6$	
	老用户	$T > 6$	

3.3 技术方案

技术方案的预研与确立指确定完成项目目标所需要使用的技术和技术实现思路。技术方案的预研与确定是项目前期的一项重点工作，项目架构师需要根据项目需求、项目目标和项目成员明确项目的技术方案，这样才能够确定广电大数据用户画像系统的整体架构，并设计出符合项目目标的大数据计算平台的具体流程。

3.3.1 技术选型

项目团队成员如下：一名项目架构师（兼任多个项目）、一名数据分析师、一名程序开发人员、若干测试和实施人员（测试和实施人员可以在需要的时候抽调）。项目团队成员主要以项目架构师、数据分析师、程序开发人员为主。数据分析师对数据分析、数据挖掘建模比较熟悉，会进行基础的 SQL 代码编写。程序开发人员能使用 Java 技术，熟悉 Spring 相关 Java Web 开发，对大数据 Hadoop、Spark 有一定了解。

针对本项目，项目架构师根据项目需求和项目团队成员提出了以下技术选型。

1. 使用 Elasticsearch 集群存储业务数据

广电公司使用 Elasticsearch 集群来存储广电用户数据，因此数据源没有选型的需求，直接使用客户（甲方）提供的数据存储技术即可。

2. 使用 Spark SQL 传输 Elasticsearch 数据到 Hive

后续的数据预处理、用户画像和 SVM 预测用户是否挽留等操作都需要用到 Elasticsearch 中的数据，为了提升程序的执行效率，需提前将 Elasticsearch 数据传输到 Hive 表中，后续操作直接使用 Spark SQL 读取 Hive 表数据。Elasticsearch 数据传输到 Hive 的方式有多种，例如，直接使用 Hive 读取 Elasticsearch 数据；使用 Spark 读取 Elasticsearch 数据得到 RDD，并将 RDD 转换成 DataFrame，再把 DataFrame 保存到 Hive 中；使用 Spark SQL 直接读取 Elasticsearch 数据为 DataFrame 并保存到 Hive 中。考虑到使用 Spark 相关技术读取 Elasticsearch 数据相比使用 Hive 读取 Elasticsearch 数据效率更高，而 Spark SQL 可直接将 Elasticsearch 数据转换为 DataFrame，因此本项目采用 Spark SQL 技术将 Elasticsearch 数据传输到 Hive 中。

3. 使用 Spark SQL 进行用户画像

用户画像的核心工作就是给用户标注标签，挖掘用户标签的工作侧重于数据分析。考虑到项目的数据分析师已经具备 SQL 基础，并且对 Spark SQL 有所了解，因此用户画像模块采用 Spark SQL 技术。

4. 采用 Kafka+Spark Streaming+Redis 统计数据流实时

比较常用的流式大数据实时处理技术有 Storm、Spark Streaming。因为本项目的开发人员对 Storm 技术了解较少，而对 Spark Streaming 比较熟悉，所以采用 Spark Streaming 实时统计订单信息。Spark Streaming 支持的外部数据源有 Flume、Kafka、ZeroMQ、TCP Socket 等。其中，Kafka 是一个分布式的、高吞吐量的、易于扩展的基于主题发布/订阅的消息系统，常作为流计算系统的数据源，因此选择 Spark Streaming 与 Kafka 结合。另外，因为 Redis 是一个基于内存的高性能键值对数据库，适用于少量存储数据与实时更新的场景，所以将数据的处理结果保存到 Redis 中。

5. 采用 Spark MLlib 为数据挖掘建模

Spark 提供了机器学习算法库，即 MLlib，用户可以直接调用 MLlib 提供的算法进行建模。为方便后续程序开发人员对相关代码进行系统嵌入，本项目决定采用 Scala API 的 MLlib 技术进行数据挖掘建模。

6. 采用 Hive 存储中间结果数据

这里的中间结果数据指的是数据预处理之后的数据，可以直接使用 HDFS 来存储。但是考虑到后续的用户画像使用的是 Spark SQL 技术，即进行用户画像之前需要使用 Spark SQL 读取数据预处理之后的数据，为了增加数据的可读性，决定采用 Hive 来存储中间结果数据。

7. 采用 MySQL 存储用户画像标签结果

项目一期计划实现 60 个标签，用户数量大概 300 万，如果每个用户的每个标签有 5 个子标签，那么标签表总的记录数为 60 × 3000000 × 5=900000000 条。此外，存储标签结果的数据表只有 3 个字段，即用户 ID、标签名称和父标签名称。因此，标签结果数据总体的数据量不会太大，选择 MySQL 存储标签结果数据可以提高读取的效率。

8. 采用 Spring RESTful Web Service 作为用户画像可视化接口

因为程序开发人员对 Spring 相关技术比较熟悉，所以采用 Spring RESTful Web Service 作为用户画像可视化接口。

9. 定时任务框架

任务调度框架可选用简单的 Linux Cron、Java Quartz 等来进行定时调度，但是系统要求调度任务可以自定义定时、编辑、监控任务状态。考虑到项目组人员不足，如果要再次进行开发（包括前台展示界面、后台调用操作界面等），工作量非常大，因此考虑选择一个可以直接拿来使用的、可以进行二次开发的开源项目。经过网上择优筛选，选取 GitHub 上的 XXL-JOB 项目作为定时任务框架。

3.3.2 系统架构

系统的整体架构如图 3-12 所示。系统主要分为 2 个子系统，分别为广电业务系统和广电大数据用户画像系统。

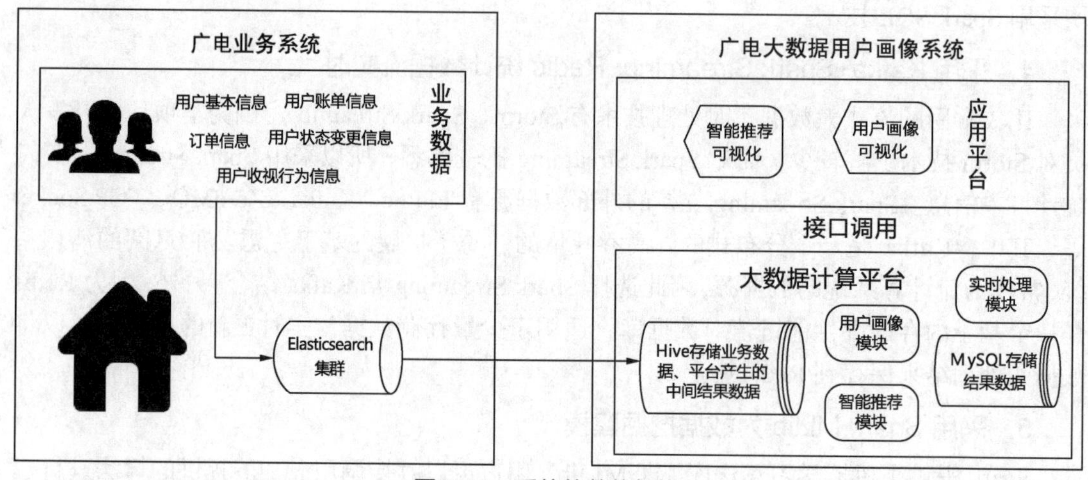

图 3-12　系统的整体架构

在广电业务系统中，用户注册、支付月租、购买产品或者观看电视节目等一系列行为所产生的数据都会被收集到广电公司的业务系统中，且业务人员会把用户画像需要用到的数据抽取到 Elasticsearch 集群中。

广电大数据用户画像系统包括大数据计算平台和应用平台。大数据计算平台从 Elasticsearch 集群中抽取业务数据并将其保存到 Hive 中，且大数据计算平台所产生的中间结果数据（如数据预处理之后的数据）也都保存在 Hive 中。用户画像模块和智能推荐模块从 Hive 中读取数据进行计算，计算得到的结果保存到 MySQL 中。同时，大数据计算平台需要实时统计订单信息。最后，应用平台通过接口调用从 MySQL 中读取用户画像的结果数据和智能推荐的结果数据并进行可视化展示。

在图 3-12 所示的两个子系统中，广电大数据用户画像系统的功能实现由乙方完成，其中，大数据计算平台的实现是整个系统的重点工作。基于目前大数据计算平台架构，对大数据计算平台模块进行细化，设计出符合项目目标的大数据计算平台的具体流程。因为在项目一期中只实现用户画像，智能推荐模块在项目二期实现，所以此处设计的流程不需要包括智能推荐模块。大数据计算平台系统流程如图 3-13 所示。

大数据计算平台系统流程主要分为以下 7 个方面。

（1）实际项目中数据已经存储在 Elasticsearch 集群中，乙方项目人员只需要直接从 Elasticsearch 集群中读取数据进行计算即可。项目前期广电公司提供了 5 份 CSV 格式的数据供乙方项目人员编写代码测试，分别为用户基本信息数据、用户状态信息变更数据、用户账单信息数据、订单信息数据、用户收视行为信息数据。为模仿实际的架构，乙方项目人员自己搭建了一个 Elasticsearch 集群，并将 5 份数据导入 Elasticsearch 集群中。

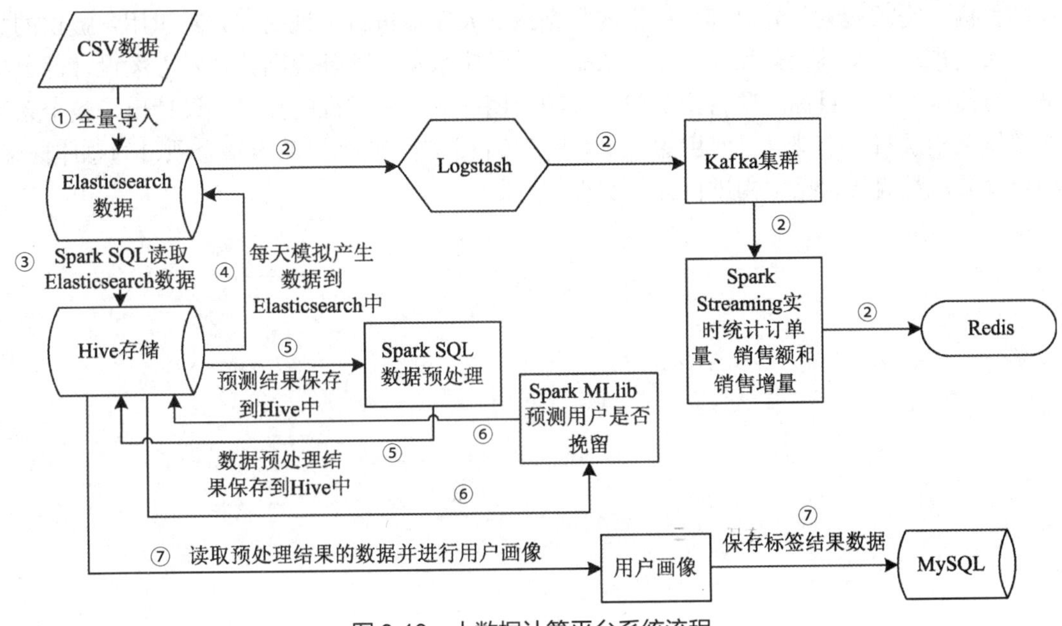

图 3-13 大数据计算平台系统流程

（2）为观察广电公司的产品的实时销售情况，使用 Logstash（开源的服务器端数据处理管道）将 Elasticsearch 集群中的订单信息数据发送至 Kafka 集群的对应主题中，再使用 Spark Streaming 获取对应主题中的数据，实时统计订单信息（如订单量、销售额和销售增量等），并将统计结果保存到 Redis 中。

（3）数据预处理或用户画像的计算都需要使用 Elasticsearch 集群中的数据。如果每次进行数据预处理或每计算一个标签都要从 Elasticsearch 集群中读取数据，那么无疑是浪费资源且效率低下。为避免每次计算都读取 Elasticsearch 集群中的数据，这里使用 Spark SQL 一次性把后面计算要用到的数据都读取并导入 Hive。

（4）为了保证模拟的环境和实际的生产环境一致，每天需要定时模拟产生数据到 Elasticsearch 集群中。

（5）从 Elasticsearch 集群中读取到的 Hive 的数据并不能直接用于用户画像，在进行用户画像之前需要进行数据预处理，并把数据预处理的结果保存到 Hive 中。

（6）用户画像中有一个标签是预测用户是否为值得挽留用户，这个标签的计算需要使用 Spark MLlib 中的 SVM 算法来实现，因此，在进行用户画像之前要先使用 Spark MLlib 中的 SVM 算法预测用户是否值得挽留，并将预测结果保存到 Hive 中。

（7）根据业务需求探索得出每个标签的计算规则，使用 Spark SQL 读取 Hive 中预处理之后的数据，根据标签规则对用户进行用户画像，将计算得到的用户标签结果保存至 MySQL 中。

小结

本章介绍了广电大数据用户画像项目的项目前期工作内容，包括项目需求分析、需求探索和技术方案设计。项目需求分析部分介绍了该项目的背景、客户的要求以及最终要实现的目标；需求探索部分对客户提供的数据意义及字段进行了说明，并基于用户提供的数据对数据进行了基础探索和业务需求探索，总结出数据的处理规则；技术方案设计部分根据项目需求、项目目标、项目团队成员及项目团队成员所掌握的技术，设计出了整个系统的整体架构及每一个部分所使用的技术。完成项目前期工作即可奠定整个项目的总体框架，项目团队成员直接根据架构进行分工实现即可。

第4章 广电大数据用户画像——数据采集与预处理

在广电公司实际的生产环境中，用户的基本信息、每个月用户的账单信息、用户的订单信息、用户状态变更信息和用户收视行为信息等数据是用户在实际使用服务中所产生的数据，是在时刻变动的。为了贴合实际生产环境，项目将模拟这种数据变化的特征，以及验证应对这种数据变化时，广电大数据用户画像系统的稳定性。本章将根据数据的来源，模拟产生数据到 Elasticsearch 集群中的过程，并通过 Spark SQL 实现 Elasticsearch 集群到 Hive 的数据传输和到 MySQL 的数据存储，再根据 3.2 节的需求探索得到的清洗规则对数据进行处理。

学习目标

（1）熟悉使用 Logstash 进行数据采集，将 CSV 格式的数据采集到 Elasticsearch 集群中。

（2）熟悉模拟数据产生的具体过程。

（3）掌握 Elasticsearch 集群与 Hive 之间的数据传输方法。

（4）掌握使用 Spark SQL 技术将 DataFrame 类型的数据保存到 MySQL 数据库中的方法。

（5）掌握根据数据处理规则实现广电大数据的预处理的方法。

（6）掌握广电大数据预处理代码的封装实现的方法。

4.1 业务数据

广电公司的业务数据是每天增量同步至 Elasticsearch 集群中的，而提供给乙方人员的业务数据是 CSV 格式的文件数据。在实际的环境中，业务数据是在时刻发生变化的，为了更加贴合实际业务环境，需要将 CSV 数据导入 Elasticsearch 集群，并将 Elasticsearch 集群中的数据同步到 Hive 表中，再将 Hive 表中的数据以增量的模式导入 Elasticsearch 集群，以此模拟业务数据的产生。

4.1.1 生产数据来源

在实际的生产环境中，用户的业务系统记录了用户的基本信息、每个月用户的账单信息、用户的订单信息、用户状态变更的信息和用户收视行为的信息。系统每天会定时将前一天业务系统产生的数据增量同步到 Elasticsearch 集群中，以服务其他业务系统，因此，在本项目中是需要从 Elasticsearch 集群读取数据的。

在用户的 Elasticsearch 集群中，用户基本信息表、用户状态信息变更表、账单信息表、订单信息表的数据在 Elasticsearch 集群中各自用一张表来存储，而用户收视行为信息表的数据较多，因此将用户收视行为信息表的数据以周为单位进行存储，如图 4-1 所示，media-index201834 表示存储的是 2018 年第 34 周的数据。

media-index201834	37.4Gi/74.9Gi 325M
media-index201835	42.0Gi/83.9Gi 358M
media-index201836	51.5Gi/103Gi 449M
media-index201837	52.9Gi/106Gi 455M
media-index201838	45.9Gi/91.8Gi 407M
media-index201839	38.7Gi/77.4Gi 345M

图 4-1　用户收视行为信息表的数据以周为单位进行存储

4.1.2　模拟产生业务数据

为了还原真实项目的环境，需要模拟产生业务数据。因为业务数据中的变化主要发生在账单信息表、订单信息表和用户收视行为信息表中，所以本小节主要介绍模拟这 3 张表的数据更新的具体步骤。

1. Elasticsearch 安装配置

因为项目中需要用到 Elasticsearch 集群来存储数据，所以需要安装配置 Elasticsearch 集群。根据 1.2.1 小节介绍的硬件环境的 Elasticsearch 集群节点规划，确定各节点的角色，各节点的角色如表 4-1 所示。

表 4-1　Elasticsearch 集群各节点的角色

节点	IP 地址	角色
node1	192.168.111.75	既作为 data 节点也能够被选举为 master 节点
node2	192.168.111.76	既作为 data 节点也能够被选举为 master 节点
node3	192.168.111.77	既作为 data 节点也能够被选举为 master 节点

选择 Elasticsearch 的版本为 6.3.2，Elasticsearch 安装配置的具体步骤如下。

（1）在 node1 节点上下载并解压 elasticsearch-6.3.2.tar.gz，并修改 elasticsearch-6.3.2/config/elasticsearch.yml 文件，其内容如代码 4-1 所示。

代码 4-1　elasticsearch.yml 文件内容

```
cluster.name: tipdm-es          #Elasticsearch 集群名称
node.name: es-node1             #Elasticsearch 节点名称，每个节点的名称不能相同
node.master: true               #指定该节点是否有资格被选举成为 master，默认为 true
node.data: true                 #指定该节点是否存储索引数据，默认为 true
network.host: 192.168.111.75    #节点的 IP 地址
#设置集群中 master 节点的初始列表，可以通过这些节点来自动发现新加入集群的节点
discovery.zen.ping.unicast.hosts: ["192.168.111.75", "192.168.111.76", "192.168.111.77"]
#设置这个参数以保证集群中的节点可以知道其他 N 个有 master 资格的节点。默认值为 1，对于大的集群
#来说，可以设置大一点的值（2～4）
discovery.zen.minimum_master_nodes: 2
#如果要使用 head，那么需要设置以下 2 个参数，使 head 插件可以访问 Elasticsearch
http.cors.enabled: true
```

```
http.cors.allow-origin: "*"
#这是因为 CentOS 6 不支持 SecComp，而 Elasticsearch 默认 bootstrap.system_call_filter 为 true 进行检测，所以导致检测失败，失败后直接导致 Elasticsearch 无法启动
bootstrap.memory_lock: false
bootstrap.system_call_filter: false
```

（2）为避免 Elasticsearch 启动报错，需要调整系统设置。打开/etc/security/limits.conf 文件，其添加的内容如代码 4-2 所示。

代码 4-2 limits.conf 文件添加的内容

```
* soft nofile 65536
* hard nofile 131072
* soft nproc 2048
* hard nproc 4096
```

修改 limits.conf 文件后，还需要修改/etc/security/limits.d/90-nproc.conf 文件，将“* soft nproc 1024”修改为“* soft nproc 4096”。此外，需要修改/etc/sysctl.conf 文件，在文件末尾添加“vm.max_map_count=655360”，并在命令行中执行命令“sysctl -p”使其生效。

（3）因为 root 用户启动 Elasticsearch 会报错，所以需要创建一个新的用户，此处创建一个 elasticsearch 用户，并更改 elasticsearch-6.3.2 目录的所属用户和所属组为 elasticsearch，如代码 4-3 所示。

代码 4-3 创建 elasticsearch 用户并更改 elasticsearch-6.3.2 目录的属性

```
useradd elasticsearch
passwd elasticsearch
chown -R elasticsearch:elasticsearch elasticsearch-6.3.2
```

（4）从 root 用户切换为 elasticsearch 用户，再启动 Elasticsearch，如代码 4-4 所示。

代码 4-4 启动 Elasticsearch

```
su elasticsearch
cd elasticsearch-6.3.2
./bin/elasticsearch
```

执行代码 4-4 所示的命令后，通过浏览器的地址栏访问“http://192.168.111.75:9200”，若能进入图 4-2 所示的界面，则表示 Elasticsearch 已经安装并启动成功。

192.168.111.75:9200

不安全 | 192.168.111.75:9200


```
{
  "name" : "es-node1",
  "cluster_name" : "tipdm-es",
  "cluster_uuid" : "_na_",
  "version" : {
    "number" : "6.3.2",
    "build_flavor" : "default",
    "build_type" : "tar",
    "build_hash" : "053779d",
    "build_date" : "2018-07-20T05:20:23.451332Z",
    "build_snapshot" : false,
    "lucene_version" : "7.3.1",
    "minimum_wire_compatibility_version" : "5.6.0",
    "minimum_index_compatibility_version" : "5.0.0"
  },
  "tagline" : "You Know, for Search"
}
```


图 4-2 Elasticsearch 启动成功界面

至于其他节点的安装，可以复制当前节点的整个 elasticsearch-6.3.2 目录到其他节点中，只需要修改 elasticsearch.yml 的 node.name 和 network.host 的值即可，其他步骤是相同的。

Elasticsearch 集群的 head 插件是一种界面化的集群操作和管理工具，用来辅助管理 Elasticsearch 集群，因此建议读者自行查阅相关资料安装 Elasticsearch 的 head 插件，此处不做介绍。Elasticsearch 集群的 head 插件安装在 node1 节点上，使用默认的端口 9100。

安装好并成功启动 Elasticsearch 集群及 Elasticsearch 的 head 插件后，通过浏览器访问 head 插件的地址 "http://192.168.111.75:9100"，可以进入图 4-3 所示的界面，从图 4-3 中可以浏览 Elasticsearch 集群的基本信息。

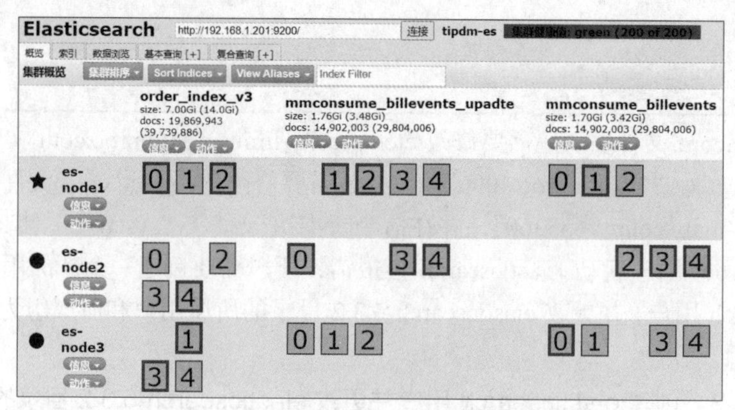

图 4-3　Elastisearch 集群管理界面

2. 数据导入 Elasticsearch 集群

因为乙方人员拿到的数据是 CSV 格式的文件数据，所以需要先将 CSV 格式的文件数据导入 Elasticsearch 集群，可以使用 Logstash 将文件数据导入 Elasticsearch 集群。Logstash 是开源的服务器端数据处理管道，能够同时从多个数据来源采集数据、转换数据，并将数据发送到指定的数据存储库中。选择与 Elasticsearch 集群对应的 Logstash 版本 6.3.2。Logstash 的安装比较简单，解压即可使用，此处在 server1（IP 地址为 192.168.111.73）上安装了 Logstash。

Logstash 运行时需要指定一个配置文件，配置文件中可以指定输入（input）、过滤（filter）及输出（output）这 3 个组件的插件类型。Logstash 在输入、过滤和输出管道中都支持多种类型的插件，需要根据实际需求来选择合适的插件。因为给定的数据源是 CSV 文件格式的，并且最终数据需要存储到 Elasticsearch 集群中，所以选择输入插件的类型为 file，过滤插件的类型为 CSV，输出插件的类型为 Elasticsearch。CSV 数据导入 Elasticsearch 集群的流程如图 4-4 所示。

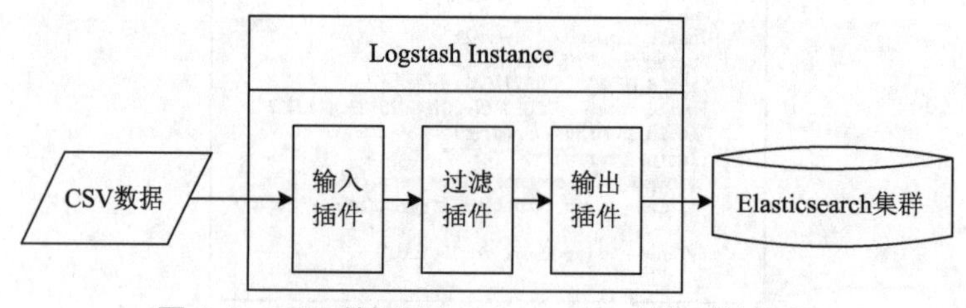

图 4-4　CSV 数据导入 Elasticsearch 集群的流程

对输入（input）、过滤（filter）、输出（output）插件的部分参数说明如下。

（1）input 的 file 插件的部分参数说明如代码 4-5 所示。

代码 4-5　input 的 file 插件的部分参数说明

```
input {
   file {
      codec=>...                      #可选项，默认是 plain，可通过这个参数设置编码方式
      discover_interval=>...          #可选项，Logstash 每隔多久检查一次被监听的 path 下是
#否有新文件，默认值是 15s
      exclude=>...                    #可选项，不想被监听的文件可以被排除，这里和 path 一样
#支持 glob 展开
      sincedb_path=>...               #可选项，跟踪受监听日志文件的当前位置，默认情况下会将
#sincedb 文件写入$path.data/plugins/inputs/file，注意，它必须是文件路径而不是目录路径
      stat_interval=>...              #可选项，Logstash 每隔多久检查一次被监听文件状态（是
#否有更新），默认值是 1s
      start_position=>...             #可选项，Logstash 从什么位置开始读取文件数据，默认是
#结束位置，即 Logstash 会以类似 tail -f 的形式运行。如果要导入一个完整的文件，则可以将这
#个设定改为"beginning"，Logstash 会从文件开头读取
      path=>...                       #必选项，处理的文件的路径，可以定义多个路径
      tags=>...                       #可选项，在数据处理的过程中，由具体的插件来添加或者删
#除的标记
      type=>...                       #可选项，自定义将要处理的事件类型，可以自己随便定义，
#如处理的是 Linux 操作系统日志，则可以将其定义为"syslog"
   }
}
```

（2）filter 的 CSV 插件的部分参数说明如代码 4-6 所示。

代码 4-6　filter 的 CSV 插件的部分参数说明

```
filter {
   csv {
      columns =>...                   #可选项，设置字段的名称
      separator =>...                 #可选项，数据分隔符，默认是逗号
      remove_field=>...               #可选项，删除不需要的列
   }
}
```

（3）output 的 Elasticsearch 插件的部分参数说明如代码 4-7 所示。

代码 4-7　output 的 Elasticsearch 插件的部分参数说明

```
output {
   elasticsearch {
      host =>...                      #必选项，Elasticsearch 的地址
      index=>...                      #在 Elasticsearch 中创建索引，设置索引名称
   }
}
```

各个业务数据表文件导入 Elasticsearch 集群的具体操作步骤如下。

（1）用户基本信息表导入

在 Logstash 的安装目录中，创建 custom_config 目录，在 custom_config 目录下创建 Logstash 的配置文件 mediamatch_usermsg.conf，文件的内容如代码 4-8 所示。导入 mediamatch_usermsg.csv 数据到 Elasticsearch 集群中。

代码 4-8　mediamatch_usermsg.conf 文件的内容

```
input{
   file{
      path => ["/opt/data/usermsg/mediamatch_usermsg.csv"]
      start_position =>"beginning"
      sincedb_path => "/data/logstash-6.3.2/sincedb_path/userevent.sdp"
      }
}
filter{
   csv{
      autodetect_column_names => false
      separator => ";"
      columns => ["terminal_no","phone_no","sm_name","run_name","sm_code",
"owner_name","owner_code","run_time","address oj","estate_name","open_time",
"force"]
      remove_field => ["message","path","host","@version","@timestamp"]
      }
}
output{
   elasticsearch{
      index => "mediamatch_usermsg"
      hosts => ["192.168.111.75:9200","192.168.111.76:9200","192.168.111.77:9200"]
   }
}
```

在执行命令前需要将 mediamatch_usermsg.csv 文件上传到 server1（IP 地址为 192.168.111.73）的/opt/data/usermsg 目录下。此外，需要在 Logstash 的安装目录下创建 sincedb_path 目录，并在 sincedb_path 目录下创建 userevent.sdp 文件，用于保存读取文件进度的记录。编写完 mediamatch_usermsg.conf 文件后，执行代码 4-9 所示的命令，启动 Logstash 进程进行数据传输。

代码 4-9　启动 Logstash 将 mediamatch_usermsg.csv 导入 Elasticsearch 集群

```
#进入 Logstash 的安装目录
cd /data/logstash-6.3.2/
#创建/opt/path_data/mediamatch_usermsg
mkdir -p /opt/path_data/mediamatch_usermsg
#后台运行模式
nohup  bin/logstash  -f  custom_config/mediamatch_usermsg.conf  --path.data=
/opt/path_data/mediamatch_usermsg  >>usermsg_es.log  &
```

执行代码 4-9 所示的命令后，Logstash 会不断地将数据更新到 Elasticsearch 集群的 mediamatch_usermsg 索引中。访问 Elasticsearch 集群管理页面，选择“索引”选项卡，可以看到 mediamatch_usermsg 索引的空间大小（Size）和记录数（Docs）会随着数据的不断输入而增大，直到这两个字段的值不再增长，说明此时数据传输已完成，因为这个过程比较耗时，所以要在后台运行。选择“基本查询[+]”选项卡，可以查看 Elasticsearch 的用户基本信息表的记录，如图 4-5 所示。

Elasticsearch http://192.168.1.201:9200/ 连接 tipdm-es 集群健康值: green (190 of 190) 信息

概览 索引 数据浏览 基本查询 [+] 复合查询 [+]

搜索 mediamatch_usermsg (3309539 个文档) 的文档，查询条件:

must match_all + -

搜索 返回格式: Table 显示数量: 10 显示查询语句

查询 5 个分片中用的 5 个. 3309539 命中. 耗时 0.135 秒

ode	force	terminal_no	sm_code	estate_name	owner_name	open_time	phone_no	sm_name	run_name	run_time	addressoj
	NULL	1410046265	d1	NULL	HC级	2017-07-31 15:19:49	4559217	数字电视	正常	2018-01-25 12:40:09	南沙区东涌镇西樵村樵兴南街***号之一
	NULL	2400159451	d1	NULL	HC级	2017-07-26 22:55:15	4550500	数字电视	正常	2017-08-30 11:13:05	南沙区榄核镇墩塘村*队三沙街东巷*号
	NULL	40002763	d1	NULL	EE级	2017-08-31 20:18:02	4583050	数字电视	主动暂停	2018-03-29 16:20:10	南沙区东涌镇官坦村富佳超市楼上四楼
	NULL	1410070515	d1	NULL	HC级	2017-09-04 10:26:51	4583315	数字电视	正常	2017-09-04 10:26:51	南沙区榄核镇榄核居委镇南路**号
	NULL	1410068647	d1	NULL	HC级	2017-09-03 10:43:37	4583411	数字电视	正常	2017-09-03 10:43:37	南沙区大岗镇镇南社区长堤路**号
	NULL	1410075696	d1	NULL	HC级	2017-09-19 17:20:51	4590334	数字电视	正常	2017-09-19 17:20:51	南沙区榄核镇雁沙村*队海傍西街*巷**号
	NULL	1410074829	d1	NULL	HC级	2017-09-25 15:24:51	4590506	数字电视	正常	2018-01-21 10:46:39	南沙区榄核镇大坳村**队中坊街**号
	宽带生效	2400177175	b0	万地雅涛阁	HC级	2017-09-21 22:24:46	4592084	珠江宽频	正常	2017-09-21 22:24:46	南沙区万顷沙镇万地雅涛阁**-***
	NULL	1410075976	d1	NULL	HC级	2017-09-28 11:36:17	4592714	数字电视	正常	2017-09-28 11:36:17	南沙区大岗镇东隆村东成**路**号
	NULL	1410071235	d1	NULL	HC级	2017-09-02 13:44:17	4582237	数字电视	主动暂停	2017-12-23 14:36:17	南沙区大岗镇岭东村沙仙街**号***房

图 4-5 查看用户基本信息表的记录

（2）用户状态信息变更表导入

在 Logstash 安装目录的 custom_config 目录下创建 Logstash 的配置文件 mediamatch_userevent.conf，文件的内容如代码 4-10 所示。导入 mediamatch_userevent.csv 数据到 Elasticsearch 集群中。

代码 4-10 mediamatch_userevent.conf 文件的内容

```
input{
    file{
        path => ["/opt/data/userevent/mediamatch_userevent.csv"]
        start_position =>"beginning"
        sincedb_path => "/data/logstash-6.3.2/sincedb_path/userevent.sdp"
        }
}
filter{
    csv{
        autodetect_column_names => false
        separator => ";"
        columns => \
["phone_no","owner_name","run_name","run_time","sm_name","owner_code","ope
n_time"]
        remove_field => ["message","path","host","@version","@timestamp"]
        }
}
output{
    elasticsearch{
        index => "mediamatch_userevent"
hosts => ["192.168.111.75:9200","192.168.111.76:9200","192.168.111.77:9200"]
}
}
```

在执行命令前需要将 mediamatch_userevent.csv 文件上传到 server1（IP 地址为 192.168.111.73）的 /opt/data/userevent 目录下，同时需要在 sincedb_path 目录下创建 userevent.sdp 文件，用于保存读取文件进度的记录。执行代码 4-11 所示的命令，启动 Logstash

进程，将数据导入 Elasticsearch 集群的 mediamatch_userevent 索引。

代码 4-11　启动 Logstash 将 mediamatch_userevent.csv 导入 Elasticsearch 集群

```
#进入 Logstash 的安装目录
cd /data/logstash-6.3.2/
#创建/opt/path_data/mediamatch_userevent
mkdir -p /opt/path_data/mediamatch_userevent
#后台运行模式
nohup bin/logstash -f custom_config/mediamatch_userevent.conf --path.data=
/opt/path_data/mediamatch_userevent >>userevent_es.log  &
```

导入成功后，查看 Elasticsearch 集群中用户状态信息变更表的记录，如图 4-6 所示。

Elasticsearch　http://192.168.1.201:9200/　连接　tipdm-es　集群健康值: green (190 of 190)

概览　索引　数据浏览　基本查询 [+]　复合查询 [+]

搜索 mediamatch_userevent (3401497 个文档) 的文档，查询条件:

must　match_all　+　-

搜索　返回格式: Table　显示数量: 10　显示查询语句

查询 5 个分片中用的 5 个. 3401497 命中. 耗时 0.025 秒

_index	_type	_id	_score ▲	run_name	sm_name	owner_code	owner_name	run_time	phone_no
mediamatch_userevent	doc	VAYrv3sBnKRP0oUyIt87	1	2012-07-25 09:07:32	00	2011-04-15 17:48:49	正常	HC级	2857743
mediamatch_userevent	doc	VgYrv3sBnKRP0oUyIt87	1	2017-02-22 06:20:40	00	2010-12-02 15:44:40	正常	HC级	2841506
mediamatch_userevent	doc	VwYrv3sBnKRP0oUyIt87	1	2013-12-17 15:53:04	00	2011-03-13 16:25:30	主动暂停	HC级	2854191
mediamatch_userevent	doc	YwYrv3sBnKRP0oUyIt87	1	2015-09-17 10:17:18	00	2010-10-26 11:57:23	欠费暂停	HC级	2835928
mediamatch_userevent	doc	ZQYrv3sBnKRP0oUyIt87	1	2010-10-27 10:31:55	00	2010-10-27 10:31:55	正常	HC级	2836220
mediamatch_userevent	doc	ZgYrv3sBnKRP0oUyIt87	1	2016-03-23 10:09:24	00	2010-10-28 14:57:35	欠费暂停	HC级	2836442
mediamatch_userevent	doc	cQYrv3sBnKRP0oUyIt87	1	2015-11-21 11:15:05	00	2010-11-09 15:05:29	正常	EE级	2838548
mediamatch_userevent	doc	cwYrv3sBnKRP0oUyIt87	1	2010-11-13 19:38:39	00	2010-11-13 19:38:39	正常	HC级	2838894
mediamatch_userevent	doc	egYrv3sBnKRP0oUyIt87	1	2011-05-19 12:34:11	00	2011-05-19 12:34:11	正常	HC级	2862044
mediamatch_userevent	doc	fQYrv3sBnKRP0oUyIt87	1	2011-05-19 10:25:06	00	2011-05-19 10:25:06	正常	HC级	2862326

图 4-6　查看用户状态信息变更表的记录

（3）账单信息表导入

在 Logstash 安装目录的 custom_config 目录下创建 Logstash 的配置文件 mm_billevents.conf，文件的内容如代码 4-12 所示。导入 mmconsume_billevents.csv 数据到 Elasticsearch 集群中。

代码 4-12　mm_billevents.conf 文件的内容

```
input{
   file{
      path => ["/opt/data/billevents/mmconsume_billevents.csv"]
      start_position =>"beginning"
      sincedb_path => "/data/logstash-6.3.2/sincedb_path/billevents.sdp"
   }
}
filter{
   csv{
      autodetect_column_names => false
      separator => ";"
      columns => ["terminal_no","phone_no","fee_code","year_month","owner_
name","owner_code","sm_name","should_pay","favour_fee"]
      remove_field => ["message","path","host","@version","@timestamp"]
      }
}
output{
```

```
    elasticsearch{
        index => "mmconsume_billevents"
        hosts  => ["192.168.111.75:9200","192.168.111.76:9200","192.168.111.
77:9200"]
    }
}
```

在执行命令前需要将 mmconsume_billevents.csv 文件上传到 server1（IP 地址为 192.168.111.73）的/opt/data/billevents 目录下，同时需要在 sincedb_path 目录下创建 billevents.sdp 文件，用于保存读取文件进度的记录。执行代码 4-13 所示的命令，启动 Logstash 进程，将数据导入 Elasticsearch 集群的 mmconsume_billevents 索引。

代码 4-13　启动 Logstash 将 mmconsume_billevents.csv 导入 Elasticsearch 集群

```
#进入 Logstash 的安装目录
cd /data/logstash-6.3.2/
#创建/opt/path_data/mm_billevents
mkdir -p /opt/path_data/mm_billevents
#后台运行模式
nohup bin/logstash -f custom_config/mm_billevents.conf --path.data=/opt/
path_data/mm_billevents >>billevents_es.log &
```

导入成功后，查看 Elasticsearch 集群中账单信息表的记录，如图 4-7 所示。

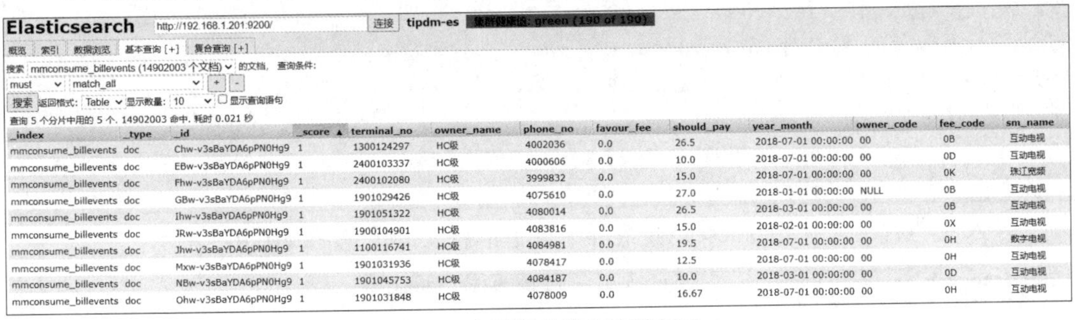

Elasticsearch　http://192.168.1.201:9200/　连接　tipdm-es　集群健康值: green (190 of 190)

概览　索引　数据浏览　基本查询 [+]　复合查询 [+]

搜索 mmconsume_billevents (14902003 个文档) 的文档，查询条件:

must　match_all　+　-

搜索　返回格式: Table　显示数量: 10　显示查询语句

查询 5 个分片中用的 5 个. 14902003 命中. 耗时 0.021 秒

_index	_type	_id	_score ▲	terminal_no	owner_name	phone_no	favour_fee	should_pay	year_month	owner_code	fee_code	sm_name
mmconsume_billevents	doc	Chw-v3sBaYDA6pPN0Hg9	1	1300124297	HC级	4002036	0.0	26.5	2018-07-01 00:00:00	00	0B	互动电视
mmconsume_billevents	doc	EBw-v3sBaYDA6pPN0Hg9	1	2400103337	HC级	4000606	0.0	10.0	2018-07-01 00:00:00	00	0D	互动电视
mmconsume_billevents	doc	Fhw-v3sBaYDA6pPN0Hg9	1	2400102080	HC级	3999832	0.0	15.0	2018-07-01 00:00:00	00	0K	珠江宽频
mmconsume_billevents	doc	GBw-v3sBaYDA6pPN0Hg9	1	1901029426	HC级	4075610	0.0	27.0	2018-01-01 00:00:00	NULL	0B	互动电视
mmconsume_billevents	doc	Ihw-v3sBaYDA6pPN0Hg9	1	1901051322	HC级	4080014	0.0	26.5	2018-03-01 00:00:00	00	0B	互动电视
mmconsume_billevents	doc	JRw-v3sBaYDA6pPN0Hg9	1	1900104901	HC级	4083816	0.0	15.0	2018-02-01 00:00:00	00	0X	互动电视
mmconsume_billevents	doc	Jhw-v3sBaYDA6pPN0Hg9	1	1100116741	HC级	4084981	0.0	19.5	2018-07-01 00:00:00	00	0H	数字电视
mmconsume_billevents	doc	Mxw-v3sBaYDA6pPN0Hg9	1	1901031936	HC级	4078417	0.0	12.5	2018-07-01 00:00:00	00	0H	互动电视
mmconsume_billevents	doc	NBw-v3sBaYDA6pPN0Hg9	1	1901045753	HC级	4084187	0.0	10.0	2018-03-01 00:00:00	00	0D	互动电视
mmconsume_billevents	doc	Ohw-v3sBaYDA6pPN0Hg9	1	1901031848	HC级	4078009	0.0	16.67	2018-07-01 00:00:00	00	0H	互动电视

图 4-7　查看账单信息表的记录

（4）订单信息表导入

在 Logstash 安装目录的 custom_config 目录下创建 Logstash 的配置文件 order_index.conf，文件的内容如代码 4-14 所示。

代码 4-14　order_index.conf 文件的内容

```
input{
    file{
        path => ["/opt/data/order_index/*.csv"]
        start_position =>"beginning"
        sincedb_path => "/data/logstash-6.3.2/sincedb_path/order.sdp"
        }
}
filter{
    csv{
        autodetect_column_names => false
        separator => ";"
```

```
        columns  =>  ["phone_no","owner_name","optdate","prodname","sm_name",
"offerid","offername","business_name","owner_code","prodprcid","prodprcnam
e","effdate","expdate","orderdate","cost","mode_time","prodstatus","run_na
me","orderno","offertype"]
        remove_field => ["message","path","host","@version","@timestamp"]
        }
}
output{
    elasticsearch{
        index => "order_index_v3"
        hosts  =>  ["192.168.111.75:9200","192.168.111.76:9200","192.168.111.
77:9200"]
    }
}
```

在执行命令前需要将 order_index_v3.csv 文件上传到 server1（IP 地址为 192.168.111.73）的/opt/data/order_index 目录下，同时需要在 sincedb_path 目录下创建 order.sdp 文件，用于保存读取文件进度的记录。执行代码 4-15 所示的命令，启动 Logstash 进程，将数据导入 Elasticsearch 集群的 order_index 集群索引。

代码 4-15　启动 Logstash 将订单数据导入 Elasticsearch 集群

```
#进入 Logstash 的安装目录
cd /data/logstash-6.3.2/
#创建/opt/path_data/order_index
mkdir -p /opt/path_data/order_index
#后台运行模式
nohup bin/logstash -f custom_config/order_index.conf --path.data=/opt/path_
data/order_index >>order_es.log  &
```

导入成功后，查看 Elasticsearch 集群中订单信息表的记录，如图 4-8 所示。

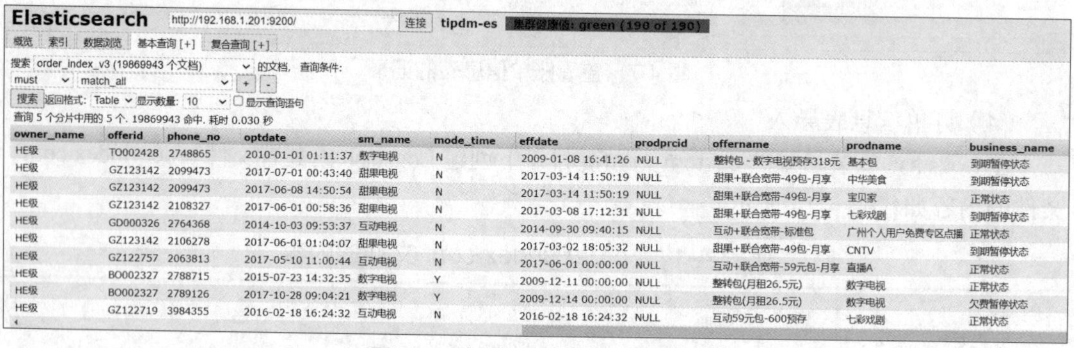

Elasticsearch　http://192.168.1.201:9200/　连接　tipdm-es　集群健康值: green (190 of 190)

概览　索引　数据浏览　基本查询 [+]　复合查询 [+]

搜索 order_index_v3 (19869943 个文档) 的文档，查询条件：
must　match_all　+　-
搜索　返回格式: Table　显示数量: 10　显示查询语句
查询 5 个分片中用的 5 个. 19869943 命中. 耗时 0.030 秒

owner_name	offerid	phone_no	optdate	sm_name	mode_time	effdate	prodprcid	offername	prodname	business_name
HE级	TO002428	2748865	2010-01-01 01:11:37	数字电视	N	2009-01-08 16:41:26	NULL	整转包 - 数字电视预存318元	基本包	到期暂停状态
HE级	GZ123142	2099473	2017-07-01 00:43:40	甜果电视	N	2017-03-14 11:50:19	NULL	甜果+联合宽带-49包-月享	中华美食	到期暂停状态
HE级	GZ123142	2099473	2017-06-08 14:50:54	甜果电视	N	2017-03-14 11:50:19	NULL	甜果+联合宽带-49包-月享	宝贝家	正常状态
HE级	GZ123142	2108327	2017-06-01 00:58:36	甜果电视	N	2017-03-08 17:12:31	NULL	甜果+联合宽带-49包-月享	七彩戏剧	到期暂停状态
HE级	GD000326	2764368	2014-10-03 09:53:37	互动电视	N	2014-09-30 09:40:15	NULL	互动+联合宽带-标准包	广州个人用户免费专区点播	正常状态
HE级	GZ123142	2106278	2017-06-01 01:04:07	甜果电视	N	2017-03-02 18:05:32	NULL	甜果+联合宽带-49包-月享	CNTV	到期暂停状态
HE级	GZ122757	2063813	2017-05-10 11:00:44	互动电视	N	2017-06-01 00:00:00	NULL	互动+联合宽带-59元包-月享	直播A	正常状态
HE级	BO002327	2788715	2015-07-23 14:32:35	数字电视	Y	2009-12-11 00:00:00	NULL	整转包(月租26.5元)	数字电视	正常状态
HE级	BO002327	2789126	2017-10-28 09:04:21	数字电视	Y	2009-12-14 00:00:00	NULL	整转包(月租26.5元)	数字电视	欠费暂停状态
HE级	GZ122719	3984355	2016-02-18 16:24:32	互动电视	N	2016-02-18 16:24:32	NULL	互动59元包-600预存	七彩戏剧	正常状态

图 4-8　查看订单信息表的记录

（5）用户收视行为信息表导入

在 Logstash 安装目录的 custom_config 目录下创建 Logstash 的配置文件 media1.conf，文件的内容如代码 4-16 所示。导入用户收视行为信息数据到 Elasticsearch 集群中。

代码 4-16　media1.conf 文件的内容

```
input{
    file{
```

```
        path => ["/opt/data/media_index/media1/*.csv"]
        start_position =>"beginning"
        sincedb_path => "/data/logstash-6.3.2/sincedb_path/media1.sdp"
        }
}
filter{
    grok{
    match => ["message","%{TIMESTAMP_ISO8601:origin_time}"]
    }
    date{
    match => ["origin_time","yyyy-MM-dd HH:mm:ss"]
    target => "@timestamp"
    }
    csv{
        quote_char => "'"
        separator => ";"
        columns => ['terminal_no','phone_no','duration','station_name','origin_time',
'end_time','owner_code','owner_name','vod_cat_tags','resolution','audio_lang',
'region','res_name','res_type','vod_title','category_name','program_title',
'sm_name','first_show_time']
        remove_field => ["message","path","host","@version"]
        }
   if "[" in [vod_cat_tags] {
   json {
    source => 'vod_cat_tags'
    target => 'vod_cat_tags'
    }}else{
    mutate { remove_field => "vod_cat_tags" }
   }
}
output{
    elasticsearch{
        index => "media_index%{+yyyyww}"
        document_type => "media"
        hosts => ["192.168.111.75:9200","192.168.111.76:9200","192.168.111.77:
9200"]
    }
}
```

在代码 4-16 中，filter 组件中使用了和前面不同的插件 grok 和 date，主要是为了使 @timestamp 的时间与 origin_time 的时间同步，从而使 Elasticsearch 集群的索引名带有和 origin_time 相同的年份和周数，实现用户收视行为信息数据在 Elasticsearch 集群中以周来存储的目标。此外，因为用户收视行为信息表中 vod_cat_tags 字段的值比较特殊，如果是直播，那么 vod_cat_tags 字段的值为空，否则为代码 4-17 所示的 JSON 结构，所以为了正确解析 vod_cat_tags 字段，在 filter 组件中引入了 JSON 插件，并使用条件语句进行判断 vod_cat_tags 是否为空。

代码 4-17　vod_cat_tags 字段的值

```
[{"level1_name":"回看","level2_name":"粤港频道","level3_name":"翡翠台",
"level4_name":"2018-6-12","level5_name":null}]
```

用户收视行为信息表的数据量比较大，可以选择将原数据文件分成多个文件，在多个节点上启动 Logstash 进程，数据导入 Elasticsearch 集群的效率会更高。配置文件只需要在 input 的 file 插件中修改对应的数据路径即可。

执行代码 4-18 所示的命令，启动 Logstash 进程进行数据传输。

代码 4-18　启动 Logstash 将用户收视行为信息数据导入 Elasticsearch 集群

```
#进入 Logstash 的安装目录
cd /data/logstash-6.3.2/
#创建/opt/path_data/media1
mkdir -p /opt/path_data/media1
#后台运行模式
nohup bin/logstash -f custom_config/media1.conf --path.data=/opt/path_data/
media1  >>media1_es.log  &
```

导入成功后，查看 Elasticsearch 集群中用户收视行为信息表的记录，如图 4-9 所示。

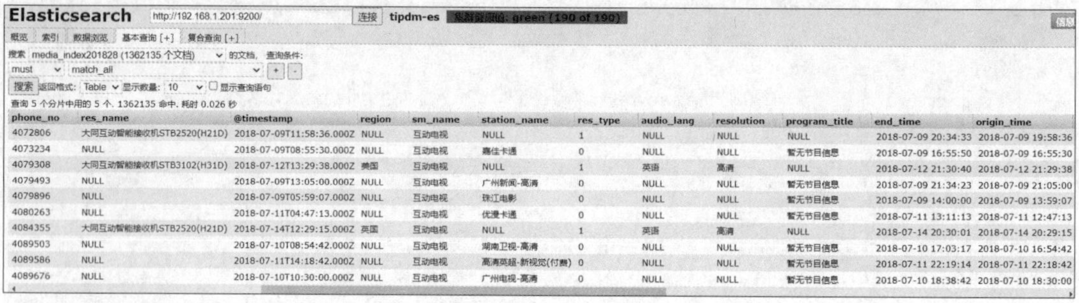

Elasticsearch　http://192.168.1.201:9200/　连接　tipdm-es

概览　索引　数据浏览　基本查询 [+]　复合查询 [+]

搜索 media_index201828 (1362135 个文档) 的文档，查询条件：

must　match_all　+　-

搜索 返回格式：Table 显示数量：10 显示查询语句

查询 5 个分片中用的 5 个. 1362135 命中. 耗时 0.026 秒

phone_no	res_name	@timestamp	region	sm_name	station_name	res_type	audio_lang	resolution	program_title	end_time	origin_time
4072806	大同互动智能接收机STB2520(H21D)	2018-07-09T11:58:36.000Z	NULL	互动电视	NULL	1	NULL	NULL	NULL	2018-07-09 20:34:33	2018-07-09 19:58:36
4073234	NULL	2018-07-09T08:55:30.000Z	NULL	互动电视	嘉佳卡通	0	NULL	NULL	暂无节目信息	2018-07-09 16:55:50	2018-07-09 16:55:30
4079308	大同互动智能接收机STB3102(H31D)	2018-07-12T13:29:38.000Z	美国	互动电视	NULL	1	英语	高清	NULL	2018-07-12 21:30:40	2018-07-12 21:29:38
4079493	NULL	2018-07-09T13:05:00.000Z	NULL	互动电视	广州新闻-高清	0	NULL	NULL	暂无节目信息	2018-07-09 21:34:23	2018-07-09 21:05:00
4079896	NULL	2018-07-09T05:59:07.000Z	NULL	互动电视	珠江电影	0	NULL	NULL	暂无节目信息	2018-07-09 14:00:00	2018-07-09 13:59:07
4080263	NULL	2018-07-11T04:47:13.000Z	NULL	互动电视	优漫卡通	0	NULL	NULL	暂无节目信息	2018-07-11 13:11:13	2018-07-11 12:47:13
4084355	大同互动智能接收机STB2520(H21D)	2018-07-14T12:29:15.000Z	英国	互动电视	NULL	1	英语	高清	NULL	2018-07-14 20:30:01	2018-07-14 20:29:15
4089503	NULL	2018-07-10T08:54:42.000Z	NULL	互动电视	湖南卫视-高清	0	NULL	NULL	暂无节目信息	2018-07-10 17:03:17	2018-07-10 16:54:42
4089586	NULL	2018-07-11T14:18:42.000Z	NULL	互动电视	高清英超-新视觉(付费)	0	NULL	NULL	暂无节目信息	2018-07-11 22:19:14	2018-07-11 22:18:42
4089676	NULL	2018-07-10T10:30:00.000Z	NULL	互动电视	广州电视-高清	0	NULL	NULL	暂无节目信息	2018-07-10 18:38:42	2018-07-10 18:30:00

图 4-9　查看用户收视行为信息表的记录

3．模拟产生数据

（1）模拟产生账单数据

① 数据产生规则。

因为账单数据是每月 1 日生成的，账单信息表中的 year_month 字段记录了账单时间，所以只需要每月更新 year_month 字段的值，得到时间不同的数据，并将其作为每月产生的新的数据即可。更改 year_month 字段的规则是先于每月 1 日计算当前年月减原始账单数据中最大的年月（2018-07）得到相差的月数 month_delta，再将 mmconsume_billevents 数据中的 year_month 字段都加上 month_delta，最后将修改后的数据更新到 Elasticsearch 集群中。

② 具体步骤。

a. 参考 4.2.1 小节数据从 Elasticsearch 集群传输到 Hive 中的方法，将 Elasticsearch 集群中的 mmconsume_billevents 数据同步到 Hive 的 user_profile 库的 mmconsume_billevents_id 表中，模拟产生的账单数据都是以 Hive 中的 mmconsume_billevents 表为基础的。

b. 编写 mmconsume_billevents_1d.sh 脚本，脚本内容如代码 4-19 所示。

代码 4-19　mmconsume_billevents_1d.sh 脚本内容

```
#!/bin/bash
#Shell 脚本开始运行的时间
start_time=$(date +%Y-%m-%d-%H:%M:%S)
```

```
log_file=/root/qwm/mmconsume/log.out
#Hive 脚本
hive_script=create_mmconsume_billevents_1d.hql
echo -e "Shell 脚本开始运行时间 : $start_time" >> $log_file
#计算当前日期与 2018-07-01 相差的月数
datatime=$1
curr_ymd=$(date +%Y-%m-%d)
curr_time2=$(($(date +%s -d $curr_ymd) - $(date +%s -d $datatime)))
month_delta=$((curr_time2/(60*60*24*30)))
echo "当前日期与 2018-07-01 相差的月数: $month_delta" >> $log_file
#在 Hive 中创建表 mmconsume_billevents_1d
echo -e "use user_profile;drop table if exists mmconsume_billevents_1d;\ncreate
table mmconsume_billevents_1d as select terminal_no,phone_no,fee_code,concat
(add_months(year_month,${month_delta}),' ','00:00:00')as year_month,owner_name,
owner_code,sm_name,should_pay,favour_fee from mmconsume_billevents;">${hive_
script}
hive -f /root/qwm/mmconsume/create_mmconsume_billevents_1d.hql
#删除 Elasticsearch 集群中的 mmconsume_billevents_test 数据
curl -XDELETE http://$2:9200/$7
#将 mmconsume_billevents_1d 表的数据导入 Elasticsearch 集群
spark-submit --class com.tipdm.scala.datasource.Hive2Elasticsearch --executor-
memory 10g --total-executor-cores 20 --master spark://$3:7077 --jars /opt/
cloudera/parcels/CDH-5.7.3-1.cdh5.7.3.p0.5/lib/spark/lib/elasticsearch-spa
rk-13_2.10-6.3.2.jar $4 user_profile.mmconsume_billevents_1d $2 $5 $6 $7 $8
end_time=$(date +%Y-%m-%d-%H:%M:%S)
echo "任务结束运行时间: $end_time" >> $log_file
```

代码 4-19 主要用于计算当前时间与 2018 年 7 月相差的月数 month_delta，即将原账单数据的 year_month 字段数据加上 month_delta，修改后的数据以覆盖的模式保存在 Hive 的 user_profile 库的 mmconsume_billevents_1d 表中，最后通过运行一个 Spark 程序将 Hive 中的 mmconsume_billevents_1d 表中的数据以覆盖的模式保存到 Elasticsearch 集群中。其中，Hive 数据导入 Elasticsearch 集群可通过代码 4-20 来实现。

代码 4-20 Hive 数据导入 Elasticsearch 集群

```
package com.tipdm.scala.chapter_3_4_3_datasource
import com.tipdm.scala.chapter_3_4_3_datasource.Elasticsearch2Hive.printUsage
import org.apache.spark.sql.hive.HiveContext
import org.apache.spark.{SparkConf, SparkContext}
import org.elasticsearch.spark.sql.EsSparkSQL
/**
  * Hive 数据导入 Elasticsearch 集群
  */
object Hive2Elasticsearch {
  def main(args: Array[String]): Unit = {
    if (args.length != 6) {
      printUsage()
      System.exit(1)
    }
```

```
    val conf = new SparkConf().setAppName("Hive2Elasticsearch")
    val sc = new SparkContext(conf)
    val sqlContext = new HiveContext(sc)
    val hiveTable = args(0)
    val esNode = args(1)
    val esPort = args(2)
    val timeColumn = args(3)
    val esIndex = args(4)
    val esType = args(5)
    val options = Map(
      ("es.nodes", esNode),
      ("es.port", esPort),
      ("es.index.auto.create", "true"),
      ("es.write.operation", "index")
    )
    if (hiveTable.contains("media")) {
      val weeks = sqlContext.sql("select distinct concat(year(" + timeColumn
+ "),'',weekofyear(" + timeColumn + ")) as week from "
        + hiveTable).rdd.map(row => row.getString(0)).collect()
      val data = sqlContext.sql("select *,concat(year(" + timeColumn + "),
'',weekofyear(" + timeColumn + ")) as week from " + hiveTable)
      val table = "media_1d" + System.currentTimeMillis()
      data.registerTempTable(table)
      for (we <- weeks) {
        val data1 = sqlContext.sql("select * from " + table + " where week=" + we)
        val df = data1.drop("week")
        df.show(3)
        println(esIndex + "" + we + "/" + esType)
        EsSparkSQL.saveToEs(df, esIndex + "" + we + "/" + esType, options)
      }
    } else {
      val data = sqlContext.sql("select * from " + hiveTable)
      EsSparkSQL.saveToEs(data, esIndex + "/" + esType, options)
    }
    sc.stop()
  }
  /**
    * 使用说明
    */
  def printUsage(): Unit = {
    buff.append("Usage : com.tipdm.scala.chapter_3_5_1_datasource.Hive2Elasticsearch").
append(" ")
      .append("<hiveTable>").append(" ")
      .append("<esNode>").append(" ")
      .append("<esPort>").append(" ")
      .append("<timeColumn>").append(" ")
      .append("<esIndex>").append(" ")
      .append("<esType>").append(" ")
    println(buff.toString())
  }
}
```

为了确保 mmconsume_billevents_1d.sh 脚本运行成功，需要先在脚本所在的节点上创建 /root/qwm/mmconsume 目录，再将代码 4-20 所示的内容打包成 user_profile_project-1.0.jar 的 JAR 包，后续模拟产生订单数据和用户收视行为数据都需要使用这个 JAR 包。

c. 为了测试 mmconsume_billevents_1d.sh 脚本能否正确运行，在 2018 年 8 月 28 日设置了一个定时策略。在 server1 节点上执行“crontab -e”，加入命令“0 10 3 1 * ? * /root/mmconsume_billevents_1d.sh 2018-07-01 192.168.111.75 server3 /root/qwm/user_profile_project-1.0.jar 9200 year_month mmconsume_billevents_update doc”。策略中设置的定时任务会在每月 1 日凌晨 3 点 10 分被触发，这样可以通过在 2018 年 9 月 1 日早上观察任务的执行结果来验证编写的脚本是否正确。在 2018 年 9 月 1 日查看的执行结果如图 4-10 所示，与 mmconsume_billevents.csv 原文件的数据记录对比可发现 year_month 字段的时间已更新。

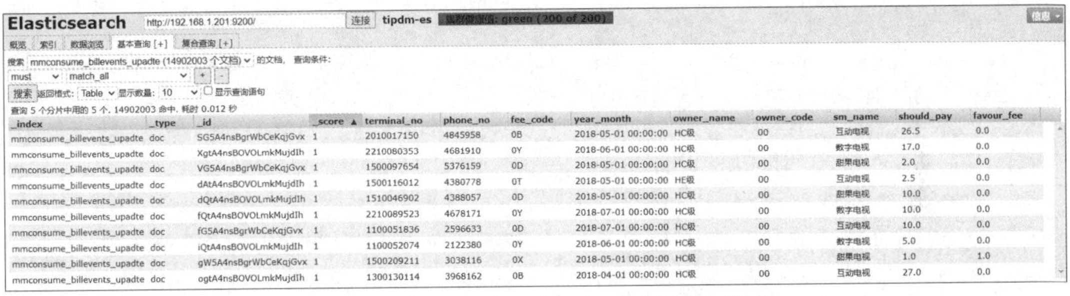

Elasticsearch http://192.168.1.201:9200/ 连接 tipdm-es 集群健康值: green (200 of 200)

概览 索引 数据浏览 基本查询 [+] 复合查询 [+]

搜索 mmconsume_billevents_upadte (14902003 个文档) 的文档，查询条件：

must match_all

搜索 返回格式：Table 显示数量：10 显示查询语句

查询 5 个分片中用的 5 个. 14902003 命中. 耗时 0.012 秒

_index	_type	_id	_score	terminal_no	phone_no	fee_code	year_month	owner_name	owner_code	sm_name	should_pay	favour_fee
mmconsume_billevents_upadte	doc	SG5A4nsBgrWbCeKqjGvx	1	2010017150	4845958	0B	2018-05-01 00:00:00	HC级	00	互动电视	26.5	0.0
mmconsume_billevents_upadte	doc	XgtA4nsBOVOLmkMujdIh	1	2210080353	4681910	0Y	2018-06-01 00:00:00	HC级	00	数字电视	17.0	0.0
mmconsume_billevents_upadte	doc	VG5A4nsBgrWbCeKqjGvx	1	1500097654	5376159	0D	2018-05-01 00:00:00	HC级	00	甜果电视	2.0	0.0
mmconsume_billevents_upadte	doc	dAtA4nsBOVOLmkMujdIh	1	1500116012	4380778	0T	2018-05-01 00:00:00	HE级	00	互动电视	2.5	0.0
mmconsume_billevents_upadte	doc	dQtA4nsBOVOLmkMujdIh	1	1510046902	4388057	0D	2018-05-01 00:00:00	HC级	00	甜果电视	10.0	0.0
mmconsume_billevents_upadte	doc	fQtA4nsBOVOLmkMujdIh	1	2210089523	4678171	0Y	2018-07-01 00:00:00	HC级	00	数字电视	10.0	0.0
mmconsume_billevents_upadte	doc	fG5A4nsBgrWbCeKqjGvx	1	1100051836	2596633	0D	2018-07-01 00:00:00	HC级	00	互动电视	10.0	0.0
mmconsume_billevents_upadte	doc	iQtA4nsBOVOLmkMujdIh	1	1100052074	2122380	0Y	2018-06-01 00:00:00	HC级	00	数字电视	5.0	0.0
mmconsume_billevents_upadte	doc	gW5A4nsBgrWbCeKqjGvx	1	1500209211	3038116	0X	2018-05-01 00:00:00	HC级	00	甜果电视	1.0	1.0
mmconsume_billevents_upadte	doc	ogtA4nsBOVOLmkMujdIh	1	1300130114	3968162	0B	2018-04-01 00:00:00	HC级	00	互动电视	27.0	0.0

图 4-10 在 2018 年 9 月 1 日查看的执行结果

（2）模拟产生订单数据

① 数据产生规则。

每天都会产生订单数据的记录，因此需要模拟每天产生的新数据。选择订单信息表中 optdate≥2018-01-01 00:00:00 的记录，使用这些数据来模拟产生订单数据，具体的规则如下。

a. 以 2018-10-10 作为基准 task_start_time，计算该日期与 2018-01-01 间隔的天数 delta。

b. 计算当前日期与 task_start_time 间隔的天数 curr_delta。

c. 每天从订单信息表（optdate≥2018-01-01 00:00:00）中选择满足（optdate 字段的值+delta）=（当前日期−curr_delta）的数据，并计算当前日期与 optdate 间隔的天数 delta1，同时将 orderdate、expdate、effdate 字段的值分别加上 delta1，再将修改后的数据增量更新到 Elasticsearch 集群中。

② 具体步骤。

a. 参考 4.2.1 小节数据从 Elasticsearch 集群传输到 Hive 中的方法，将 Elasticsearch 集群中的 order_index_v3 数据同步到 Hive 的 user_profile 库的 order_index_v3 表中，模拟产生的订单数据是以 Hive 的 order_index_v3 表中的 optdate 字段≥2018-01-01 00:00:00 的数据为基础的。

b. 编写 order_index_1d.sh 脚本，脚本内容如代码 4-21 所示。

代码 4-21 order_index_1d.sh 脚本内容

```
#!/bin/bash
#任务启动的日期
task_start_time=$1
#Hive 脚本
hive_script=$2
#Shell 脚本开始运行的时间
start_time=$(date +%Y-%m-%d-%H:%M:%S)
#日志保存的路径
log_file=/root/qwm/order_index/log.out
echo -e "\nstart time : $start_time" >> $log_file
echo "program run day: $task_start_time" >> $log_file
#任务启动的日期与数据的开始日期（2018-01-01）间隔的天数
datatime=$3
time1=$(($(date +%s -d $task_start_time) - $(date +%s -d $datatime)))
delta=$((time1/(60*60*24)))
echo "任务启动的日期与数据的开始日期（2018-01-01）间隔的天数:$delta" >> $log_file
#当前日期与任务启动的日期间隔的天数
curr_ymd=$(date +%Y-%m-%d)
curr_time2=$(($(date +%s -d $curr_ymd) - $(date +%s -d $task_start_time)))
curr_delta=$((curr_time2/(60*60*24)))
#当前日期与数据开始日期间隔的天数
curr_time3=$(($(date +%s -d $curr_ymd) - $(date +%s -d $datatime)))
curr_delta1=$((curr_time3/(60*60*24)))
echo "当前日期与任务启动的日期${task_start_time}间隔的天数:$curr_delta" >>
$log_file
echo "当前日期与数据开始日期2018-01-01间隔的天数:$curr_delta1" >> $log_file
#在 Hive 中创建表 order_index_1d，若今天(task_start_time)是 2018-10-09，则
#order_index_1d中的数据是原始数据中optdate字段为2018-01-01的数据，且计算当前日期
#与optdate字段的时间差delta，orderdate、expdate、effdate字段分别加上时间差delta
echo -e "use user_profile;drop table if exists order_index_1d;\ncreate table
order_index_1d as select phone_no,owner_name,concat(CURRENT_DATE,' ',from_
unixtime(unix_timestamp(optdate),\"HH:mm:ss\")) as optdate,prodname,sm_name,
offerid,offername,business_name,owner_code,prodprcid,prodprcname,concat(da
te_add(effdate,datediff(CURRENT_DATE,optdate)),' ',from_unixtime(unix_timestamp
(effdate),\"HH:mm:ss\")) as effdate,concat(date_add(expdate,datediff(CURRENT_
DATE,optdate)),' ',from_unixtime(unix_timestamp(expdate),\"HH:mm:ss\")) as
expdate,concat(date_add(orderdate,datediff(CURRENT_DATE,optdate)),' ',from_
unixtime(unix_timestamp(orderdate),\"HH:mm:ss\")) as orderdate,cost,mode_time,
prodstatus,run_name,orderno,offertype from order_index_v3 where date_add
(optdate,${delta}) = date_sub(CURRENT_DATE,${curr_delta});">${hive_script}
hive -f /root/qwm/order_index/${hive_script}
#将order_index_1d表的数据导入Elasticsearch集群
spark-submit --class com.tipdm.scala.datasource.Hive2Elasticsearch --executor-
memory 10g --total-executor-cores 20 --master spark://$4:7077 --jars /opt/cloudera/
parcels/CDH-5.7.3-1.cdh5.7.3.p0.5/lib/spark/lib/elasticsearch-spark-13_2.1
0-6.3.2.jar $5 user_profile.order_index_1d $6 $7 $8 $9 ${10}
end_time=$(date +%Y-%m-%d-%H:%M:%S)
```

```
echo "end time : $end_time" >> $log_file
```

代码 4-21 的主要功能是每天从 Hive 的 order_index_v3 表中提取符合模拟产生订单数据规则的数据，并按规则修改 orderdate、expdate 和 effdate 这 3 个字段的值，以覆盖的模式保存到 Hive 的 order_index_1d 表中，最后通过 Spark 程序将 order_index_1d 表中的数据以增量的模式导入 Elasticsearch 集群。

c. 为了测试 order_index_1d.sh 脚本能否正确执行，需要设置一个临时的定时任务来测试脚本。在 server1 节点上执行“crontab -e”，加入命令“0 10 2 * * ? * /root/order_index_1d.sh 2018-10-10 create_order_index_1d.hql 2018-01-01 server3 /root/qwm/user_profile_project-1.0.jar 192.168.111.75 9200 optdate order_index_update doc”。定时任务是在 2021 年 9 月 4 日设置的，order_index_1d.sh 脚本在 2021 年 9 月 15 日凌晨 2 点 10 分会被触发执行。通过在 2021 年 9 月 15 日观察任务的执行结果，验证脚本是否正确，即是否将产生的订单数据增量更新到 Elasticsearch 集群的 order_index_update 索引中。在 2021 年 9 月 15 日查看的执行结果如图 4-11 所示。

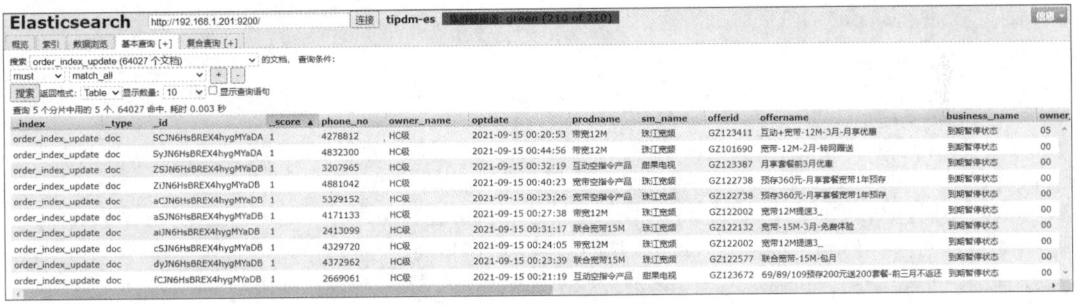

Elasticsearch http://192.168.1.201:9200/ 连接 tipdm-es green (210 of 210)

概览 索引 数据浏览 基本查询 [+] 复合查询 [+]

搜索 order_index_update (64027 个文档) 的文档，查询条件：

must match_all

搜索 返回格式：Table 显示数量：10 显示查询语句

查询 5 个分片中用的 5 个. 64027 命中. 耗时 0.003 秒

_index	_type	_id	_score ▲	phone_no	owner_name	optdate	prodname	sm_name	offerid	offername	business_name	owner_
order_index_update	doc	SCJN6HsBREX4hygMYaDA	1	4278812	HC级	2021-09-15 00:20:53	带宽12M	珠江宽频	GZ123411	互动+宽带-12M-3月-月享优惠	到期暂停状态	05
order_index_update	doc	SyJN6HsBREX4hygMYaDA	1	4832300	HC级	2021-09-15 00:44:56	带宽12M	珠江宽频	GZ101690	宽带-12M-2月-转网赠送	到期暂停状态	00
order_index_update	doc	ZSJN6HsBREX4hygMYaDB	1	3207965	HC级	2021-09-15 00:32:19	互动空指令产品	甜果电视	GZ123387	月享套餐前3月优惠	到期暂停状态	00
order_index_update	doc	ZiJN6HsBREX4hygMYaDB	1	4881042	HC级	2021-09-15 00:40:23	宽带空指令产品	珠江宽频	GZ122738	预存360元-月享套餐宽带1年预存	到期暂停状态	00
order_index_update	doc	aCJN6HsBREX4hygMYaDB	1	5329152	HC级	2021-09-15 00:23:26	宽带空指令产品	珠江宽频	GZ122738	预存360元-月享套餐宽带1年预存	到期暂停状态	00
order_index_update	doc	aSJN6HsBREX4hygMYaDB	1	4171133	HC级	2021-09-15 00:27:38	带宽12M	珠江宽频	GZ122002	宽带12M提速3_	到期暂停状态	00
order_index_update	doc	aiJN6HsBREX4hygMYaDB	1	2413099	HC级	2021-09-15 00:31:17	联合宽带15M	珠江宽频	GZ122132	宽带-15M-3月-免费体验	到期暂停状态	00
order_index_update	doc	cSJN6HsBREX4hygMYaDB	1	4329720	HC级	2021-09-15 00:24:05	带宽12M	珠江宽频	GZ122002	宽带12M提速3_	到期暂停状态	00
order_index_update	doc	dyJN6HsBREX4hygMYaDB	1	4372962	HC级	2021-09-15 00:23:39	联合宽带15M	珠江宽频	GZ122577	联合宽带-15M-包月	到期暂停状态	00
order_index_update	doc	fCJN6HsBREX4hygMYaDB	1	2669061	HC级	2021-09-15 00:21:19	互动空指令产品	甜果电视	GZ123672	69/89/109预存200元送200套餐-前三月不返还	到期暂停状态	00

图 4-11　在 2021 年 9 月 15 日查看的执行结果

从图 4-11 中可以看到，表中 optdate 字段的值为 2021-09-15，说明 order_index_1d.sh 脚本是可以模拟产生订单数据的。

（3）模拟产生用户收视行为数据

① 数据产生规则。

每天都会产生用户收视行为数据，因此需要模拟每天产生的新数据。使用 2018 年 5 月—2018 年 7 月这 3 个月的用户收视行为数据作为基准，根据如下规则模拟产生数据。

a. 选择某个日期为 task_start_time，如选择 2018-10-08 作为 task_start_time，计算该日期与用户收视行为数据的最小日期（2018-05-02）间隔的天数，即 delta=（2018-10-08）-（2018-05-02）。

b. 计算当前日期与 task_start_time 间隔的天数 curr_delta。

c. 每天从用户收视行为数据中选择满足（origin_time 字段的值+delta）=（当前日期-curr_delta）条件的数据，同时将时间字段（origin_time、end_time）修改成当前日期，并将修改后的数据更新到 Elasticsearch 集群中。

② 具体步骤。

a. 参考 4.2.1 小节数据从 Elasticsearch 集群传输到 Hive 中的方法，将 Elasticsearch 集群中的用户收视行为数据同步到 Hive 的 user_profile 库的 media_index_3m 表中。

b. 编写 media_index_1d.sh 脚本，脚本内容如代码 4-22 所示。

代码 4-22 media_index_1d.sh 脚本内容

```
#!/bin/bash
#设置任务启动的日期
task_start_time=$1
#Hive 脚本
hive_script=$2
#Shell 脚本开始运行的时间
start_time=$(date +%Y-%m-%d-%H:%M:%S)
log_file=/root/qwm/log.out
echo -e "\nstart time : $start_time" >> $log_file
echo "program run day: $task_start_time" >> $log_file
#任务启动的日期与数据的开始日期（2018-05-02）间隔的天数
datatime=$3
time1=$(($(date +%s -d $task_start_time) - $(date +%s -d $datatime)))
delta=$((time1/(60*60*24)))
echo "任务启动的日期与数据的开始日期（2018-05-02）间隔的天数:$delta" >> $log_file
#当前日期与任务启动的日期间隔的天数
curr_ymd=$(date +%Y-%m-%d)
curr_time2=$(($(date +%s -d $curr_ymd) - $(date +%s -d $task_start_time)))
curr_delta=$((curr_time2/(60*60*24)))
echo "当前日期与任务启动的日期${task_start_time}间隔的天数:$curr_delta" >>
$log_file
#在 Hive 中创建表 media_1d，若 task_start_time 是 2018-09-18，则 media_1d 中的数据是
#原始数据中 2018-05-02 的数据，且其中 origin_time 和 end_time 的日期都改为 2018-09-18，
#若到了 2018-09-19，则 media_1d 中的数据是 2018-05-03 的数据，且其中 origin_time 和
#end_time 的日期都改为 2018-09-19
echo -e "use user_profile;drop table if exists media_1d;\ncreate table media_1d
as select terminal_no,phone_no,duration,station_name,concat(CURRENT_DATE,'
',from_unixtime(unix_timestamp(origin_time),\"HH:mm:ss\")) as origin_time,concat
(CURRENT_DATE,' ',from_unixtime(unix_timestamp(end_time),\"HH:mm:ss\")) as
end_time,owner_code,owner_name,vod_cat_tags,resolution,audio_lang,region,r
es_name,res_type,vod_title,category_name,program_title,sm_name,first_show_
time from media_index_3m where date_add(origin_time,${delta}) = date_sub
(CURRENT_DATE,${curr_delta});">${hive_script}
hive -f /root/qwm/${hive_script}
#将 media_1d 表的数据导入 Elasticsearch 集群
spark-submit --class com.tipdm.scala.datasource.Hive2Elasticsearch --executor-
memory 10g --total-executor-cores 20 --master spark://$4:7077 --jars /opt/cloudera/
parcels/CDH-5.7.3-1.cdh5.7.3.p0.5/lib/spark/lib/elasticsearch-spark-13_2.1
0-6.3.2.jar $5 user_profile.media_1d $6 $7 $8 $9 ${10}
#Shell 脚本结束运行的时间
end_time=$(date +%Y-%m-%d-%H:%M:%S)
echo "end time : $end_time" >> $log_file
```

代码 4-22 的主要功能是从用户收视行为数据中筛选出时间符合规则的数据，并且将 origin_time 字段和 end_time 字段的日期修改为当前日期，将修改后的数据以覆盖的模式保存到 Hive 的 user_profile 库的 media_1d 表中，最后通过 Spark 程序将 media_1d 表中的数据以增量的模式更新到 Elasticsearch 集群中。

c. 为了测试 media_index_1d.sh 脚本是否正确，需要设置定时策略来测试脚本。在

server1 节点上输入“crontab -e”，加入命令“0 10 1 * * ? * /root/media_index_1d.sh 2018-10-10 create_media_1d.hql 2018-05-02 server3 /root/qwm/user_profile_project-1.0.jar 192.168.111.75 9200 origin_time media_index media”。定时任务是在 2018 年 10 月 15 日设置的，media_index_1d.sh 脚本会在 2018 年 10 月 16 日凌晨 1 点 10 分被触发执行，将数据增量更新到 Elasticsearch 集群的 media_indexyyyyww 索引中，其中，yyyyww 代表年份与周数，即以周为单位存储用户收视行为数据。在 2018 年 10 月 16 日查看的执行结果如图 4-12 所示。

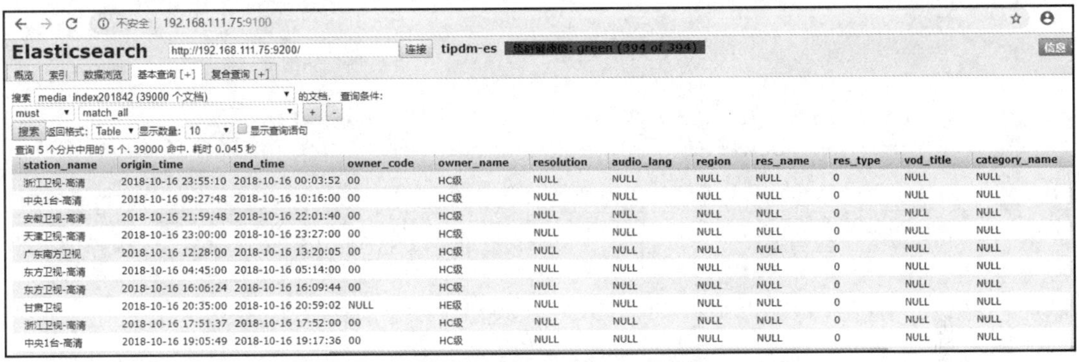

station_name	origin_time	end_time	owner_code	owner_name	resolution	audio_lang	region	res_name	res_type	vod_title	category_name
浙江卫视-高清	2018-10-16 23:55:10	2018-10-16 00:03:52	00	HC级	NULL	NULL	NULL	NULL	0	NULL	NULL
中央1台-高清	2018-10-16 09:27:48	2018-10-16 10:16:00	00	HC级	NULL	NULL	NULL	NULL	0	NULL	NULL
安徽卫视-高清	2018-10-16 21:59:48	2018-10-16 22:01:40	00	HC级	NULL	NULL	NULL	NULL	0	NULL	NULL
天津卫视-高清	2018-10-16 23:00:00	2018-10-16 23:27:00	00	HC级	NULL	NULL	NULL	NULL	0	NULL	NULL
广东南方卫视	2018-10-16 12:28:00	2018-10-16 13:20:05	00	HC级	NULL	NULL	NULL	NULL	0	NULL	NULL
东方卫视-高清	2018-10-16 04:45:00	2018-10-16 05:14:00	00	HC级	NULL	NULL	NULL	NULL	0	NULL	NULL
东方卫视-高清	2018-10-16 16:06:24	2018-10-16 16:09:44	00	HC级	NULL	NULL	NULL	NULL	0	NULL	NULL
甘肃卫视	2018-10-16 20:35:00	2018-10-16 20:59:02	NULL	HE级	NULL	NULL	NULL	NULL	0	NULL	NULL
浙江卫视-高清	2018-10-16 17:51:37	2018-10-16 17:52:00	00	HC级	NULL	NULL	NULL	NULL	0	NULL	NULL
中央1台-高清	2018-10-16 19:05:49	2018-10-16 19:17:36	00	HC级	NULL	NULL	NULL	NULL	0	NULL	NULL

图 4-12　在 2018 年 10 月 16 日查看的执行结果

从图 4-12 中可以看到，media_index201842 索引出现了 origin_time 为 2018-10-16 的记录，说明 media_index_1d.sh 的编写是正确的。

4.2 数据存储与传输

广电大数据用户画像项目采用了 Spark SQL 技术，根据需求探索环节得到的数据处理规则进行数据预处理，并根据用户画像规则进行标签计算，将画像标签结果保存到 MySQL 中。如果将数据存储在 Hive 中，那么 Spark 读取与存储数据更加方便。因此，需要对数据进行传输，读取 Elasticsearch 集群中的数据并将其导入 Hive 表；测试是否可将用户画像标签结果数据存储到 MySQL 中。

4.2.1 Elasticsearch 数据传输到 Hive

Spark 可以读取 Elasticsearch 数据并将其转换成 RDD，也可以直接将 Elasticsearch 数据读取为 DataFrame，考虑最终读取的数据需要保存到 Hive 中，本项目使用 Spark SQL 直接读取 Elasticsearch 数据为 DataFrame。Spark SQL 提供了 esDF(resource: String, query: String, cfg: scala.collection.Map[String, String])方法读取 Elasticsearch 数据，esDF()方法中的参数解释如表 4-2 所示。

表 4-2　esDF()方法中的参数解释

参数	解释
resource	Elasticsearch 数据的 index 和 type，如 mediamatch_usermsg/doc
query	查询语句
cfg	Elasticsearch 集群的相关配置，如 Elasticsearch 节点 IP 地址、端口号等

读取 mediamatch_usermsg 数据，并选取 phone_no 为 5143217 的数据，如代码 4-23 所示。

代码 4-23 读取 mediamatch_usermsg 数据

```
scala> import org.elasticsearch.spark.sql._
import org.elasticsearch.spark.sql._
scala> val defaultQuery: String = "?q=phone_no:5143217"
defaultQuery: String = ?q=phone_no:5143217
scala> val options =
     |   Map(
     |     ("es.nodes", "192.168.111.75"),
     |     ("es.port", "9200"),
     |     ("es.read.metadata", "false"),
     |     ("es.mapping.date.rich", "false")
     |   )
options: scala.collection.immutable.Map[String,String] = Map(es.nodes -> 
192.168.111.75, es.port -> 9200, es.read.metadata -> false, es.mapping.
date.rich -> false)
scala> val esTable="mediamatch_usermsg/doc"
esTable: String = mediamatch_usermsg/doc
scala> val esDf = sqlContext.esDF(esTable, defaultQuery, options)
esDf: org.apache.spark.sql.DataFrame = [@timestamp: string, @version: string,
addressoj: string, estate_name: string, force: string, host: string, message:
string, open_time: string, owner_code: string, owner_name: string, path:
string, phone_no: string, run_name: string, run_time: string, sm_code: string,
sm_name: string, tags: string, terminal_no: string]
scala>
esDf.select("phone_no","owner_name","owner_code","run_name","run_time").show()
+--------+----------+----------+--------+-------------------+
|phone_no|owner_name|owner_code|run_name|           run_time|
+--------+----------+----------+--------+-------------------+
| 5143217|      HC级|        00|    正常|2013-01-19 16:35:13|
+--------+----------+----------+--------+-------------------+
```

根据项目需求可知，本项目并非读取 Elasticsearch 集群中的所有数据，而是选取某个时间段的数据。例如，用户收视行为数据是选择 media_index 数据中当前时间之前 3 个月的数据，mediamatch_usermsg 则是选取当前时间之前 50 年的数据。因此，读取 Elasticsearch 数据时需要根据数据的时间字段选取指定时间范围的数据。

下面以 mediamatch_usermsg 数据为例，介绍如何从 Elasticsearch 集群中读取指定时间范围内的数据。先从 Elasticsearch 集群中读取 mediamatch_usermsg 中的所有数据为 DataFrame，再将 DataFrame 注册成为临时表，如代码 4-24 所示。

代码 4-24 读取 mediamatch_usermsg 中的所有数据

```
scala> val defaultQuery: String = "?q=*:*"
defaultQuery: String = ?q=*:*
scala> val esTable="mediamatch_usermsg/doc"
esTable: String = mediamatch_usermsg/doc
scala> val options =
     |   Map(
```

```
     |     ("es.nodes", "192.168.111.75"),
     |     ("es.port", "9200"),
     |     ("es.read.metadata", "false"),
     |     ("es.mapping.date.rich", "false")
     |  )
options:  scala.collection.immutable.Map[String,String]  =  Map(es.nodes  ->
192.168.111.75,  es.port  ->  9200,  es.read.metadata  ->  false,  es.mapping.
date.rich -> false)
scala> val esDf = sqlContext.esDF(esTable, defaultQuery, options)
esDf: org.apache.spark.sql.DataFrame = [@timestamp: string, @version: string,
addressoj: string, estate_name: string, force: string, host: string, message:
string,  open_time:  string,  owner_code:  string,  owner_name:  string,  path:
string, phone_no: string, run_name: string, run_time: string, sm_code: string,
sm_name: string, tags: string, terminal_no: string]
scala> val sparkTable: String = "tmp" + System.currentTimeMillis()
sparkTable: String = tmp1547517925190
scala> esDf.registerTempTable(sparkTable)
```

得到临时表后，可以通过一条SQL语句从临时表中过滤指定时间范围的数据，如选取当前时间之前50年的数据，通过“select * from sparkTable where run_time>当前时间-50年”的SQL语句实现，其中，“当前时间-50年”的计算可以通过自定义方法来实现，如代码4-25所示。定义getBeforeTime(timePattern:String,timeRangeValue:Int,timeRangePattern:String,startTime:String)方法，该方法中有4个参数，各个参数的解释如表4-3所示。

表4-3　getBeforeTime()方法中的参数解释

参数	解释
timePattern	指时间数据的格式，如“yyyy-MM-dd HH:mm:ss”
timeRangeValue、timeRangePattern	指时间跨度，如若为50年，则timeRangeValue为50，timeRangePattern为“Y”
startTime	指从哪个时间开始往前截取数据，如截取“2018-08-01 00:00:00”之前的数据，则startTime为“2018-08-01 00:00:00”

如果startTime为当前时间，那么可以通过自定义的getNowDate(timePattern:String)方法获取当前的时间。getBeforeTime()方法返回的类型是Date类型，而SQL语句中的过滤条件“run_time>当前时间-50年”中的“当前时间-50年”的结果需要为时间字符串类型，因此，自定义getBeforeTimeStr(timePattern: String, timeRangeValue:Int,timeRangePattern:String,startTime:String)方法，将getBeforeTime()方法返回的Date类型转换为String类型，如代码4-25所示。

代码4-25　通过自定义方法获取某个时间之前的时间

```
import java.text.SimpleDateFormat
import java.util.{Calendar, Date, Properties}
scala> def getBeforeTime(timePattern: String, timeRangeValue: Int, timeRangePattern:
String, startTime: String): Date = {
     |   val dateFormat: SimpleDateFormat = new SimpleDateFormat(timePattern)
     |   val cal: Calendar = Calendar.getInstance()
```

```
    |   cal.setTime(dateFormat.parse(startTime))
    |   timeRangePattern match {
    |     case "Y" => cal.add(Calendar.YEAR, -timeRangeValue)
    |     case "M" => cal.add(Calendar.MONTH, -timeRangeValue)
    |     case "D" => cal.add(Calendar.DATE, -timeRangeValue)
    |     case _ => throw new Exception("timeRangePattern not found!
timeRangePattern:" + timeRangePattern)
    |   }
    |   cal.getTime
    | }
getBeforeTime: (timePattern: String, timeRangeValue: Int, timeRangePattern:
String, startTime: String)java.util.Date
scala> def getBeforeTimeStr(timePattern: String, timeRangeValue: Int,
timeRangePattern: String, startTime: String): String = {
    |   val dateFormat: SimpleDateFormat = new SimpleDateFormat(timePattern)
    |   dateFormat.format(getBeforeTime(timePattern, timeRangeValue,
timeRangePattern, startTime))
    | }
getBeforeTimeStr: (timePattern: String, timeRangeValue: Int, timeRangePattern:
String, startTime: String)String
scala> def getNowDate(timePattern: String): String = {
    | val now: Date = new Date()
    | val dateFormat: SimpleDateFormat = new SimpleDateFormat(timePattern)
    | val date = dateFormat.format(now)
    | return date
    | }
getNowDate: (timePattern: String)String
scala> getBeforeTimeStr("yyyy-MM-dd HH:mm:ss",3,"M",getNowDate("yyyy-MM-dd
HH:mm:ss"))
res8: String = 2018-10-15 10:38:03
```

通过代码 4-25 定义的方法，从代码 4-24 中注册的临时表中选取当前时间之前 50 年的数据，并将选中的数据保存到 Hive 的 user_profile 库的 mediamatch_usermsg 中，如代码 4-26 所示。

代码 4-26　选取指定时间的数据并将其保存到 Hive 中

```
scala> val hiveTable="user_profile.mediamatch_usermsg"
hiveTable: String =user_profile. mediamatch_usermsg
scala> val selectedCols= \
"addressoj,estate_name,force,host,message,open_time,owner_code,owner_name,
phone_no, run_name,run_time, sm_code,sm_name,tags,terminal_no"
selectedCols: String = addressoj,estate_name,force,host,message,open_time,
owner_code,owner_name,phone_no, run_name,run_time, sm_code,sm_name,tags,terminal_no
scala> val timeColName="run_time"
timeColName: String = run_time
scala> val sql = "CREATE TABLE " + hiveTable + " as  select " + selectedCols +
" from " + sparkTable +" where " + timeColName + " > '" + getBeforeTimeStr
("yyyy-MM-dd HH:mm:ss",50,"Y",getNowDate("yyyy-MM-dd HH:mm:ss")) + "'"
```

```
sql: String = CREATE TABLE user_profile. mediamatch_usermsg as  select addressoj,
estate_name,force,host,message,open_time,owner_code,owner_name,phone_no,  run_
name,run_time, sm_code,sm_name,tags,terminal_no from tmp1547517925190 where
run_ time > '1969-01-15 11:09:11'
scala>sqlContext.sql(sql)
```

在本项目提供的数据中，mediamatch_usermsg、mediamatch_userevent、order_index 和 mmconsume_billevents 数据都是一份数据保存到一个表中，Spark 从 Elasticsearch 集群中读取这 4 份数据的操作可参考代码 4-26 所示的内容实现。

media_index 与其他 4 份数据不一样，media_index 是根据时间，将每周的数据保存到一张表（一个 index）中，如图 4-1 所示。因此，读取 media_index 数据的方式也与 mediamatch_usermsg 等数据稍有不同。读取 media_index 数据需要将指定时间范围内的所有与 media_index 相关的 index/type 找到，如读取 2018-08-01 之前 10 天的数据，需要匹配到 media_index201830/media 和 media_index201831/media。为了能够根据指定的时间范围获取 media_index 的所有 index/type，需要自定义 getRangeDays()方法和 getBeforeTimeTableNames()方法来实现，如代码 4-27 所示。

代码 4-27 自定义方法获取 index/type

```
scala> import scala.collection.mutable.ArrayBuffer
import scala.collection.mutable.ArrayBuffer
scala> import java.text.SimpleDateFormat
import java.text.SimpleDateFormat
scala> import java.util.{Calendar, Date, Properties}
import java.util.{Calendar, Date, Properties}
scala> def getRangeDays(start: Date, end: Date): Array[Date] = {
     |   val calBegin = Calendar.getInstance()
     |   // 使用给定的 Date 设置此 Calendar 的时间
     |   calBegin.setTime(start);
     |   val calEnd = Calendar.getInstance()
     |   // 使用给定的 Date 设置此 Calendar 的时间
     |   calEnd.setTime(end)
     |   val arr = new ArrayBuffer[Date]()
     |   while (end.after(calBegin.getTime())) {
     |     arr.append(calBegin.getTime)
     |     calBegin.add(Calendar.DATE, 1) // day
     |   }
     |   arr.toArray
     | }
getRangeDays: (start: java.util.Date, end: java.util.Date)Array[java.util.Date]
scala> def getBeforeTimeTableNames(timePattern: String, timeRangeValue: Int,
timeRangePattern: String,esIndexPre: String, esIndexType: String, esType:
String, startTime: String): List[String] = {
     |   val start = getBeforeTime(timePattern, timeRangeValue, timeRangePattern,
startTime)
     |   val df: SimpleDateFormat = new SimpleDateFormat(timePattern)
     |   val end = df.parse(startTime)
     |   val dateFormat = new SimpleDateFormat(esIndexType)
```

```
    |   val days = for (d <- getRangeDays(start, end)) yield dateFormat.
format(d)
    |   days.toSet.map((x: String) => esIndexPre + x + "/" + esType).toList
    | }
getBeforeTimeTableNames:  (timePattern:  String,  timeRangeValue:  Int,
timeRangePattern: String, esIndexPre: String, esIndexType: String, esType:
String, startTime: String)List[String]
scala> getBeforeTimeTableNames("yyyy-MM-dd HH:mm:ss",10,"D","media_index",
"yyyyww","media","2018-08-01 00:00:00")
res7: List[String] = List(media_index201830/media, media_index201831/media)
```

在代码 4-27 中，getRangeDays()方法用于获取指定时间范围内的所有日期。如获取 2018-08-01 之前 10 天的日期，使用 getRangeDays()方法获取的即 2018-07-22 至 2018-07-31 这 10 天的日期。getBeforeTimeTableNames()方法有 7 个参数，各个参数的解释如表 4-4 所示。

表 4-4　getBeforeTimeTableNames()方法中的参数解释

参数	解释
timePattern	指时间数据的格式，如“yyyy-MM-dd HH:mm:ss”
timeRangeValue、timeRangePattern	指时间跨度，若为 50 年，则 timeRangeValue 为 50，timeRangePattern 为“Y”
esIndexPre	指的是 index 的前缀，如 index 为“media_index201831”中的“media_index”
exIndexType	指的是 index 后缀的类型，media_index 每周会形成一个 index，因此 esIndexType 为“yyyyww”
esType	指的是 media_index 数据的 type，即“media”
startTime	指从哪个时间开始往前截取数据，如要截取“2018-08-01 00:00:00”之前的数据，则 startTime 为“2018-08-01 00:00:00”

虽然代码 4-27 获取了 media_index 数据指定时间范围内的所有 index/type，但是获取到的 index/type 并不一定真实存在于 Elasticsearch 数据中，现实中可能会由于某些原因（如甲方人员收集数据失误）而导致数据缺失，如果有一周的数据缺失，那么 Elasticsearch 数据中不会有这一周对应的 index/type，而读取不存在的 index/type 数据会导致程序运行失败，因此，为了提升程序的健壮性，需要对 getBeforeTimeTableNames()方法得到的 index/type 列表进行进一步筛选，筛选出 Elasticsearch 数据中存在的 index/type。为判断 index/type 是否存在于 Elasticsearch 数据中并获取到判断结果，需要使用 RestClient 类与 Elasticsearch 集群进行通信，通过 RestClient 类中的 typeExists(String index,String type)方法判断 index/type 是否存在，如代码 4-28 所示。

代码 4-28　判断 index/type 是否存在于 Elasticsearch 数据中

```
scala> import org.elasticsearch.hadoop.rest.RestClient
import org.elasticsearch.hadoop.rest.RestClient
scala> import org.elasticsearch.spark.cfg.SparkSettingsManager
import org.elasticsearch.spark.cfg.SparkSettingsManager
```

```
scala> import org.elasticsearch.spark.sql._
import org.elasticsearch.spark.sql._
scala> import scala.collection.JavaConverters._
import scala.collection.JavaConverters._
scala> val options = Map(
     |   ("es.nodes", "192.168.111.75"),
     |   ("es.port", "9200"),
     |   ("es.read.metadata", "false"),
     |   ("es.mapping.date.rich", "false"),
     |   ("es.read.field.as.array.include", "vod_cat_tags")
     | )
options: scala.collection.immutable.Map[String,String] = Map(es.read.field.
as.array.include -> vod_cat_tags, es.mapping.date.rich -> false, es.port ->
9200, es.read.metadata -> false, es.nodes -> 192.168.111.75)
scala> val settings = new SparkSettingsManager().load(sqlContext.sparkContext.
getConf).merge(options.asJava)
settings: org.elasticsearch.hadoop.cfg.Settings = org.elasticsearch.spark.cfg.
SparkSettings@1b599d06
scala> val client = new RestClient(settings)
client: org.elasticsearch.hadoop.rest.RestClient = org.elasticsearch.hadoop.
rest.RestClient@59114cd5
scala> val allExistTables = allTables.filter { x => val s = x.split("/");
client.typeExists(s(0), s(1)) }
allExistTables: List[String] = List(media_index201830/media, media_index201831/
media)
scala> client.close()
scala> println("allExistTables : " + allExistTables)
allExistTables : List(media_index201830/media, media_index201831/media)
```

代码 4-28 得到的 allExistTables 列表即所有满足条件的 index/type，然后通过列表的 map 操作，使用 esDF()方法获取 index/type 的数据并将其转换为 DataFrame，得到一个包含多个 DataFrame 元素的列表。因为最后需要将获取到的所有数据存储到同一张 Hive 表中，所以可以先将列表中的多个 DataFrame 元素合并成一个 DataFrame，如代码 4-29 所示。参考代码 4-26 将获取到的 media_index 数据保存到 Hive 表中。

代码 4-29　获取所有 index/type 数据为 DataFrame

```
scala> val selectedCols= \
"audio_lang,category_name,duration,end_time,first_show_time,origin_time,owner_
code,owner_name,phone_no,program_title,region,res_name,res_type,resolution,
sm_name,station_name,terminal_no,vod_cat_tags,vod_title"
selectedCols: String = \
audio_lang,category_name,duration,end_time,first_show_time,origin_time,owner_
code,owner_name,phone_no,program_title,region,res_name,res_type,resolution,
sm_name,station_name,terminal_no,vod_cat_tags,vod_title
scala> val selectedColsArr = selectedCols.split(",")
selectedColsArr: Array[String] = Array(audio_lang, category_name, duration,
end_time, first_show_time, origin_time, owner_code, owner_name, phone_no,
program_title, region, res_name, res_type, resolution, sm_name, station_name,
terminal_no, vod_cat_tags, vod_title)
```

```
scala> val firstCol = selectedColsArr(0).trim
firstCol: String = audio_lang
scala> val tailCols = selectedColsArr.slice(1, selectedColsArr.length).map(_.trim)
tailCols: Array[String] = Array(category_name, duration, end_time, first_
show_time, origin_time, owner_code, owner_name, phone_no, program_title,
region, res_name, res_type, resolution, sm_name, station_name, terminal_no,
vod_cat_tags, vod_title)
scala> val esDf = allExistTables.map(x => sqlContext.esDF(x, default_query,
options)).reduce((x1, x2) => x1.select(firstCol, tailCols: _*).unionAll
(x2.select(firstCol, tailCols: _*)))
esDf: org.apache.spark.sql.DataFrame = [audio_lang: string, category_name:
string, duration: string, end_time: string, first_show_time: string,
origin_time: string, owner_code: string, owner_name: string, phone_no: string,
program_title: string, region: string, res_name: string, res_type: string,
resolution: string, sm_name: string, station_name: string, terminal_no: string,
vod_cat_tags:
array<struct<level1_name:string,level2_name:string,level3_name:string,leve
l4_name:string,level5_name:string>>, vod_title: string]
```

因为后续需要定时将 Elasticsearch 数据传输到 Hive 中，为了方便任务的调度，所以对 Elasticsearch 数据传输到 Hive 中的代码进行封装。Elasticsearch 中的数据有两种类型：一种是一张表的数据，如 mediamatch_usermsg；另一种是多张表的数据，如 media_index。要将这两种类型的数据传输到 Hive 中，需要封装两个类。

针对将一张表的数据传输到 Hive 中的封装类 Elasticsearch2Hive，设置的参数如下。

（1）selectedCols：Elasticsearch 资源同步的列字段名。

（2）timeColName：Elasticsearch 资源时间列名称。

（3）timeColPattern：Elasticsearch 资源时间列格式，如 yyyyMMdd HH:mm:ss。

（4）timeRangeValue：Elasticsearch 资源同步时间段值，如同步 1 个月的数据，则该参数值需要设置为 1。

（5）timeRangeType：Elasticsearch 资源同步时间段类型（Y|M|D），如同步 1 个月的数据，则该参数值需要设置为 M。

（6）esTable：Elasticsearch 资源名，即 index/type，如 mediamatch_usermsg/doc。

（7）startTime：同步设置的某个时间之前的数据，如提供的业务数据是 2018-08-01 之前的数据，则该参数值可以设置为 2018-08-01。

封装类 Elasticsearch2Hive 代码的具体实现如代码 4-30 所示。

代码 4-30　封装类 Elasticsearch2Hive 代码的具体实现

```
package com.tipdm.scala.chapter_3_5_1_datasource
import com.tipdm.scala.util.SparkUtils
import org.apache.spark.sql.hive.HiveContext
import org.apache.spark.{SparkConf, SparkContext}
import org.elasticsearch.spark.sql._
/**
  * hiveTable:Hive 表
  * selectedCols: Elasticsearch 资源同步的列字段名
```

```
  * timeColName: Elasticsearch 资源时间列名称
  * timeColPattern: Elasticsearch 资源时间列格式，如 yyyyMMdd HH:mm:ss
  * timeRangeValue (required): Elasticsearch 资源同步时间段值
  * timeRangeType (required): Elasticsearch 资源同步时间段类型，Y|M|D
  * esTable: Elasticsearch 资源名，index/type
  * startTime:同步设置的某个时间之前的数据
  */
object Elasticsearch2Hive {
  val defaultQuery: String = "?q=*:*"
  def main(args: Array[String]): Unit = {
    if (args.length != 9) {
      printUsage()
      System.exit(1)
    }
    val hiveTable = args(0)
    val selectedCols = args(1)
    val timeColName = args(2)
    val timeColPattern = args(3)
    val timeRangeValue = args(4).toInt
    val timeRangeType = args(5)
    val esTable = args(6)
    val startTime = args(7)
    val db=args(8)
    val options =
      Map(
        ("es.nodes", "192.168.111.75"),
        ("es.port", "9200"),
        ("es.read.metadata", "false"),
        ("es.mapping.date.rich", "false")
      )
    val conf = new SparkConf().setAppName("Elasticsearch2Hive")
    val sc = new SparkContext(conf)
    val sqlContext = new HiveContext(sc)
    sqlContext.sql("set spark.sql.caseSensitive=true")
    val sparkTable: String = "tmp" + System.currentTimeMillis()
    val esDf = sqlContext.esDF(esTable, defaultQuery, options)
    esDf.registerTempTable(sparkTable)
    val sql = "CREATE TABLE " + db+"."+hiveTable + " as  select " + selectedCols
+ " from " + sparkTable +
      " where " + timeColName + " > '" + SparkUtils.getBeforeTimeStr
(timeColPattern, timeRangeValue, timeRangeType, startTime) + "'"
    println("Hive 中是否存在"+":"+SparkUtils.exists(sqlContext, db, hiveTable))
    if (SparkUtils.exists(sqlContext, db, hiveTable)) {
      SparkUtils.dropTable(sqlContext, db,hiveTable)
      println("Hive 中是否存在"+db+"."+hiveTable+":"+SparkUtils.exists(sqlContext,
db, hiveTable))
      sqlContext.sql(sql)
    } else {
      sqlContext.sql(sql)
    }
```

```
    sc.stop()
  }
  /**
    * 使用说明
    */
  def printUsage(): Unit = {
    val buff = new StringBuilder
    buff.append("Usage : com.tipdm.scala.chapter_3_5_1_datasource.
Elasticsearch2Hive"). append(" ")
      .append("<hiveTable>").append(" ")
      .append("<selectedCols>").append(" ")
      .append("<timeColName>").append(" ")
      .append("<timeColPattern>").append(" ")
      .append("<timeRangeValue>").append(" ")
      .append("<timeRangeType>").append(" ")
      .append("<esTable>").append(" ")
      .append("<startTime>").append(" ")
      .append("<db>").append(" ")
    println(buff.toString())
  }
}
```

针对代码 4-30，编写相关的测试代码，如代码 4-31 所示。

代码 4-31　Elasticsearch2Hive 类的测试代码

```
package com.tipdm.java.chapter_3_5_1_datasource;
import com.tipdm.engine.SparkYarnJob;
import com.tipdm.engine.model.Args;
import com.tipdm.engine.model.SubmitResult;
public class Elasticsearch2Hive {
   private static String className = "com.tipdm.scala.chapter_3_5_1_
datasource.Elasticsearch2Hive";
   private static String applicationName = "Elasticsearch2Hive";
   public static void main(String[] args) throws Exception {
      String[] arguments = new String[9];
      arguments[0]="mediamatch_usermsg_test";
arguments[1]="terminal_no,phone_no,sm_name,run_name,sm_code,owner_name,
owner_code,run_time,addressoj,estate_name,open_time,force";
      arguments[2]="run_time";
      arguments[3]="yyyy-MM-dd HH:mm:ss";
      arguments[4]="1";
      arguments[5]="Y";
      arguments[6]="mediamatch_usermsg/doc";
      arguments[7]="2018-08-01 00:00:00";
      arguments[8]="user_profile";
      Args innerArgs = Args.getArgs(applicationName,className,arguments);
      SubmitResult submitResult = SparkYarnJob.run(innerArgs);
      SparkYarnJob.monitor(submitResult);
      System.out.println("任务运行成功");
   }
}
```

执行代码 4-31，并查看 Elasticsearch2Hive 类的测试运行结果，如图 4-13 所示。

driver-20190301155124-0012	Fri Mar 01 15:51:24 CST 2019	worker-20190301145033-192.168.111.77-7078	FINISHED	1	1024.0 MB	com.tipdm.scala.chatper_3_5_1_datasource.Elasticsearch2Hive		
app-20190301155128-0015	Elasticsearch2Hive	20	8.0 GB		2019/03/01 15:51:28	spark	FINISHED	19 min

图 4-13　Elasticsearch2Hive 类的测试运行结果

若任务状态如图 4-13 所示，则表示任务运行成功。任务运行成功后，查看 Hive 中的数据存储情况，如代码 4-32 所示。

代码 4-32　Elasticsearch2Hive 类测试代码运行成功后查看 Hive 中的数据存储情况

```
0: jdbc:hive2://server3:10000> use user_profile;
No rows affected (0.023 seconds)
0: jdbc:hive2://server3:10000> show tables;
+--------------------------+--+
|        tab_name          |
+--------------------------+--+
| mediamatch_usermsg_test  |
+--------------------------+--+
5 rows selected (0.031 seconds)0: jdbc:hive2://server3:10000> select
terminal_no,phone_no,sm_name,run_name from mediamatch_usermsg_test limit 5;
+--------------+-----------+----------+-----------+--+
| terminal_no  | phone_no  |sm_name   | run_name  |
+--------------+-----------+----------+-----------+--+
| 2000014626   | 5138414   |互动电视  | 正常      |
| 2000255020   | 5160854   |数字电视  | 主动暂停  |
|  20000131    | 5149463   |甜果电视  | 正常      |
| 2000227048   | 5149941   |互动电视  | 欠费暂停  |
|  20000043    | 5156820   |数字电视  | 正常      |
+--------------+-----------+----------+-----------+--+
5 rows selected (1.602 seconds)
```

根据代码 4-32 的结果，可以判断 Elasticsearch2Hive 类的代码封装是正确的，能根据实际需求从 Elasticsearch 集群中读取数据并将其保存到 Hive 表中。

针对将多张表的数据传输到 Hive 中的封装类 ElasticsearchMulti2Hive，设置的参数如下。

（1）selectedCols：Elasticsearch 资源同步的列字段名。

（2）tsColsName：Elasticsearch 资源时间列名称。

（3）tsColPattern：Elasticsearch 资源时间列格式，如 yyyyMMdd HH:mm:ss。

（4）timeRangeValue：Elasticsearch 资源同步时间段值，如同步 1 个月的数据，则该参数值需要设置为 1。

（5）timeRangeType：Elasticsearch 资源同步时间段类型（Y|M|D），如同步 1 个月的数据，则该参数值需要设置为 M。

（6）esIndexType：Elasticsearch 资源名 index 后缀类型（yyyyMMdd|yyyyMM|yyyyww），yyyyMMdd 表示每天一个 index，yyyyMM 表示每月一个 index，yyyyww 表示每周一个 index。

（7）esIndexPre：Elasticsearch 资源名 index 前缀，如 index 为 media_index201819，则

其前缀为 media_index。

（8）esType：Elasticsearch 资源名的类型，设置为“media”。

（9）startTime：同步设置的某个时间之前的数据，如提供的业务数据是 2018-08-01 之前的数据，则该参数值可以设置为 2018-08-01。

封装类 ElasticsearchMulti2Hive 代码的具体实现，如代码 4-33 所示。

代码 4-33　封装类 ElasticsearchMulti2Hive 代码的具体实现

```
package com.tipdm.scala.chapter_3_5_1_datasource
import com.tipdm.scala.util.SparkUtils
import org.apache.spark.{SparkConf, SparkContext}
import org.elasticsearch.hadoop.rest.RestClient
import org.elasticsearch.spark.cfg.SparkSettingsManager
import org.elasticsearch.spark.sql._
import scala.collection.JavaConverters._
/**
  * hiveTable:Hive 表
  * selectedCols: Elasticsearch 资源同步的列字段名
  * tsColName: Elasticsearch 资源时间列名称
  * tsColPattern: Elasticsearch 资源时间列格式，如 yyyyMMdd HH:mm:ss
  * timeRangeValue (required): Elasticsearch 资源同步时间段值
  * timeRangeType (required): Elasticsearch 资源同步时间段类型，Y|M|D
  * esIndexType: Elasticsearch 资源名 index 后缀类型，yyyyMMdd|yyyyMM|yyyyww
  * esIndexPre: Elasticsearch 资源名 index 前缀，index prefix
  * esType: Elasticsearch 资源名的类型
  * startTime:同步设置的某个时间之前的数据
  */
object ElasticsearchMulti2Hive {
  val default_query: String = "?q=*:*"
  def main(args: Array[String]): Unit = {
      if (args.length != 11) {
      printUsage()
      System.exit(1)
    }
    val hiveTable = args(0)
    val selectedCols = args(1)
    val tsColName = args(2)
    val tsColPattern = args(3)
    val timeRangeValue = args(4).toInt
    val timeRangeType = args(5)
    val esIndexType = args(6)
    val esIndexPre = args(7)
    val esType = args(8)
    val startTime = args(9)
    val db=args(10)
    val options = Map(
      ("es.nodes", "192.168.111.75"),
      ("es.port", "9200"),
      ("es.read.metadata", "false"),
```

```
      ("es.mapping.date.rich", "false"),
      ("es.read.field.as.array.include", "vod_cat_tags")
    )
    val conf = new SparkConf().setAppName("ElasticsearchMulti2Hive")
    val sc = new SparkContext(conf)
    val sqlContext = new org.apache.spark.sql.hive.HiveContext(sc)
    sqlContext.sql("set spark.sql.caseSensitive=true")
    val sparkTable: String = "tmp" + System.currentTimeMillis()
    val sql = "CREATE TABLE " + db+"."+hiveTable + " as  select " +
selectedCols + " from " + sparkTable + " where " + tsColName + " >
'" + SparkUtils.getBeforeTimeStr(tsColPattern, timeRangeValue,
timeRangeType, startTime) + "'"
    val selectedColsArr = selectedCols.split(",")
    val firstCol = selectedColsArr(0).trim
    val tailCols = selectedColsArr.slice(1, selectedColsArr.length)
.map(_.trim)
    val allTables = SparkUtils.getBeforeTimeTableNames(tsColPattern,
 timeRangeValue, timeRangeType, esIndexPre, esIndexType, esType, startTime)
    println("allTables:" + allTables)
    // 先判断表是否存在
    val settings = new SparkSettingsManager().load(sqlContext.sparkContext.
getConf).merge(options.asJava)
    val client = new RestClient(settings)
    val allExistTables = allTables.filter { x => val s = x.split("/");
client.typeExists(s(0), s(1)) }
    client.close()
    println("allExistTables : " + allExistTables)
    val esDf = allExistTables.map(x => sqlContext.esDF(x, default_query,
options))
      .reduce((x1, x2) => x1.select(firstCol, tailCols: _*).unionAll
(x2.select(firstCol, tailCols: _*)))
    esDf.registerTempTable(sparkTable)
    if (SparkUtils.exists(sqlContext, db, hiveTable)) {
      SparkUtils.dropTable(sqlContext, db,hiveTable)
      sqlContext.sql(sql)
    } else {
      sqlContext.sql(sql)
    }
    sc.stop()
  }
  /**
    * 使用说明
    */
  def printUsage(): Unit = {
    val buff = new StringBuilder
    buff.append("Usage : com.tipdm.scala.chapter_3_5_1_datasource.
ElasticsearchMulti2Hive").append(" ")
      .append("<hiveTable>").append(" ")
      .append("<selectedCols>").append(" ")
      .append("<timeColName>").append(" ")
```

```
        .append("<timeColPattern>").append(" ")
        .append("<timeRangeValue>").append(" ")
        .append("<timeRangeType>").append(" ")
        .append("<esIndexType>").append(" ")
        .append("<esIndexPre>").append(" ")
        .append("<esType>").append(" ")
        .append("<startTime>").append(" ")
        .append("<db>").append(" ")
      println(buff.toString())
    }
}
```

针对代码 4-33，编写相关的测试代码，如代码 4-34 所示。

代码 4-34　ElasticsearchMulti2Hive 类的测试代码

```
package com.tipdm.java.com.tipdm.java.chapter_3_5_1_datasource;
import com.tipdm.engine.SparkYarnJob;
import com.tipdm.engine.model.Args;
import com.tipdm.engine.model.SubmitResult;
public class ElasticsearchMulti2Hive {
   private static String className = "com.tipdm.scala.chapter_3_5_1_
datasource.ElasticsearchMulti2Hive";
   private static String applicationName = "ElasticsearchMulti2Hive";
   public static void main(String[] args) throws Exception {
      String[] arguments = new String[11];
      arguments[0] = "media_index_3m_test";
      arguments[1] = "terminal_no,phone_no,duration,station_name,origin_
time,end_time,owner_code,owner_name,vod_cat_tags,resolution,audio_lang,
region,res_name,res_type,vod_title,category_name,program_title,sm_name,
first_show_time";
      arguments[2] = "origin_time";
      arguments[3] = "yyyy-MM-dd HH:mm:ss";
      arguments[4] = "1";
      arguments[5] = "D";
      arguments[6] = "yyyyww";
      arguments[7] = "media_index";
      arguments[8] = "media";
      arguments[9] = "2018-08-01 00:00:00";
      arguments[10]="user_profile";
      Args innerArgs = Args.getArgs(applicationName,className,arguments);
      SubmitResult submitResult = SparkYarnJob.run(innerArgs);
      SparkYarnJob.monitor(submitResult);
      System.out.println("任务运行成功");
   }
}
```

执行代码 4-34，并查看 ElasticsearchMulti2Hive 类的测试运行结果，如图 4-14 所示。

driver-20190301160954-0013	Fri Mar 01 16:09:54 CST 2019	worker-20190301145033-192.168.111.77-7078	FINISHED	1	1024.0 MB	com.tipdm.scala.chatper_3_5_1_datasource.ElasticsearchMulti2Hive

app-20190301160957-0016	media_index_3m	12	8.0 GB	2019/03/01 16:09:57	spark	FINISHED	17 s

图 4-14　ElasticsearchMulti2Hive 类的测试运行结果

若任务状态如图 4-14 所示，则表示任务运行完成。任务运行成功后，查看 Hive 中的数据存储情况，如代码 4-35 所示。

代码 4-35　ElasticsearchMulti2Hive 类测试代码运行成功后查看 Hive 中的数据存储情况

```
0: jdbc:hive2://server3:10000> use user_profile;
No rows affected (0.023 seconds)
0: jdbc:hive2://server3:10000> show tables;
+--------------------------+--+
|       tab_name           |
+--------------------------+--+
| media_index_3m_test      |
| mediamatch_usermsg_test  |
+--------------------------+--+
5 rows selected (0.031 seconds)
0: jdbc:hive2://server3:10000> select terminal_no,phone_no,sm_name,owner_name,
program_title from media_index_3m_test limit 5;
+--------------+-----------+----------+------------+---------------+--+
| terminal_no  | phone_no  | sm_name  | owner_name | program_title    |
+--------------+-----------+----------+------------+---------------+--+
| 2000361347   | 1452711   | 互动电视 | HC 级      | 城市话题         |
| 2000015154   | 1312616   | 互动电视 | HC 级      | 铁梨花(35)       |
| 2000430876   | 1319543   | 互动电视 | HC 级      | 广州早晨         |
| 2000364012   | 1940936   | 互动电视 | HC 级      | 宣传片           |
| 1200067394   | 1985705   | 互动电视 | HC 级      | 铁梨花(36)       |
+--------------+-----------+----------+------------+---------------+--+
5 rows selected (1.064 seconds)
```

根据代码 4-35 的结果，可以判断 ElasticsearchMulti2Hive 类的代码封装是正确的。可以根据实际需求从 Elasticsearch 集群中读取具有相同前缀的不同 index 数据并将其保存到 Hive 表中。

4.2.2　用户画像标签结果保存到 MySQL

3.3.1 小节中提到使用 Spark SQL 技术实现用户画像，并将用户画像的结果保存在 MySQL 中。Spark SQL 实现用户画像得到的是 DataFrame 类型的数据，包含 3 个字段，分别为 phone_no（用户编号）、label（标签）、parent_label（父标签）。要想将 DataFrame 类型的数据保存到 MySQL 中，需要在 MySQL 数据库中设计并创建一张标签表 user_label，用于保存用户画像标签结果数据，如代码 4-36 所示。

代码 4-36　创建标签表 user_label

```
create database user_profile;
use user_profile;
CREATE TABLE 'user_label' (
  'phone_no' text,
  'label' text,
  'parent_label' text NOT NULL
) ENGINE=MyISAM DEFAULT CHARSET=utf8;
```

将 DataFrame 类型的数据保存到 MySQL 数据库中时，需要先将其通过 write()方法转换为 DataFrameWriter，DataFrameWriter 通过 jdbc()方法将数据传输到 MySQL 中，jdbc()方法的定义如图 4-15 所示。

```
def jdbc(url: String, table: String, connectionProperties: Properties): Unit
```

Saves the content of the DataFrame to an external database table via JDBC. In the case the table already exists in the external database, behavior of this function depends on the save mode, specified by the mode function (default to throwing an exception).

Don't create too many partitions in parallel on a large cluster; otherwise Spark might crash your external database systems.

You can set the following JDBC-specific option(s) for storing JDBC:

- truncate (default false): use TRUNCATE TABLE instead of DROP TABLE.

In case of failures, users should turn off truncate option to use DROP TABLE again. Also, due to the different behavior of TRUNCATE TABLE among DBMS, it's not always safe to use this. MySQLDialect, DB2Dialect, MsSqlServerDialect, DerbyDialect, and OracleDialect supports this while PostgresDialect and default JDBCDirect doesn't. For unknown and unsupported JDBCDirect, the user option truncate is ignored.

url JDBC database url of the form jdbc:subprotocol:subname

table Name of the table in the external database.

connectionProperties JDBC database connection arguments, a list of arbitrary string tag/value. Normally at least a "user" and "password" property should be included. "batchsize" can be used to control the number of rows per insert. "isolationLevel" can be one of "NONE", "READ_COMMITTED", "READ_UNCOMMITTED", "REPEATABLE_READ", or "SERIALIZABLE", corresponding to standard transaction isolation levels defined by JDBC's Connection object, with default of "READ_UNCOMMITTED".

图 4-15 jdbc()方法的定义

jdbc()方法中有 3 个参数，url 参数指的是连接至 MySQL 数据库的 URL，如 jdbc:mysql://192.168.111.75:3306/user_profile；table 参数指的是要存放数据的表名；connectionProperties 参数指的是连接数据库的配置，如用户名和密码。

为测试是否可以将 DataFrame 数据保存到 MySQL 中，通过代码 4-37 所示的内容，计算得到品牌名称的标签数据 data。

代码 4-37 电视入网程度标签数据 data

```
scala> val data=sqlContext.sql("select t1.phone_no,case when T>8 then '老用户' when T>4 and T<=8 then '中等用户' when T<=4 then '新用户' end as label,'电视入网程度' as parent_label from(select \
phone_no,max(datediff(current_date(),open_time)/365) as T from user_profile.mediamatch_usermsg_process where sm_name like '%电视%' and open_time is not NULL group by phone_no) t1")
scala> data.show(5)
+--------+-----+------------+
|phone_no|label|parent_label|
+--------+-----+------------+
| 1055997|  老用户|       电视入网程度|
| 1111129|  老用户|       电视入网程度|
| 1317671|  老用户|       电视入网程度|
| 1321504|  老用户|       电视入网程度|
```

```
| 1433825|  老用户|      电视入网程度|
+--------+-----+------------+
only showing top 5 rows
```

将 data 保存到 MySQL 中，设置 url 为“jdbc:mysql://192.168.111.75:3306/user_profile”，数据库的用户名为“root”，密码为“root”，保存模式为“append”。先通过 write()方法将数据转换为 DataFrameWriter，再调用 jdbc()方法将 data 数据保存到 MySQL 中，如代码 4-38 所示。

代码 4-38　将 data 数据保存到 MySQL 中

```
scala> val saveMode = "append"
saveMode: String = append
scala> val outputTable="user_label"
outputTable: String = user_label
scala> val url = "jdbc:mysql://192.168.111.75:3306/user_profile"
url: String = jdbc:mysql://192.168.111.75:3306/user_profile
scala> import java.util.Properties
import java.util.Properties
scala> val connectionProperties = new Properties()
connectionProperties: java.util.Properties = {}
scala> connectionProperties.setProperty("user","root")
res5: Object = root
scala> connectionProperties.setProperty("password","root")
res6: Object = root
scala> data.write.mode(saveMode).jdbc(url, outputTable, connectionProperties)
```

查询保存到 MySQL 中的 user_label 的前 10 条记录，结果如图 4-16 所示。

```
1  select * from user_label limit 10;
```

信息　结果1　概况　状态

phone_no	label	parent_label
1062678	老用户	电视入网程度
1356032	老用户	电视入网程度
1431434	老用户	电视入网程度
1063611	老用户	电视入网程度
1452260	老用户	电视入网程度
1462223	老用户	电视入网程度
1109737	老用户	电视入网程度
1462836	老用户	电视入网程度
1476985	老用户	电视入网程度
1503605	老用户	电视入网程度

图 4-16　user_label 前 10 条记录的查询结果

图 4-16 所示的结果说明 DataFrame 数据可以保存到 MySQL 中，后续实现用户画像时可将画像标签结果数据保存到 MySQL 中。

4.3 基础数据预处理

根据 3.2.2 小节中的异常数据探索和 3.2.3 小节中的用户收视行为无效数据的探索结果，得到业务数据表预处理规则如表 4-5 所示。

表 4-5　业务数据表预处理规则

表名称	表预处理规则
用户状态信息变更表 账单信息表 订单信息表 用户基本信息表 用户收视行为信息表	1. 数据去重。 2. 删除 owner_name='EA 级', 'EB 级', 'EC 级', 'ED 级', 'EE 级'的记录。 3. 删除 owner_code=02,09,10 的记录。 4. 保留 sm_name='珠江宽频', '数字电视', '互动电视', '甜果电视'的记录
用户状态信息变更表 订单信息表 用户基本信息表	保留 run_name='正常','主动暂停','欠费暂停','主动销户'的记录
用户基本信息表	每个用户只保留 run_time 时间最大的记录
用户收视行为信息表	1. 保留收视时长 20s≤duration≤5h 的记录。 2. 删除用户收视行为信息表中 res_type=0 时 origin_time 和 end_time 中秒单位为 00 的记录。 3. 删除用户编号为 5401487 的记录

从表 4-5 中可以发现，这 5 张表的数据预处理规则都有共同的地方。此外，用户基本信息表与用户收视行为信息表还有其他处理规则。根据此特点，可以封装一个函数来实现数据处理过程，以达到代码重用的目的。此函数需要考虑的参数问题如下。

（1）因为需要在 Hive 中进行读取和写入的操作，所以需要传入一个 HiveContext 实例。

（2）需要指定从 Hive 中读取数据的表名称。

（3）因为不同的表的处理逻辑不一样，所以需要使用一个标记参数以区分输入的表。

（4）数据预处理完毕后，需要指定数据存储在 Hive 中的表名称。

根据以上业务逻辑和参数封装，进行数据预处理功能代码的封装，如代码 4-39 所示。

代码 4-39　封装数据预处理功能

```
package com.tipdm.scala.chapter_3_6_processing
import org.apache.spark.sql.hive.HiveContext
import org.apache.spark.{SparkConf, SparkContext}
import org.apache.spark.sql._
import org.apache.spark.sql.functions._
/**
  * 数据预处理功能
  */
object DataProcess {
  def main(args: Array[String]): Unit = {
    if(args.length!=10){
      printUsage()
      System.exit(1)
    }
    val conf = new SparkConf().setAppName("DataProcess")
```

```
    val sc = new SparkContext(conf)
    val sqlContext = new HiveContext(sc)
    // media_index 数据预处理
    val originMediaIndexTable = args(0)
    val processMediaIndexTable = args(1)
    dataProcessing(sqlContext,originMediaIndexTable,"media",processMediaIndex
Table)
    // mediamatch_userevent 数据预处理
    val originMediamatchUsereventTable = args(2)
    val processMediamatchUsereventTable = args(3)
 dataProcessing(sqlContext,originMediamatchUsereventTable,"userevent",
processMediamatchUsereventTable)
    //mediamatch_usermsg 数据预处理
    val originalMediamatchUsermsgTable = args(4)
    val processMediamatchUsermsgTable = args(5)
  dataProcessing(sqlContext,originalMediamatchUsermsgTable,"usermsg",
processMediamatchUsermsgTable)
    //mmconsume_billevents 数据预处理
    val originalMMConsumeBilleventsTable = args(6)
    val processMMConsumeBilleventsTable = args(7)
  dataProcessing(sqlContext,originalMMConsumeBilleventsTable,"bill",
processMMConsumeBilleventsTable)
    //order_index 数据预处理
    val originalOrderIndexTable = args(8)
    val processOrderIndexTable = args(9)
    dataProcessing(sqlContext,originalOrderIndexTable,"order",
processOrderIndexTable)
    sc.stop()
  }
  def dataProcessing(sqlContext: HiveContext,inputTable:String,flag:String,
outputTable:String): Unit ={
    val df = sqlContext.sql("select * from "+inputTable)
    //数据去重，并去除政企用户的数据
    val commonDF = df.distinct().filter("owner_name!='EA 级' and owner_name!=
'EB 级' and owner_name!='EC 级' and owner_name!='ED 级' and owner_name!='EE 级
'")
    //去除特殊线路的用户数据
      .filter("owner_code!='02' and owner_code!='09' and owner_code!='10'")
    //保留 sm_name='珠江宽频','数字电视' ,'互动电视' ,'甜果电视'的数据
      .filter("sm_name='珠江宽频' or sm_name='数字电视' or sm_name='互动电视' or
sm_name='甜果电视'")
    val resultDF = if(flag.equals("media")){
      //用户收视行为记录保留 duration>=20s 且 duration<=5h
      commonDF.filter("duration>=20000 and duration<=18000000")
          //过滤用户编号为 5401487 的记录
          .filter("phone_no!=5401487")
        //过滤 res_type!=0 or origin_time not rlike '00$' or end_time not rlike
'00$'的记录
```

```
        .filter(col("res_type").notEqual(0) or !col("origin_time").rlike("00$")
or !col("end_time").rlike("00$"))
    }else if(flag.equals("usermsg")){
      val maxTimeUsermsg = commonDF.groupBy("phone_no").agg(max("run_time").
alias("run_time"))
      commonDF.join(maxTimeUsermsg,Seq("phone_no","run_time"))
        .filter("run_name='正常' or run_name='主动暂停' or run_name='欠费暂停' or
run_name='主动销户'")
    }else if(flag.endsWith("order") || flag.equals("userevent")){
      //保留 run_name='正常','主动暂停','欠费暂停','主动销户'的记录
      commonDF.filter("run_name='正常' or run_name='主动暂停' or run_name='欠费
暂停' or run_name='主动销户'")
    }else{
      commonDF
    }
    //输出到 Hive 表中
    resultDF.write.mode(SaveMode.Overwrite).saveAsTable(outputTable)
  }
  /**
    * 使用说明
    */
  def printUsage(): Unit = {
    val buff = new StringBuilder
    buff.append("Usage : com.tipdm.scala.chapter_3_6_processing.DataProcess").append(" ")
      .append("<originMediaIndexTable>").append(" ")
      .append("<processMediaIndexTable>").append(" ")
      .append("<originMediamatchUsereventTable>").append(" ")
      .append("<processMediamatchUsereventTable>").append(" ")
      .append("<originalMediamatchUsermsgTable>").append(" ")
      .append("<processMediamatchUsermsgTable>").append(" ")
      .append("<originalMMConsumeBilleventsTable>").append(" ")
      .append("<processMMConsumeBilleventsTable>").append(" ")
      .append("<originalOrderIndexTable>").append(" ")
      .append("<processOrderIndexTable>").append(" ")
    println(buff.toString())
  }
}
```

针对代码 4-39，编写测试代码，如代码 4-40 所示。

代码 4-40　数据预处理的测试代码

```
package com.tipdm.java. chapter_3_6_processing;
import com.tipdm.engine.SparkYarnJob;
import com.tipdm.engine.model.Args;
import com.tipdm.engine.model.SubmitResult;
/**
 * 数据预处理的调用
 */
public class DataProcess {
   private static String className = "com.tipdm.scala.chapter_3_6_processing.
DataProcess";
```

```
    private static String applicationName = "DataProcess";
    public static void main(String[] args) throws Exception {
        String[] arguments = new String[10];
        arguments[0] = "user_profile.media_index_3m";
        arguments[1] = "user_profile.media_index_3m_process";
        arguments[2] = "user_profile.mediamatch_userevent";
        arguments[3] = "user_profile.mediamatch_userevent_process";
        arguments[4] = "user_profile.mediamatch_usermsg";
        arguments[5] = "user_profile.mediamatch_usermsg_process";
        arguments[6] = "user_profile.mmconsume_billevents";
        arguments[7] = "user_profile.mmconsume_billevent_process";
        arguments[8] = "user_profile.order_index_v3";
        arguments[9] = "user_profile.order_index_process";
        Args innerArgs = Args.getArgs(applicationName,className,arguments);
        SubmitResult submitResult = SparkYarnJob.run(innerArgs);
        SparkYarnJob.monitor(submitResult);
        System.out.println("任务运行成功");
    }
}
```

执行代码4-40后，将数据预处理任务提交到Spark集群中运行，在Spark中查看该任务状态。任务运行完成后，查看Spark监控，数据预处理任务运行状态如图4-17所示。

app-20190226154019-0111	DataProcess	20	8.0 GB	2019/02/26 15:40:19	spark	FINISHED	22 min

driver-20190226154012-0086	Tue Feb 26 15:40:12 CST 2019	worker-20190226134040-192.168.111.78-7078	FINISHED	1	1024.0 MB	com.tipdm.scala.chapter_3_6_processing.DataProcess

图4-17　数据预处理任务运行状态

确认预处理任务运行成功后，查看用户基本信息表在Hive中的数据存储情况，如代码4-41所示。

代码4-41　用户基本信息表在Hive中的数据存储情况

```
scala> val originUsermsg = sqlContext.sql("select * from user_profile.
mediamatch_usermsg")
originUsermsg: org.apache.spark.sql.DataFrame = [terminal_no: string, phone_no:
string, sm_name: string, run_name: string, sm_code: string, owner_name: string,
owner_code: string, run_time: string, addressoj: string, estate_name: string,
open_time: string, force: string]
scala> val processedUsermsg = sqlContext.sql("select * from user_profile.
mediamatch_usermsg_process")
processedUsermsg: org.apache.spark.sql.DataFrame = [phone_no: string, run_time:
string, terminal_no: string, sm_name: string, run_name: string, sm_code: string,
owner_name: string, owner_code: string, addressoj: string, estate_name: string,
open_time: string, force: string]
scala> originUsermsg.count
res2: Long = 5401493
scala> processedUsermsg.count
res3: Long = 2811801
scala> processedUsermsg.select("phone_no").distinct.count
res4: Long = 2811801
```

```
scala> processedUsermsg.select("sm_name").distinct.show
+--------+
|sm_name|
+--------+
|   互动电视|
|   数字电视|
|   甜果电视|
|   珠江宽频|
+--------+
scala> processedUsermsg.select("run_name").distinct.show
+---------+
|run_name|
+---------+
|    主动销户|
|    主动暂停|
|      正常|
|    欠费暂停|
+---------+
scala> processedUsermsg.select("owner_name").distinct.show
+----------+
|owner_name|
+----------+
|       HE 级|
|       HC 级|
|       HB 级|
|       HA 级|
+----------+
scala> processedUsermsg.select("owner_code").distinct.show
+----------+
|owner_code|
+----------+
|        00|
|        01|
|        04|
|        05|
|        06|
|        07|
|        08|
|        15|
|      NULL|
|        31|
+----------+
```

根据代码 4-41 的运行结果，可以发现根据用户基本信息表的数据预处理规则过滤后，该表的记录数几乎减少了一半，表中的记录数和用户数符合设置的规则（每个用户有且只有一条记录），且对其他字段的过滤如 owner_name、owner_code、sm_name、run_name 等达到了预期的效果。基于用户基本信息表的数据预处理效果，说明数据预处理的代码是正确的。

小结

本章主要介绍了广电大数据用户画像项目的数据来源、传输与处理的过程。为了贴合实际生产环境，本项目通过 Logstash 将 CSV 格式的数据采集到 Elasticsearch 集群中，通过 Spark 读取 Elasticsearch 数据，将其导入 Hive 并进行必要的转换后存储到 Elasticsearch 集群中，模拟新数据产生的过程。为了使后续数据处理更加高效，通过 Spark SQL 读取 Elasticsearch 集群中的数据并存入 Hive 后，根据探索环节得到的数据清洗规则实现了广电大数据的预处理，为进一步分析与建模准备了质量更好的数据集。

第 5 章 广电大数据用户画像——实时统计订单信息

目前，实时流数据处理在大数据处理领域已经占据着越来越重要的地位。谁拥有准确、及时的数据，谁就拥有更多的话语权。如何对海量数据进行快速有效的采集和分析已经成为大数据分析与应用领域中亟待解决的重要问题。本章将介绍如何通过 Kafka 获取实时订单数据流，结合 Spark Streaming 进行实时流式数据处理，完成广电大数据用户画像项目中订单信息实时统计任务。

学习目标

（1）熟悉 CDH 集群中 Kafka 的安装和配置。

（2）掌握使用 Kafka 模拟产生实时数据流的方法。

（3）掌握使用 Spark Streaming 实现数据的实时流式处理的方法。

5.1 实时统计目标

为了实时掌握订单信息的具体情况，需要使用实时处理相关技术来统计订单信息。具体的统计信息要求如下。

（1）要求每 30min 统计一次新增营业额、新增订单数、新增有效订单数，因为有一些订单是免费的（如 cost 字段为空），所以这些订单是无效订单。此外，需要统计总订单数、有效订单总数、总营业额。

（2）将这些指标实时更新到 Redis 中。此外，要保存整点的总订单数、有效订单总数、总营业额的对应数值。

（3）避免重复统计数据，保证结果的准确性。

考虑到项目框架的整体相关性及开发技术的成本，选择 Spark Streaming 和 Kafka 实现实时订单信息统计。

5.2 Kafka 安装和配置

因为广电大数据用户画像项目中使用的是 CDH 集群，所以本节主要介绍如何在 CDH 集群中安装 Kafka。因为 CDH 的 parcel 包没有包含 Kafka，所以需要单独加载 Kafka 的 parcel 包到 Cloudera Manager 中，parcel 包的安装步骤如下。

（1）从 Cloudera 官网上下载 Kafka 的 parcel 相关资源，需要注意的是，Kafka 的版本要与 CDH 的版本相对应。根据项目中的集群环境，下载 KAFKA-2.2.0-1.2.2.0.p0.68-el6.parcel、KAFKA-2.2.0-1.2.2.0.p0.68-el6.parcel.sha1、manifest.json。此外，需要下载 KAFKA-1.2.0.jar 文件。

（2）在 Cloudera Manager 的 server 节点上配置相关资源。因为本项目中 CDH 集群的 Cloudera Manager 是安装在 server1 上的，所以在 server1 上把 KAFKA-2.2.0-1.2.2.0.p0.68-el6.parcel、KAFKA-2.2.0-1.2.2.0.p0.68-el6.parcel.sha1、manifest.json 文件添加到/opt/cloudera/parcel-repo 目录下，并把 KAFKA-2.2.0-1.2.2.0.p0.68-el6.parcel.sha1 重新命名为 KAFKA-2.2.0-1.2.2.0.p0.68-el6.parcel.sha，覆盖/opt/cloudera/parcel-repo/目录下的 manifest.json。另外，需要把 KAFKA-1.2.0.jar 添加到/opt/cloudera/csd 目录下。

（3）登录 Cloudera Manager 的管理界面，选择“主机”→“Parcel”选项，在新的界面中单击“检查新 Parcel”按钮，此时弹出 2.2.0-1.2.2.0.p0.68 未分配的提示，单击“分配”按钮，分配完成后单击“激活”按钮，Kafka parcel 包激活成功后的界面如图 5-1 所示。

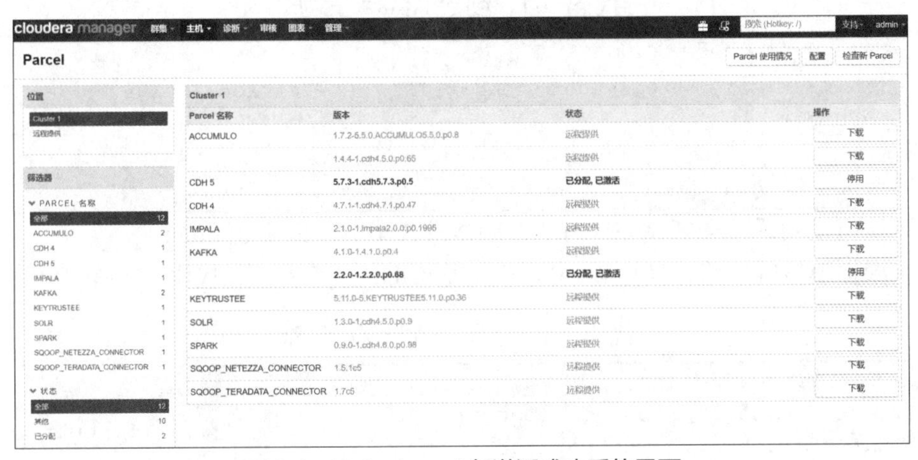

图 5-1 Kafka parcel 包激活成功后的界面

（4）Kafka 的 parcel 包添加成功后，需要在 Cloudera Manager 中安装 Kafka 集群。单击“添加服务”按钮，选择“Kafka”选项，单击“继续”按钮，选择将 Kafka Broker 安装在节点 node2、node3、node4 上。因为此处用不到 Kafka MirrorMaker，所以不用安装，单击“继续”按钮即可完成 Kafka 集群的安装。如图 5-2 所示，Kafka 集群已成功安装在 node2、node3、node4 上。

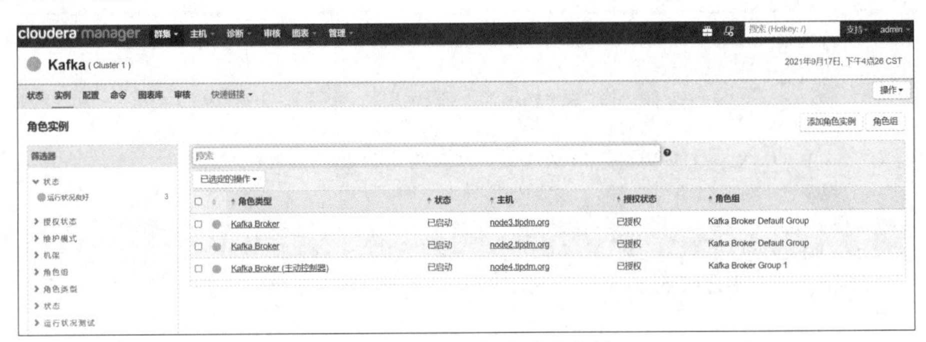

图 5-2 Kafka 集群成功安装

5.3 实时统计订单信息

在实际项目中，甲方的业务系统中已存在实时订单数据流（Kafka 集群），使用 Spark Streaming 读取 Kafka 订单主题即可实现订单统计。但是因为甲方的业务人员只提供了一份约 7GB 的订单静态数据文件，所以为了实现实时统计订单的功能，需要使用这份数据文件来模拟产生实时的订单数据流，并通过 Kafka 和 Spark Streaming 进行处理。

5.3.1 模拟产生订单实时数据流

模拟订单实时数据流的主要思路是通过使用 Kafka 生产者逐行读取订单数据文件，并向订单主题发送数据。为了模拟正常的订单产生速率，在生产者每次发送完消息后，线程随机睡眠 0～10s。模拟产生订单实时数据流的具体过程如下。

（1）创建一个 Kafka 主题，其名称为 order，在 node2 上执行代码 5-1 所示的命令。

代码 5-1　创建 Kafka 主题

```
#创建主题
kafka-topics --create --ZooKeeper node3:2181 --replication-factor 3 --partitions
3 --topic order
#查看主题
kafka-topics --ZooKeeper node3:2181 --list
```

代码执行成功后，会创建一个有 3 个副本及 3 个分区的主题 order。

（2）根据模拟的思路，使用 Java 语言实现 Kafka 生产者，如代码 5-2 所示。

代码 5-2　实现 Kafka 生产者

```
package com.tipdm.java.chapter_3_7_streaming;
import kafka.javaapi.producer.Producer;
import kafka.producer.KeyedMessage;
import kafka.producer.ProducerConfig;
import org.slf4j.Logger;
import org.slf4j.LoggerFactory;
import java.io.*;
import java.util.Properties;
/**
 * Kafka 生产者
 * 程序接收一个参数：订单文件
 */
public class KafkaProducer {
    private static final Logger log = LoggerFactory.getLogger(KafkaProducer.
class);
    public static void main(String[] args) {
        Properties properties = new Properties();
        InputStream in = KafkaProducer.class.getResourceAsStream("/sysconfig/
kafka.properties");
        BufferedReader reader = null;
        try {
            Properties pro = new Properties();
            pro.load(new BufferedInputStream(in));
```

```
        String meta = pro.getProperty("kafka.brokers");
        String topic = pro.getProperty("kafka.topics");
        properties.put("metadata.broker.list",meta);
        properties.put("request.required.acks", "1");
        properties.put("serializer.class", "kafka.serializer.StringEncoder");
        Producer<String, String> producer = new Producer<String, String>(new
ProducerConfig(properties));
        // 订单文件
        File file = new File(args[0]);
        reader = new BufferedReader(new FileReader(file));
        String line = null;
        int index =1;
        while((line = reader.readLine())!=null) {
            producer.send(new KeyedMessage<String, String>(topic,index+"",
line));
            if(index%1000==0){
                log.info("index:"+index+"   context:"+line);
            }
            index +=1;
            Thread.sleep((int)(1+Math.random()*10000));
        }
    } catch (Exception e) {
        e.printStackTrace();
    }finally {
        try {
            reader.close();
            in.close();
        } catch (Exception e) {
            e.printStackTrace();
        }
    }
  }
}
```

将代码5-2编译成user_profile_project-1.0.jar包并上传到CDH集群的node1节点的/root目录下，并执行代码5-3所示的命令以启动Kafka生产者。

代码5-3 启动Kafka生产者

```
java -cp \
.:/root/user_profile_project-1.0.jar \
-Djava.ext.dirs=/opt/cloudera/parcels/CDH-5.7.3-1.cdh5.7.3.p0.5/ \
jars/ com.tipdm.java.streaming.KafkaProducer /data/order.csv
```

为了检验程序是否可以正常运行，需要启动消费者消费order主题，如代码5-4所示。

代码5-4 启动消费者消费order主题

```
kafka-console-consumer --ZooKeeper node3:2181 --topic order --from-beginning
```

成功启动消费者后，当order主题产生数据时，消费者会实时从该主题中获取数据，不断地在屏幕上输出订单记录，消费者消费内容如图5-3所示。

```
2526958;HE级;2015-08-21 13:48:11;整转赠送-高尔夫;数字电视;BO002327;整转包(月租26.5元);正常状态;NULL;NULL;NULL;2008-06-12
00:00:00;2008-07-13 09:33:27;2008-06-12 09:33:27;NULL;N;BY;正常;1000002927687;0
2526958;HE级;2015-08-21 13:48:11;整转赠送-CHC;数字电视;BO002327;整转包(月租26.5元);正常状态;NULL;NULL;NULL;2008-06-12 00:
00:00;2008-07-13 09:33:27;2008-06-12 09:33:27;NULL;N;BY;正常;1000002927687;0
1394054;HE级;2012-12-10 12:58:32;基本包;数字电视;BO002327;整转包(月租26.5元);报停状态;NULL;NULL;NULL;2008-06-12 00:00:00;
2050-01-01 00:00:00;2008-06-12 09:34:20;NULL;N;YG;主动暂停;1000002927832;0
2647977;HE级;2014-04-15 08:03:27;整转赠送-文广包2;数字电视;BO002327;整转包(月租26.5元);到期暂停状态;NULL;NULL;NULL;2008-0
6-12 00:00:00;2008-07-13 09:36:47;2008-06-12 09:36:47;NULL;N;YD;正常;1000002927835;0
2595840;HE级;2015-08-20 10:40:00;整转赠送-鼎视及国防军事;数字电视;BO002327;整转包(月租26.5元);欠费暂停状态;NULL;NULL;NULL
;2008-06-12 00:00:00;2008-07-13 09:37:52;2008-06-12 09:37:52;NULL;N;DB;欠费暂停;1000002927836;0
2595840;HE级;2015-08-20 10:40:00;整转赠送-文广包2;数字电视;BO002327;整转包(月租26.5元);欠费暂停状态;NULL;NULL;NULL;2008-0
6-12 00:00:00;2008-07-13 09:37:52;2008-06-12 09:37:52;NULL;N;DB;欠费暂停;1000002927836;0
3015082;HE级;2014-04-14 21:14:53;整转赠送-CHC;数字电视;BO002340;购机购卡 (月租5元);到期暂停状态;NULL;NULL;NULL;2008-06-12
 00:00:00;2008-07-13 09:38:31;2008-06-12 09:38:31;NULL;N;YD;正常;1000002927837;0
3015082;HE级;2014-04-14 21:14:53;整转赠送-鼎视及国防军事;数字电视;BO002340;购机购卡 (月租5元);到期暂停状态;NULL;NULL;NULL
;2008-06-12 00:00:00;2008-07-13 09:38:31;2008-06-12 09:38:31;NULL;N;YD;正常;1000002927837;0
```

图 5-3　消费者消费内容

当看到类似图 5-3 所示的内容时，即说明模拟产生订单实时数据流的思路是可行的。

5.3.2　Spark Streaming 实时统计订单信息

基于实时统计订单信息的业务需求，给出实时统计订单的完整计算代码，如代码 5-5 所示。

代码 5-5　实时统计订单

```
package com.tipdm.scala.chapter_3_7_streaming
import java.io.{BufferedInputStream, InputStream}
import java.text.SimpleDateFormat
import java.util.{Date, Properties}
import com.tipdm.scala.util.InternalRedisClient
import kafka.common.TopicAndPartition
import kafka.message.MessageAndMetadata
import kafka.serializer.StringDecoder
import org.apache.spark.sql.SQLContext
import org.apache.spark.streaming.kafka.{HasOffsetRanges, KafkaUtils, OffsetRange}
import org.apache.spark.streaming.{Seconds, StreamingContext}
import org.apache.spark.{Accumulator, SparkConf}
import redis.clients.jedis.Pipeline
object KafkaStream {
  def main(args: Array[String]): Unit = {
    val conf = new SparkConf()
    conf.setAppName("kafkaStream")
    // 模式运行
     conf.setMaster("local[*]")
    // 将窗口时间设置为 30min
    val ssc = new StreamingContext(conf, Seconds(60*30))
    val sqlContext = new SQLContext(ssc.sparkContext)
    val properties: Properties = new Properties()
    val inputStream: InputStream = getClass.getResourceAsStream("/sysconfig/
kafka.properties")
    properties.load(new BufferedInputStream(inputStream))
    // Kafka 主题
    val topic = properties.getProperty("kafka.topics")
    // Kafka 的分区数
    val partitions = properties.getProperty("kafka.num.partitions").toInt
    val topics = Set(topic)
    val brokers = properties.getProperty("kafka.brokers")
```

```
val kafkaParams: Map[String, String] = Map[String, String](
  "metadata.broker.list" -> brokers,
  "group.id" -> "exactly-once",
  "enable.auto.commit" -> "false"
)
inputStream.close()
// 从 Redis 中读取主题各分区的 offerSet
val jedis = InternalRedisClient.getJedis()
var fromOffsets: Map[TopicAndPartition, Long] = Map()
for (i <- 0 until partitions) {
  var offerSet = 0l
  val tmp = jedis.get(topic + "_" + i)
  if (null != tmp) {
    offerSet = tmp.toLong
  }
  fromOffsets += (new TopicAndPartition(topic, i) -> offerSet)
}
// 定义订单总金额
val totalCost: Accumulator[Double] = ssc.sparkContext.accumulator(0.0)
// 定义有效订单总数
val validOrders: Accumulator[Int] = ssc.sparkContext.accumulator(0)
val historyTotal = jedis.get("totalcost")
val historyValidOrders = jedis.get("validOrders")
val historyOrders = jedis.get("totalOrders")
jedis.close()
// 定义新增有效订单数
val increaseValidOrders = ssc.sparkContext.accumulator(0)
// 定义新增订单金额
val increaseCost: Accumulator[Double] = ssc.sparkContext.accumulator(0.0)
var sum: Long = 0
if (null != historyOrders) {
  sum = historyOrders.toInt
}
if (null != historyValidOrders) {
  validOrders.add(historyValidOrders.toInt)
}
if (null != historyTotal) {
  totalCost.add(historyTotal.toDouble)
}
val messageHandler: MessageAndMetadata[String, String] => (String, String) =
  (mmd: MessageAndMetadata[String, String]) => (mmd.key(), mmd.message())
val offsetRanges = Array[OffsetRange]()
// 创建一个直接从 Kafka 代理获取消息的输入流
val kafkaStream = KafkaUtils.createDirectStream[String, String, StringDecoder,
StringDecoder, (String, String)](ssc, kafkaParams, fromOffsets, messageHandler)
// 在此处理每一批传送过来的数据
kafkaStream.foreachRDD(
  rdd => {
    val offsetRanges = rdd.asInstanceOf[HasOffsetRanges].offsetRanges
```

```
        // 新增订单每一批次重置为 0
        increaseValidOrders.setValue(0)
        // 新增订单营业额重置为 0
        increaseCost.setValue(0.0)
        // 按业务逻辑处理每一行数据
        rdd.foreach(row=>{
          val fields = row._2.split(";")
          val cost = fields(14)
          // 订单有效判断
          if (null == cost || cost.equalsIgnoreCase("null") || cost.startsWith
("YH")) {
          } else {
            // 有效订单统计
            validOrders.add(1)
            increaseValidOrders.add(1)
            // 每批次的总营业额
            increaseCost.add(cost.toDouble)
            // 总营业额
            totalCost.add(cost.toDouble)
          }
        })
        // 记录总数（即订单总数）
        sum = sum + rdd.count()
        println("新增订单数: " + rdd.count())
        println("新增有效订单数: " + increaseValidOrders)
        println("有效订单总数:  " + validOrders)
        println("总订单数: " + sum)
        println("新增订单营业额: " + increaseCost)
        println("订单总营业额: " + totalCost)
        val jedis = InternalRedisClient.getJedis()
        val pipeLine: Pipeline = jedis.pipelined()
        pipeLine.multi() // 开启事务
        pipeLine.set("totalcost", totalCost.toString())
        // 每小时统计一次总营业额、总订单数、有效订单总数并保存到 Redis 中，key 以时间开头
(如 2018082413_xxx)
        val key = nowTime()
        if (key.endsWith("00")) {
          val totalKey = key.substring(0, 10) + "_totalcost"
          pipeLine.set(totalKey, totalCost.toString())
          pipeLine.set(key.substring(0, 10) + "_totalorders", sum.toString)
          pipeLine.set(key.substring(0,  10)  +  "_validorders",  validOrders.
toString())
        }
        // 每一批次都更新统计指标
        pipeLine.set("increase_cost", increaseCost.toString())
        pipeLine.set("increase_valid_order", increaseValidOrders.toString)
        pipeLine.set("totalValidOrders", validOrders.toString())
        pipeLine.set("totalOrders", sum.toString)
        pipeLine.set("increase_order", rdd.count().toString)
```

```
        // 保存Spark Streaming消费Kafka order主题的分区中的offset，以便重启后继续
// 从上次的消费位置消费
        offsetRanges.foreach { offsetRange =>
          println("partition : " + offsetRange.partition + " fromOffset:  " +
offsetRange.fromOffset + " untilOffset: " + offsetRange.untilOffset)
          val topic_partition_key = offsetRange.topic + "_" + offsetRange.
partition
          pipeLine.set(topic_partition_key, offsetRange.untilOffset + "")
        }
        pipeLine.exec(); // 提交事务
        pipeLine.sync(); // 关闭pipeLine
        jedis.close()
      }
    )
    ssc.start()
    ssc.awaitTermination()
  }
  /**
    * 获取当前时间，返回的格式为yyyyMMddHHmm
    * @return
    */
  def nowTime(): String = {
    val now: Date = new Date()
    val dataFormate: SimpleDateFormat = new SimpleDateFormat("yyyyMMddHHmm")
    val date = dataFormate.format(now)
    return date
  }
}
```

为了避免出现重复消费的问题，代码 5-5 中使用 Redis 来保存消费位置。程序在启动时，先从 Redis 中读取上一次的消费位置，每次成功消费后，将消费记录的下标更新到 Redis 中。

为了减少在 Spark 集群中配置外部依赖包，代码 5-5 编译封装 user_profile_project-1.0.jar 的 JAR 包时，将 Redis 依赖添加到 JAR 包中，如图 5-4 所示。

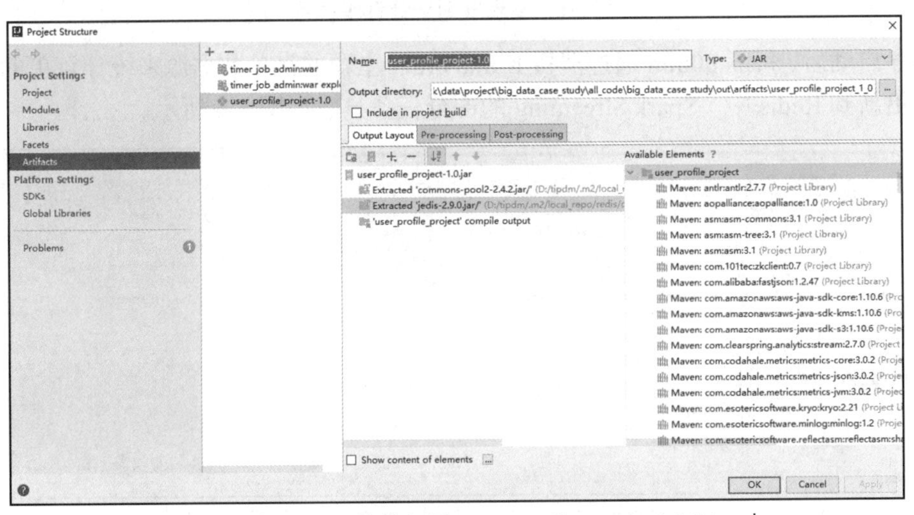

图 5-4　添加 Redis 依赖包到 user_profile_project-1.0.jar 中

针对代码 5-5 编写测试代码，如代码 5-6 所示。

代码 5-6　实时统计订单的测试代码

```
package com.tipdm.java.chapter_3_7_streaming;
import com.tipdm.engine.SparkYarnJob;
import com.tipdm.engine.model.Args;
import com.tipdm.engine.model.SubmitResult;
import org.slf4j.Logger;
import org.slf4j.LoggerFactory;
/**
 * 订单流的调用
 */
public class OrderStreaming {
    private static final Logger logger = LoggerFactory.getLogger(OrderStreaming.
class);
    private static String className = "com.tipdm.scala.chapter_3_7_streaming.
KafkaStream";
    private static String applicationName = "Order Spark Streaming";
    public static void main(String[] args) throws Exception {
        String[] arguments = new String[1];
        arguments[0] = "test";
        Args innerArgs = Args.getArgs(applicationName,className,arguments);
        SubmitResult submitResult = SparkYarnJob.run(innerArgs);
        SparkYarnJob.monitor(submitResult);
        logger.info("运行成功");
    }
}
```

在执行代码 5-6 的同时，需要启动模拟生产订单数据源程序。启动程序后，可以在 Spark 监控界面中看到提交的任务 ID，即查看实时统计订单任务，如图 5-5 所示。

Application ID	Name	Cores	Memory per Node	Submitted Time	User	State	Duration
app-20190227100137-0127 (kill)	kafkaStream	4	1024.0 MB	2019/02/27 10:01:37	spark	RUNNING	1.0 h

Submission ID	Submitted Time	Worker	State	Cores	Memory	Main Class
driver-20190227100131-0108 (kill)	Wed Feb 27 10:01:31 CST 2019	worker-20190226230934-192.168.111.77-7078	RUNNING	1	1024.0 MB	com.tipdm.scala.chapter_3_7_streaming.KafkaStream

图 5-5　查看实时统计订单任务

程序启动后，每隔 30min 通过消费 Kafka order 主题的订单数据源来统计订单的相关指标，并更新到 Redis 中。Spark Streaming 的部分任务日志如图 5-6 所示。

```
Bytes 0 - 45972 of 45972
2019-02-27 11:30:00,126 INFO [org.apache.spark.storage.BlockManagerInfo] - Added broadcast_2_piece0 in memory on node3:51438 (size: 1862.0 B, free: 530.0 MB)
2019-02-27 11:30:00,150 INFO [org.apache.spark.scheduler.TaskSetManager] - Finished task 2.0 in stage 2.0 (TID 7) in 59 ms on node4 (1/3)
2019-02-27 11:30:00,151 INFO [org.apache.spark.scheduler.TaskSetManager] - Finished task 0.0 in stage 2.0 (TID 8) in 60 ms on node3 (2/3)
2019-02-27 11:30:00,375 INFO [org.apache.spark.storage.BlockManagerInfo] - Added broadcast_2_piece0 in memory on node2:33554 (size: 1862.0 B, free: 530.0 MB)
2019-02-27 11:30:00,434 INFO [org.apache.spark.scheduler.TaskSetManager] - Finished task 1.0 in stage 2.0 (TID 6) in 352 ms on node2 (3/3)
2019-02-27 11:30:00,434 INFO [org.apache.spark.scheduler.TaskSchedulerImpl] - Removed TaskSet 2.0, whose tasks have all completed, from pool
2019-02-27 11:30:00,435 INFO [org.apache.spark.scheduler.DAGScheduler] - ResultStage 2 (foreach at KafkaStream.scala:93) finished in 0.352 s
2019-02-27 11:30:00,435 INFO [org.apache.spark.scheduler.DAGScheduler] - Job 2 finished: foreach at KafkaStream.scala:93, took 0.379331 s
新增订单数: 352
新增有效订单数: 53
有效订单总数:  728
总订单数: 18945
新增订单营业额: 1746.0
订单总营业额: 252756.55
partition : 1 fromOffset:  6200 untilOffset: 6318
partition : 2 fromOffset:  6197 untilOffset: 6314
partition : 0 fromOffset:  6196 untilOffset: 6313
2019-02-27 11:30:00,439 INFO [org.apache.spark.streaming.scheduler.JobScheduler] - Finished job streaming job 1551238200000 ms.0 from job set of time 1551238
200000 ms
```

图 5-6　Spark Streaming 的部分任务日志

从图 5-6 中可以看到，该日志记录了 2019 年 02 月 27 日 11 时 30 分的统计信息及消费订单主题的每个分区的位置。

任务运行一段时间后，查看 Redis 订单统计信息及消费 Kafka order 主题的 offset，如图 5-7 所示。

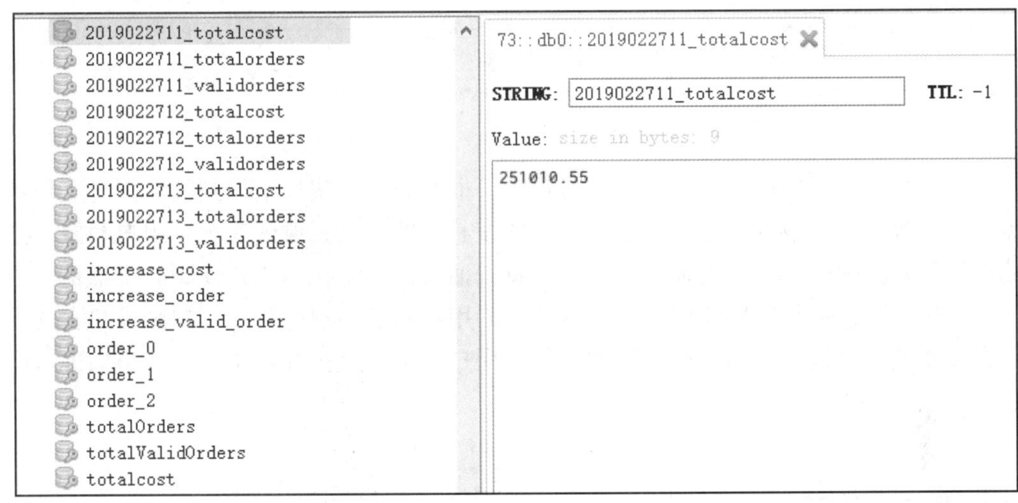

图 5-7　Redis 订单统计信息及消费 Kafka order 主题的 offset

从图 5-7 中可以看到，Redis 中出现了 4 种类型的键，其中，以日期开头的键是每小时的整点记录的此刻的订单的总营业额、有效订单总数及总订单数；以 increase 为前缀的键记录了每次统计的新增的营业额、订单数和有效订单数，这 3 个值随着每次统计而改变；以 order 为前缀的键记录了 Spark Streaming 消费 Kafka order 主题的 offset，由于 order 主题有 3 个分区，因此用 3 个键来记录每次处理的各分区的消费位置，这样有效避免了重复消费或消息丢失的问题，且这 3 个值也随着每次统计而改变；以 total 为前缀的键记录了累计的总订单数、有效订单总数和总营业额，这 3 个值也随着每次统计而改变。

从以上统计结果可以看到，实时统计订单数据的需求基本实现了。在真实的项目中，统计实时订单的指标会更加多样和复杂，并会要求 Redis 的统计指标在界面中实时展示出来。

小结

本章的目的是实现数据的实时流式处理，先通过 Java 程序读取静态订单数据模拟订单数据的产生过程，再通过 Kafka 接收订单数据到主题中，最后通过 Spark Streaming 作为消费者消费 Kafka 主题中的数据，实现订单信息的实时统计，并将结果写入 Redis。Kafka、Spark Streaming 组合是目前常用的实时处理方式，实时性和准确性高，准确、实时的订单信息能够帮助经营者了解用户的订购需求，及时调整销售策略，提高企业效益。

第 6 章 广电大数据用户画像——用户标签计算与可视化

完成广电大数据传输、数据存储等过程后，根据需求探索环节得到的数据清洗规则对数据进行了预处理。数据预处理后进入项目的核心，即用户画像的实现。用户画像是指根据用户的属性、偏好、生活习惯、行为等信息，抽象出标签化的用户模型。本章将介绍广电大数据用户画像项目中期阶段完成的内容，即用户画像实现过程。用户画像的实现将根据甲方提供的计算规则，结合已有数据给每个用户定义标签，并对画像标签结果进行可视化展示。

学习目标

（1）了解用户画像的核心实现过程。

（2）熟悉使用 SVM 算法预测用户是否值得挽留的实现过程。

（3）熟悉用户画像标签的计算规则及对应的计算实现过程。

（4）掌握 SVM 模型的构建与预测方法、用户画像计算的工程封装及测试的方法。

（5）掌握简化版的用户画像可视化工程实现的方法。

6.1 SVM 预测用户是否值得挽留

用户画像中的一个标签为用户是否值得挽留。该标签的计算规则比较复杂，并非通过统计用户的数据获得，而是需要建立模型并根据指定的特征进行预测。

6.1.1 SVM 算法

SVM 是一种二分类的模型。SVM 可以分为线性和非线性两大类。SVM 算法的主要思想如下：找到空间中的一个能够对所有数据样本都进行划分的超平面，并使得样本集中所有数据到这个超平面的距离最短。举个简单的例子，图 6-1 所示为一个 SVM 二维平面，平面上有两类不同的数据，分别用圆圈和方块表示。可以很简单地找到一条直线使得两类数据正好能够完全分开。但是能将数据完全划分开的直线不止一条。对于图 6-1 中的 3 条直线，从直观上看，L_2 划分的效果要更好一些。SVM 的目标就是寻找这样的直线，使得距离这条直线最近的点到这条直线的距离最短。

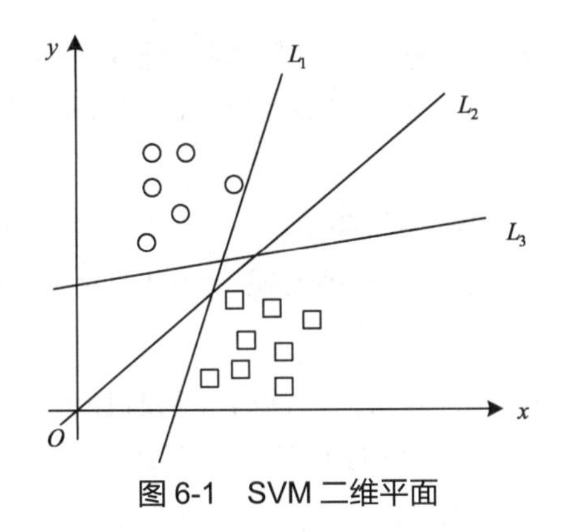

图 6-1 SVM 二维平面

在高维空间中，这样的“直线”称为超平面，而那些距离这个超平面最近的点即支持向量（Support Vector）。实际上，只要确定了支持向量，就确定了这个超平面。如图 6-2 所示，距离分离超平面最近的两个不同类别的样本点称为支持向量，两个支持向量构成了两条平行于分离超平面的长带，两者之间的距离称为 margin。margin 越大，分类正确的确信度越高。

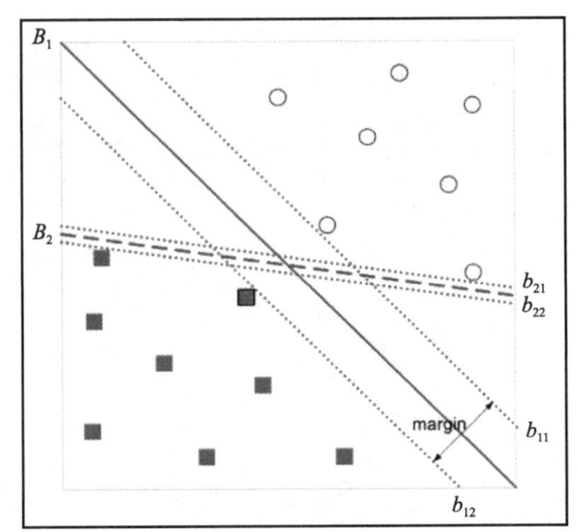

图 6-2 SVM 分离超平面

6.1.2 构建特征列和标签列数据

既然 SVM 是一个二分类模型，那么用于训练模型的数据就需要有特征列和标签列。但是业务数据中并没有一份既包含特征列又包含标签列的数据，因此需要利用业务数据来构建一份适用于 SVM 建模的数据。

预测用户是否值得挽留时，首先需要了解具备哪些特征的用户才值得被挽留。一般情况下，商家会更倾向于挽留那些经常消费的老客户。对于电视用户来说，如果用户的开户时间比较久，并且经常看电视或经常消费电视产品，那么这类用户往往会被广电公司作为挽留对象。因此，可以将电视消费水平、电视依赖度和电视入网时长这 3 个特征作为判断用户是否值得挽留的特征。

1. 用户的电视消费水平

电视消费水平需计算用户平均每个月购买电视产品的消费金额，计算这个特征值时，可根据 mmconsume_billevents 数据统计用户总的消费金额并除以 3；之所以除以 3，是因为每次计算都是取 mmconsume_billevents 数据中当前时间之前 3 个月的数据。

2. 用户的电视依赖度

电视依赖度指用户是否经常观看电视节目，或用户在看电视上花费的时间有多长，可通过 media_index 数据统计每个用户平均每天观看电视的时长来体现用户的电视依赖度。

3. 用户的电视入网时长

电视入网时长是用户从开户的时间到当前时间的时长，可通过 mediamatch_userevent 数据计算当前时间与用户开户的时间的差值来体现用户的电视入网时长。mediamatch_userevent 数据的 run_time 字段为用户的状态变更时间，因此用户的开户时间为每个用户对应的 run_time 字段的最大值。

对每个用户的电视消费水平、电视依赖度和电视入网时长这 3 个特征值进行计算，如代码 6-1 所示。

代码 6-1 构建特征列

```
// 统计每个用户的月均消费金额
scala> val billevents = sqlContext.sql("select phone_no, sum(should_pay)/3
consume from user_profile.mmconsume_billevent_process where sm_name not like
'%珠江宽频%' group by phone_no")
billevents: org.apache.spark.sql.DataFrame = [phone_no: string, consume:
double]
// 统计每个用户的入网时长 max(当前时间-run_time)
scala> val userevents = sqlContext.sql("select phone_no,max(months_between
(current_date(),run_time)/12) join_time from user_profile.mediamatch_
userevent_process group by phone_no")
userevents: org.apache.spark.sql.DataFrame = [phone_no: string, join_time:
double]
// 统计每个用户平均每次看多少小时电视
scala> val media_index = sqlContext.sql("select phone_no,(sum(media.duration)
/(1000*60*60))/count(1) as count_duration from user_profile.media_index_
3m_process media group by phone_no")
media_index: org.apache.spark.sql.DataFrame = [phone_no: string, count_
duration: double]
scala> val billevents_userevents_media = billevents.join(userevents, Seq
("phone_no")).join(media_index, Seq("phone_no"))
billevents_userevents_media: org.apache.spark.sql.DataFrame = [phone_no:
string, consume: double, join_time: double, count_duration: double]
scala> billevents_userevents_media.show(5)
+--------+------------------+-----------------+------------------+
|phone_no|           consume|        join_time|    count_duration|
+--------+------------------+-----------------+------------------+
| 1007666|62.333333333333336|      10.20854008|0.4549130434782608 4|
| 1013290|62.333333333333336|10.341526750833333| 0.2507136752136752|
| 1016576| 65.66333333333334|      9.991957045|0.1967915968732551 8|
| 1017962|138.66666666666666|       11.6737173|0.3219277108433734 6|
| 1018240|191.99333333333334| 6.479970193333333|0.2875261813537675 6|
+--------+------------------+-----------------+------------------+
only showing top 5 rows
```

构建了特征列数据后，再继续构建标签列数据。标签列标识的是用户是否值得挽留，挽留是指努力留下还未离开的用户。挽留用户的首要条件是用户状态是正常的，主动暂停或主动销户的用户为非挽留用户。挽留用户除需要是正常用户之外，还应该关注那些在观看电视节目上花费时间较多的用户，因为用户观看电视节目的时间越多，说明用户购买电视产品的可能性越大，这类用户是广电公司的潜在发展用户。根据 media_index 数据计算每个用户的当前日期前一个月观看电视节目的总时长，并统计所有用户观看时长的最大值、最小值、均值、方差、30%分位数和中位数，由于目前的真实收视数据是 2018-08-01 之前 3 个月的数据，因此将 2018-08-01 作为当前日期，如代码 6-2 所示。

代码 6-2 计算用户前一个月观看电视的时长

```
scala> val mediaIndex=sqlContext.sql("select phone_no,round(sum(duration)/(1000*60*60),2) as \
total_one_hours from user_profile.media_index_3m_process where origin_time>=
add_months('2018-08-01 00:00:00',-1) group by phone_no")
mediaIndex: org.apache.spark.sql.DataFrame = [phone_no: string, total_one_hours:
double]
scala> mediaIndex.show(3)
+--------+---------------+
|phone_no|total_one_hours|
+--------+---------------+
| 4476164|         102.33|
| 2011985|          65.33|
| 2701105|         104.81|
+--------+---------------+
only showing top 3 rows
scala>  import org.apache.commons.math3.stat.descriptive.rank.Percentile
import org.apache.commons.math3.stat.descriptive.rank.Percentile
scala> import org.apache.spark.mllib.linalg.Vectors
import org.apache.spark.mllib.linalg.Vectors
scala> import org.apache.commons.math3.stat.descriptive.rank.Percentile
import org.apache.commons.math3.stat.descriptive.rank.Percentile
scala> val _30_P = new Percentile(30.0)
_30_P: org.apache.commons.math3.stat.descriptive.rank.Percentile = org.apache.
commons.math3.stat.descriptive.rank.Percentile@7dd66054
scala> val _30_udf = \
udf{(arr: scala.collection.mutable.WrappedArray[Double]) => _30_P.evaluate
(arr.sorted.toArray)}
_30_udf: org.apache.spark.sql.UserDefinedFunction = UserDefinedFunction
(<function1>,DoubleType,List(ArrayType(DoubleType,false)))
scala> val _50_P = new Percentile(50.0)
_50_P: org.apache.commons.math3.stat.descriptive.rank.Percentile = org.apache.
commons.math3.stat.descriptive.rank.Percentile@5f5a5143
scala> val _50_udf = \
udf{(arr: scala.collection.mutable.WrappedArray[Double]) => _50_P.evaluate
(arr.sorted.toArray)}
_50_udf: org.apache.spark.sql.UserDefinedFunction = UserDefinedFunction
(<function1>,DoubleType,List(ArrayType(DoubleType,false)))
```

```
scala> mediaIndex.select(max("total_one_hours").alias("max"),min("total_one_hours").alias("min"),avg("total_one_hours").alias("avg"),stddev("total_one_hours").alias("std"),_30_udf(collect_list("total_one_hours")).alias("30%"),_50_udf(collect_list("total_one_hours")).alias("median")).show
+------+----+-----------------+-----------------+-----+------+
|   max| min|              avg|              std|  30%|median|
+------+----+-----------------+-----------------+-----+------+
|376.37|0.01|38.23807645879191|37.72779922489686|12.06| 26.93|
+------+----+-----------------+-----------------+-----+------+
```

根据代码 6-2 所示的统计结果可知，数据中用户观看电视的最大时长是 376.37h，最小时长不足一个小时，平均值是 38.23807645879191，方差是 37.72779922489686，30%分位数为 12.06，中位数为 26.93。针对代码 6-2 得到的数据结果，使用图表的形式进行展示，如图 6-3 所示。

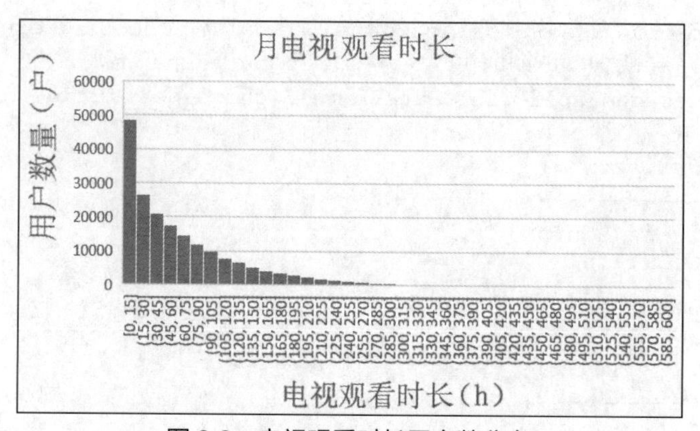

图 6-3 电视观看时长用户数分布

从图 6-3 中可以看出，电视观看时长在 0～15h 的人数是最多的。进一步对电视观看时长为 0～15h 的数据进行分析，取出观看时长小于或等于 15h 的数据，并统计这部分数据的最大值、最小值、均值、方差、30%分位数和中位数，如代码 6-3 所示。

代码 6-3 统计电视观看时长为 0～15h 的信息

```
scala> val mediaIndex_15=mediaIndex.filter("total_one_hours<=15")
mediaIndex_15: org.apache.spark.sql.DataFrame = [phone_no: string, total_one_hours: double]
scala> mediaIndex_15.select(max("total_one_hours").alias("max"),min("total_one_hours").alias("min"),avg("total_one_hours").alias("avg"),stddev("total_one_hours").alias("std"),_30_udf(collect_list("total_one_hours")).alias("30%"),_50_udf(collect_list("total_one_hours")).alias("median")).show
+----+----+-----------------+-----------------+----+------+
| max| min|              avg|              std| 30%|median|
+----+----+-----------------+-----------------+----+------+
|15.0|0.01|6.007243869064587|4.425724686005674|2.56|  5.26|
+----+----+-----------------+-----------------+----+------+
```

根据代码 6-3 的统计结果可知，对于观看时长为 0～15h 的数据，用户电视观看时长最大值为 15，最小值为 0.01，平均值为 6.007243869064587，方差为 4.425724686005674，30%分位数为 2.56，中位数为 5.26。结合实际情况以及甲方业务人员的建议，决定将 5.26 作为判断用户是否活跃的阈值。若一个月的电视观看时长小于5.26h，则说明用户平常很少看电视，也基本不会购买电视产品，因此将这类用户归为不活跃用户。对于电视观看时长大于或等于5.26h的用户，如果用户之前购买过产品，即在order_index数据中有该用户的消费记录，则将这类用户归为活跃用户，而活跃用户即可作为挽留的对象，因此，用户是否值得挽留的标签可根据如下规则得到。

（1）根据mediamatch_usermsg选择run_name字段为主动暂停或主动销户的用户并标注类别0，0代表用户为非挽留用户。

（2）根据mediamatch_usermsg选择run_name字段为正常并且是活跃的用户标注类别1，1代表用户为挽留用户。

根据探索得到的计算用户挽留标签的规则构建标签列数据，并将标签列数据和通过代码6-1计算得到的特征列数据进行合并，即可构建一份完整的SVM建模数据，如代码6-4所示。

代码6-4 构建一份完整的SVM建模数据

```
//统计活跃用户和非活跃用户
scala> val msg = \
sqlContext.sql("select distinct phone_no,0 as col1 from user_profile.mediamatch_
usermsg_process")
msg: org.apache.spark.sql.DataFrame = [phone_no: string, col1: int]
scala> val mediaIndex = sqlContext.sql("select phone_no,sum(duration) as
total_one_month_seconds from user_profile media_index_3m_process where origin_
time>=add_months('2018-08-01 00:00:00',-1) group by phone_no having total_
one_month_seconds>18936000").select("phone_no", "total_one_month_seconds")
mediaIndex: org.apache.spark.sql.DataFrame = [phone_no: string, total_one_
month_seconds: double]
scala> val orderIndexTV = sqlContext.sql("select * from user_profile. order_
index_process where run_name='正常' and offername!='废' and offername!='赠送
' and offername!='免费体验' and offername!='提速' and offername!='提价' and
offername!='转网优惠' and offername!='虚拟' and offername!='空包' and offername
not like '%宽带%'").select("phone_no").distinct()
orderIndexTV: org.apache.spark.sql.DataFrame = [phone_no: string]
scala> val media_order = mediaIndex.join(orderIndexTV, Seq("phone_no"),
"inner").selectExpr("phone_no", "1 as col2").distinct()
media_order: org.apache.spark.sql.DataFrame = [phone_no: string, col2: int]
scala> val msg_media_order = \
msg.join(media_order, Seq("phone_no"), "left_outer").na.fill(0).selectExpr("phone_
no", "col2 as col1")
msg_media_order: org.apache.spark.sql.DataFrame = [phone_no: string, col1: int]
scala> val billevents = sqlContext.sql("select phone_no, sum(should_pay)/3
consume from user_profile.mmconsume_billevent_process where sm_name not like
'%珠江宽频%' group by phone_no")
billevents: org.apache.spark.sql.DataFrame = [phone_no: string, consume: double]
scala> val userevents = sqlContext.sql("select phone_no,max(months_between
(current_date(),run_time)/12) join_time from user_profile.mediamatch_userevent_
process group by phone_no")
userevents: org.apache.spark.sql.DataFrame = [phone_no: string, join_time: double]
```

```
scala> val media_index = sqlContext.sql("select phone_no,(sum(media.duration)/(1000*60*60))/count(1) as count_duration from user_profile.media_index_3m_process media group by phone_no")
media_index: org.apache.spark.sql.DataFrame = [phone_no: string, count_duration: double]
scala> val billevents_userevents_media = billevents.join(userevents, Seq("phone_no")).join(media_index, Seq("phone_no"))
billevents_userevents_media: org.apache.spark.sql.DataFrame = [phone_no: string, consume: double, join_time: double, count_duration: double]
//标记非挽留用户
scala> val usermsg = sqlContext.sql("select * from user_profile. mediamatch_usermsg_process where run_name ='主动销户' or run_name='暂停' ")
usermsg: org.apache.spark.sql.DataFrame = [phone_no: string, run_time: string, terminal_no: string, sm_name: string, run_name: string, sm_code: string, owner_name: string, owner_code: string, addressoj: string, estate_name: string, open_time: string, force: string]
scala> val usermsg_billevents_userevents_media = usermsg.join(billevents_userevents_media, Seq("phone_no"), "inner").withColumn("label", billevents_userevents_media("consume") * 0)
usermsg_billevents_userevents_media: org.apache.spark.sql.DataFrame = [phone_no: string, run_time: string, terminal_no: string, sm_name: string, run_name: string, sm_code: string, owner_name: string, owner_code: string, addressoj: string, estate_name: string, open_time: string, force: string, consume: double, join_time: double, count_duration: double, label: double]
//标记挽留用户
scala> val activateUser = msg_media_order.where("col1=1")
activateUser: org.apache.spark.sql.DataFrame = [phone_no: string, col1: int]
scala> val billevents_userevents_activateUser = billevents_userevents_media.join(activateUser, Seq("phone_no"), "inner").withColumn("label", billevents_userevents_media("consume") * 0 + 1)
billevents_userevents_activateUser: org.apache.spark.sql.DataFrame = [phone_no: string, consume: double, join_time: double, count_duration: double, col1: int, label: double]
scala> val unionData = usermsg_billevents_userevents_media.select("phone_no", "consume", "join_time", "count_duration", "label").unionAll(billevents_userevents_activateUser.select("phone_no", "consume", "join_time", "count_duration", "label"))
unionData: org.apache.spark.sql.DataFrame = [phone_no: string, consume: double, join_time: double, count_duration: double, label: double]
scala> unionData.show(5)
+--------+------------------+------------------+-------------------+-----+
|phone_no|           consume|         join_time|     count_duration|label|
+--------+------------------+------------------+-------------------+-----+
| 1310650|62.333333333333336|11.894158265833333|0.10915343915343916|  0.0|
| 1314827|62.333333333333336|10.570727549166667|0.12366319444444444|  0.0|
| 1446137| 64.49333333333334|11.646563962499998| 0.2430332056194125|  0.0|
| 1498527|182.33333333333334|     10.4123364075| 0.4784998398975345|  0.0|
| 1498860|62.333333333333336| 8.281217829166666|0.26294074074074075|  0.0|
+--------+------------------+------------------+-------------------+-----+
only showing top 5 rows
```

执行代码 6-4 后，得到一份 DataFrame 类型的数据 unionData，数据中包含 5 个字段，分别表示用户编号、电视消费水平、电视入网时长、电视依赖度和类别。后续可利用这份数据来进行 SVM 建模。

6.1.3　建立 SVM 模型

查看 Spark 的 Scala API 文档中建立 SVM 模型的方法，如图 6-4 所示。用于建立 SVM 模型的 train()方法需要 5 个参数：input 参数是用于训练模型的数据，并且要求数据是 RDD[LabeledPoint]类型；numIterations 是运行梯度下降的迭代次数，默认值是 100；stepSize 是每次迭代梯度下降所使用的步长，默认值是 1.0；regParam 是正则化参数，默认值是 0.01；miniBatchFraction 是批处理粒度，默认值是 1.0。

```
def train(input: RDD[LabeledPoint], numIterations: Int, stepSize: Double, regParam: Double,
    miniBatchFraction: Double): SVMModel
```

Train a SVM model given an RDD of (label, features) pairs. We run a fixed number of iterations of gradient descent using the specified step size. Each iteration uses `miniBatchFraction` fraction of the data to calculate the gradient.

input	RDD of (label, array of features) pairs.
numIterations	Number of iterations of gradient descent to run.
stepSize	Step size to be used for each iteration of gradient descent.
regParam	Regularization parameter.
miniBatchFraction	Fraction of data to be used per iteration.

Annotations	@Since("0.8.0")
Note	Labels used in SVM should be {0, 1}

图 6-4　建立 SVM 模型的方法

执行代码 6-4 后，得到的 unionData 数据是 DataFrame 类型的，因此，在建模之前，需要把特征列数据转换成 RDD[LabeledPoint]类型。此外，代码 6-4 构建得到的 SVM 数据的几个特征列之间存在一定的量纲影响，为了消除量纲影响，需要对数据进行标准化处理。因为后续需要对模型进行评估，所以根据二八原则将标准化数据划分成验证集和训练集，使用训练集建立 SVM 模型，使用验证集评估模型。调用 train()方法建立模型，因为集群资源有限，迭代次数 numIterations 不适合设置为 100，所以设置 numIterations=10、stepSize=1.0、regParam=0.01、miniBatchFraction=1.0 建立模型。数据标准化处理、数据集划分和建立 SVM 模型过程如代码 6-5 所示。

代码 6-5　数据标准化处理、数据集划分和建立 SVM 模型过程

```
scala> import org.apache.spark.mllib.regression.LabeledPoint
import org.apache.spark.mllib.regression.LabeledPoint
scala> import org.apache.spark.mllib.linalg.Vectors
import org.apache.spark.mllib.linalg.Vectors
scala> import org.apache.spark.mllib.feature.StandardScaler
import org.apache.spark.mllib.feature.StandardScaler
scala> val traindata = \
unionData.select("consume", "join_time", "count_duration").rdd.zip(unionData.
select("label").rdd).map(x  =>  LabeledPoint(x._2.get(0).toString.toDouble,
Vectors.dense(x._1.toSeq.toArray.map(_.toString.toDouble))))
```

```
traindata: org.apache.spark.rdd.RDD[org.apache.spark.mllib.regression.LabeledPoint]
= MapPartitionsRDD[811] at map at <console>:59
scala> val scaler = new StandardScaler(withMean = true, withStd = true).
fit(traindata.map(x => x.features))
scaler: org.apache.spark.mllib.feature.StandardScalerModel = org.apache.spark.
mllib.feature.StandardScalerModel@28f52242
scala> val data2 = \
traindata.map(x => LabeledPoint(x.label, scaler.transform(Vectors.dense (x.features.
toArray))))
data2:
org.apache.spark.rdd.RDD[org.apache.spark.mllib.regression.LabeledPoint] =
MapPartitionsRDD[817] at map at <console>:63
scala> val data2_test = data2.map(x => (x.label, scaler.transform(Vectors.
dense(x.features.toArray))))
data2_test: org.apache.spark.rdd.RDD[(Double, org.apache.spark.mllib.linalg.
Vector)] = MapPartitionsRDD[818] at map at <console>:65
scala> val train_validate = data2.randomSplit(Array(0.8, 0.2))
train_validate: Array[org.apache.spark.rdd.RDD[org.apache.spark.mllib.regression.
LabeledPoint]] = Array(MapPartitionsRDD[819] at randomSplit at <console>:65,
MapPartitionsRDD[820] at randomSplit at <console>:65)
scala> val (train_data, validate_data) = (train_validate(0), train_validate(1))
train_data: org.apache.spark.rdd.RDD[org.apache.spark.mllib.regression.LabeledPoint]
= MapPartitionsRDD[819] at randomSplit at <console>:65
validate_data: org.apache.spark.rdd.RDD[org.apache.spark.mllib.regression.
LabeledPoint] = MapPartitionsRDD[820] at randomSplit at <console>:65
scala> train_data.cache()
res2: train_data.type = MapPartitionsRDD[335] at randomSplit at <console>:65
scala> validate_data.cache()
res3: validate_data.type = MapPartitionsRDD[336] at randomSplit at <console>:65
scala> import org.apache.spark.mllib.classification.SVMWithSGD
import org.apache.spark.mllib.classification.SVMWithSGD
scala> val model = SVMWithSGD.train(train_data, 10, 1.0, 0.01, 1.0)
model: org.apache.spark.mllib.classification.SVMModel = org.apache.spark.
mllib.classification.SVMModel: intercept = 0.0, numFeatures = 3, numClasses
= 2, threshold = 0.0
```

6.1.4 模型评估

模型评估以验证集作为输入，使用模型对验证集进行预测，根据对验证集预测的结果验证模型的效果。Spark MLlib 提供了一套指标用于评估机器学习模型，其中，受试者工作特征（Receiver Operating Characteristic，ROC）曲线、PR 曲线和曲线下面积（Area Under the Curve，AUC）常被用来评估一个二值分类器的优劣。在代码 6-5 中，根据二八原则划分了训练集和验证集，并使用训练集训练了一个模型。对于训练好的模型，使用验证集验证模型的效果，计算模型的准确率、AUROC（ROC 曲线下面积）值和 AUPRC（PR 曲线下面积）值，如代码 6-6 所示。

代码 6-6　模型评估

```
scala> import org.apache.spark.mllib.evaluation.BinaryClassificationMetrics
import org.apache.spark.mllib.evaluation.BinaryClassificationMetrics
scala> import org.apache.spark.sql.types.{DoubleType, StringType, StructField,
StructType}
```

```
import org.apache.spark.sql.types.{DoubleType, StringType, StructField, StructType}
scala> import org.apache.spark.sql.Row
import org.apache.spark.sql.Row
scala> val predictAndLabel = validate_data.map(row => {
     |        val predict = model.predict(row.features)
     |        val label = row.label
     |        (predict, label)
     |     })
predictAndLabel: org.apache.spark.rdd.RDD[(Double, Double)] = MapPartitionsRDD
[290] at map at <console>:35
scala> val validateCorrectRate = \
predictAndLabel.filter(r => r._1 == r._2).count.toDouble / validate_data.
count()
validateCorrectRate: Double = 0.46533672424612743
scala> val metrics = new BinaryClassificationMetrics(predictAndLabel)
metrics:  org.apache.spark.mllib.evaluation.BinaryClassificationMetrics  =
org.apache.spark.mllib.evaluation.BinaryClassificationMetrics@3758f3a3
scala> val schema1 = StructType(Array(
     |        StructField("param_original", StringType, false),
     |        StructField("value", DoubleType, false)))
schema1:  org.apache.spark.sql.types.StructType  =  StructType(StructField
(param_original,StringType,false), StructField(value,DoubleType,false))
scala> val rdd1 = sc.parallelize(Array(
     |        Row("correctRate", validateCorrectRate),
     |        Row("areaUnderROC", metrics.areaUnderROC()),
     |        Row("areaUnderPR", metrics.areaUnderPR())
     |     ))
rdd1: org.apache.spark.rdd.RDD[org.apache.spark.sql.Row] = ParallelCollectionRDD
[311] at parallelize at <console>:46
scala>  val evaluation = sqlContext.createDataFrame(rdd1, schema1)
evaluation: org.apache.spark.sql.DataFrame = [param_original: string, value:
double]
scala> evaluation.show()
+--------------+------------------+
|param_original|             value|
+--------------+------------------+
|   correctRate| 0.522111712130906|
|  areaUnderROC|0.6657931892455329|
|   areaUnderPR|0.9742817059516753|
+--------------+------------------+
```

根据代码 6-6 的计算结果可知，模型的准确率约为 0.522，AUROC 值约为 0.666，AUPRC 值约为 0.974。一般情况下，AUROC 的取值范围为 0.5～1，其值越大，说明分类效果越好。根据代码 6-6 的计算结果可知，模型准确率不高，且 AUROC 值只有约 0.666，模型的分类效果并不是十分理想。在集群资源充足的情况下，可通过参数寻优进行择优，从而提高模型的分类效果。

6.1.5 模型预测

训练得到一个分类效果不错的模型之后，即可使用该模型来预测用户的类别（是否值得挽留）。在此之前，需要构建一份测试数据来作为预测的输入源，测试数据是一份包含用户编号和用户编号对应的特征值（不包含类别列）的数据，即测试数据只需包含用户编号及用户的电视消费水平、用户的电视依赖度、用户的电视入网时长这 3 个特征值。执行代码 6-1 后，得到一份包含用户编号及 3 个特征列的数据 billevents_userevents_media，因此将这份数据作为测试数据即可。

查看 Spark 的 Scala API 文档，发现 SVMModel 类中有一个可用于预测的方法，即 predict() 方法。该方法的定义如图 6-5 所示。

def **predict**(testData: RDD[Vector]): RDD[Double]

Predict values for the given data set using the model trained.

testData	RDD representing data points to be predicted
returns	RDD[Double] where each entry contains the corresponding prediction

Definition Classes	GeneralizedLinearModel
Annotations	@Since("1.0.0")

图 6-5　predict()方法的定义

predict()方法需要将测试数据作为参数，同时要求测试数据的类型为 RDD[Vector]。因此，需要对 billevents_userevents_media 数据（测试数据）的数据类型进行进一步转换，将其转换成 RDD[(String,Vector)]类型，并取出 RDD 中的 Vector 数据作为 predict()方法的输入参数，如代码 6-7 所示。

代码 6-7　模型预测

```
scala> val test_data = \
billevents_userevents_media.select("phone_no").rdd.zip(billevents_usereven
ts_media.select("consume", "join_time", "count_duration").rdd).map(x =>
(x._1.get(0).toString,
Vectors.dense(x._2.toSeq.toArray.map(_.toString.toDouble))))
test_data: org.apache.spark.rdd.RDD[(String, org.apache.spark.mllib.linalg.
Vector)] = MapPartitionsRDD[317] at map at <console>:37
scala> val predictData = test_data.map(row => {
     |      val predict = model.predict(row._2)
     |      Row(row._1, row._2(0), row._2(1), row._2(2), predict)
     |    })
predictData: org.apache.spark.rdd.RDD[org.apache.spark.sql.Row] = MapPartitionsRDD[318]
at map at <console>:44
scala> val schema = StructType(Array(StructField("phone_no", StringType,
false), StructField("consume", DoubleType, false), StructField("join_time",
DoubleType, false), StructField("count_duration", DoubleType, false),
StructField("label", DoubleType, false)))
schema: org.apache.spark.sql.types.StructType = StructType(StructField(phone_
no,StringType,false), StructField(consume,DoubleType,false), StructField(join_time,
DoubleType,false), StructField(count_duration,DoubleType,false), StructField(label,
DoubleType,false))
scala>    val predictDF = sqlContext.createDataFrame(predictData, schema)
predictDF: org.apache.spark.sql.DataFrame = [phone_no: string, consume:
double, join_time: double, count_duration: double, label: double]
```

```
scala> predictDF.show(5)
+--------+------------------+------------------+-------------------+-----+
|phone_no|           consume|         join_time|     count_duration|label|
+--------+------------------+------------------+-------------------+-----+
| 1007666|62.333333333333336|10.200475564166666|0.45491304347826084|  1.0|
| 1013290|62.333333333333336|       10.333462235| 0.2507136752136752|  1.0|
| 1016576| 65.66333333333334| 9.983892529166667|0.19679159687325518|  1.0|
| 1017962|138.66666666666666|11.666666666666666|0.32192771084337346|  1.0|
| 1018240|191.99333333333334| 6.471905677500001|0.28752618135376756|  1.0|
+--------+------------------+------------------+-------------------+-----+
```

执行代码 6-7 后，得到了预测结果数据 predictDF，且数据的类型为 DataFrame，为方便后续进行用户画像，需要将预测得到的结果保存到 Hive 中。

6.1.6 整体实现及参数封装

本小节主要描述使用 SVM 算法预测用户是否值得挽留的整个过程及参数封装。综合 6.1 节中其他各小节的实现过程，可设置的参数如下。

1. 输入参数

输入参数为用于构建 SVM 模型数据的 Hive 表，包括 Hive 中 user_profile 库中的 media_index_3m_process、mediamatch_usermsg_process、mmconsume_billevent_process、mediamatch_userevent_process、order_index_process。

2. 输出参数

（1）模型评估结果 Hive 表。

（2）预测结果 Hive 表。

3. 模型参数

（1）运行梯度下降的迭代次数 numIterations。

（2）每次迭代梯度下降所使用的步长 stepSize。

（3）正则化参数 regParam。

（4）批处理粒度 miniBatchFraction。

对 SVM 预测用户是否值得挽留的过程和参数进行封装，如代码 6-8 所示。

代码 6-8 SVM 预测用户是否值得挽留的过程和参数封装

```
package com.tipdm.scala.chapter_3_8_6_svm
import com.tipdm.scala.util.SparkUtils.exists
import org.apache.spark.mllib.classification.SVMWithSGD
import org.apache.spark.mllib.evaluation.BinaryClassificationMetrics
import org.apache.spark.mllib.feature.StandardScaler
import org.apache.spark.mllib.linalg.Vectors
import org.apache.spark.mllib.regression.LabeledPoint
import org.apache.spark.sql.Row
import org.apache.spark.sql.hive.HiveContext
import  org.apache.spark.sql.types.{DoubleType,  StringType,  StructField,
StructType}
import org.apache.spark.{SparkConf, SparkContext}
```

```
object SVM {
  def main(args: Array[String]): Unit = {
    if (args.length != 12) {
      printUsage()
      System.exit(1)
    }
    val conf = new SparkConf().setAppName("SVM").setJars(Seq("/opt/cloudera/
parcels/CDH-5.7.3-1.cdh5.7.3.p0.5/lib/hive/lib/mysql-connector-java-5.1.7-
bin.jar"))
    val sc = new SparkContext(conf)
    val sqlContext = new HiveContext(sc)
    val mmconsume = args(0)
    val mediamatchUserevents = args(1)
    val tmpMediaIndex = args(2)
    val mediamatchUsermsg = args(3)
    val order_index = args(4)
    // 评估结果表
    val activate_table = args(5)
    // 预测表
    val predictionTable = args(6)
    val numIterations = args(7).toInt
    val stepSize = args(8).toDouble
    val regParam = args(9).toDouble
    val miniBatchFraction = args(10).toDouble
    val db = args(11)
    // 电视用户活跃度标签计算
    val msg = sqlContext.sql("select distinct phone_no,0 as col1 from " + db
+ "." + mediamatchUsermsg)
    val mediaIndex = sqlContext.sql("select phone_no,sum(duration) as total_
one_month_seconds from " + db + "." + tmpMediaIndex +" where origin_
time>=add_months('2018-08-01 00:00:00',-1) group by phone_no having total_one_
month_seconds>18936000").select("phone_no", "total_one_month_seconds")
    val orderIndexTV = sqlContext.sql("select * from " + db + "." + order_index
+ " where run_name='正常' and offername!='废' and " +"offername!='赠送' and
offername!='免费体验' and offername!='提速' and offername!='提价' and offername!=
'转网优惠' and offername!='测试' and offername!='虚拟' and offername!='空包' and
offername not like '%宽带%'").select("phone_ no").distinct()
    val media_order = mediaIndex.join(orderIndexTV, Seq("phone_no"), "inner").
selectExpr("phone_no", "1 as col2").distinct()
    // 电视用户活跃度，col1 为 1 时表示为活跃用户，为 0 时表示为不活跃用户
val msg_media_order = \
msg.join(media_order, Seq("phone_no"), "left_outer").na.fill(0).selectExpr
("phone_no", "col2 as col1")
    // 构建 SVM 数据
    // 统计每个用户的月均消费金额
    val billevents = sqlContext.sql("select phone_no, sum(should_pay)/3 consume
from " + db + "." + mmconsume + " where sm_name not like '%珠江宽频%' group by
phone_no")
    // 统计每个用户的入网时长 max(当前时间-run_time)
```

```
val userevents = sqlContext.sql("select phone_no,max(months_between(current_
date(),run_time)/12) join_time from " + db + "." + mediamatchUserevents + "
group by phone_no")
// 统计每个用户平均每次看多少小时电视
val media_index = sqlContext.sql("select phone_no,(sum(media.duration)/
(1000*60*60))/count(1) as count_duration from " + db + "." + tmpMediaIndex +
" media group by phone_no")
val billevents_userevents_media = billevents.join(userevents, Seq("phone_
no")).join(media_index, Seq("phone_no"))
// 从 mediamatch_usermsg 中选出离网的用户（run_name ='主动销户' or run_name='
主动//暂停'）并标记为类别 0（离网）；在正常用户中提取有活跃标签的用户并标记为类别 1（不离网）
val usermsg = sqlContext.sql("select *  from " + db + "." + mediamatchUsermsg
+ " where  run_name ='主动销户' or run_name='主动暂停' ")
// 为离网用户标记类别 0
val usermsg_billevents_userevents = usermsg.join(billevents_userevents_
media, Seq("phone_no"), "inner").withColumn("label", billevents_userevents_
media("consume") * 0)
// 在正常用户中提取有活跃标签的用户并标记为类别 1（不离网）
val activateUser = msg_media_order.where("col1=1")
val billevents_userevents_activateUser = billevents_userevents_media.
join(activateUser, Seq("phone_no"), "inner").withColumn("label", billevents_
userevents_media("consume") * 0 + 1)
val unionData = usermsg_billevents_userevents.select("phone_no", "consume",
"join_time", "count_duration", "label").unionAll(billevents_userevents_activateUser.
select("phone_no", "consume", "join_time", "count_duration", "label"))
// 训练数据为 trainData
val traindata = unionData.select("consume", "join_time", "count_duration").
rdd.zip(unionData.select("label").rdd).map(x  =>  LabeledPoint(x._2.get(0).
toString.toDouble,
   Vectors.dense(x._1.toSeq.toArray.map(_.toString.toDouble))))
// 测试数据集
val test_data = \
billevents_userevents_media.select("phone_no").rdd.zip(billevents_usereven
ts_media.select("consume", "join_time", "count_duration").rdd).
   map(x  =>  (x._1.get(0).toString,  Vectors.dense(x._2.toSeq.toArray.map
(_.toString.toDouble))))
// 归一化
val scaler = new StandardScaler(withMean = true, withStd = true).fit
(traindata.map(x => x.features))
val data2 = traindata.map(x => LabeledPoint(x.label, scaler.transform
(Vectors.dense(x.features.toArray))))
val data2_test = data2.map(x => (x.label, scaler.transform(Vectors.dense
(x.features.toArray))))
// 将数据分为训练集和验证集
val train_validate = data2.randomSplit(Array(0.8, 0.2))
val (train_data, validate_data) = (train_validate(0), train_validate(1))
train_data.cache()
validate_data.cache()
```

```
    // 建模
    val model = SVMWithSGD.train(train_data, numIterations, stepSize, regParam,
miniBatchFraction)
    //val model = SVMWithSGD.train(train_data,10, 1.0, 0.01, 1.0)
    train_data.unpersist()
    // 评估
    val predictAndLabel = validate_data.map(row => {
      val predict = model.predict(row.features)
      val label = row.label
      (predict, label)
    })
    val right_count = predictAndLabel.filter(r => r._1 == r._2).count.toDouble
    val validate_count = validate_data.count()
    val validateCorrectRate = right_count / validate_count
    val predictAndLabelNew = predictAndLabel.repartition(100)
    val metrics = new BinaryClassificationMetrics(predictAndLabelNew)
    val schema1 = StructType(Array(
      StructField("param_original", StringType, false),
      StructField("value", DoubleType, false)))
    val rdd1 = sc.parallelize(Array(
      Row("correctRate", validateCorrectRate),
      Row("areaUnderROC", metrics.areaUnderROC()),
      Row("areaUnderPR", metrics.areaUnderPR())
    ))
    val evaluation = sqlContext.createDataFrame(rdd1, schema1)
    val tmpTable = "tmp" + System.currentTimeMillis()
    evaluation.registerTempTable(tmpTable)
    if (exists(sqlContext, db, activate_table)) {
      sqlContext.sql("drop table " + db + "." + activate_table)
      sqlContext.sql("create table " + db + "." + activate_table + " as select
* from " + tmpTable)
    } else {
      sqlContext.sql("create table " + db + "." + activate_table + " as select
* from " + tmpTable)
    }
    validate_data.unpersist()
    // 预测
    val predictData = test_data.map(row => {
      val predict = model.predict(row._2)
      Row(row._1, row._2(0), row._2(1), row._2(2), predict)
    })
    val schema = StructType(Array(StructField("phone_no", StringType, false),
StructField("consume", DoubleType, false), StructField("join_time", DoubleType,
false), StructField("count_duration", DoubleType, false), StructField("label",
DoubleType, false)))
    val predictDF = sqlContext.createDataFrame(predictData, schema)
    val tmpTable1 = "tmp" + System.currentTimeMillis()
    predictDF.registerTempTable(tmpTable1)
    if (exists(sqlContext, db, predictionTable)) {
      sqlContext.sql("drop table " + db + "." + predictionTable)
```

```
      sqlContext.sql("create table " + db + "." + predictionTable + " as select
* from " + tmpTable1)
    } else {
      sqlContext.sql("create table " + db + "." + predictionTable + " as select
* from " + tmpTable1)
    }
    sc.stop()
  }
  /**
    * 使用说明
    */
  def printUsage(): Unit = {
    val buff = new StringBuilder
    buff.append("Usage : com.tipdm.scala.chapter_3_8_6_svm.SVM").append(" ")
      .append("<mmconsume>").append(" ")
      .append("<mediamatchUserevents>").append(" ")
      .append("<tmpMediaIndex>").append(" ")
      .append("<mediamatchUsermsg>").append(" ")
      .append("<order_index>").append(" ")
      .append("<activate_table>").append(" ")
      .append("<predictionTable>").append(" ")
      .append("<numIterations>").append(" ")
      .append("<stepSize>").append(" ")
      .append("<regParam>").append(" ")
      .append("<miniBatchFraction>").append(" ")
      .append("<db>").append(" ")
    println(buff.toString())
  }
}
```

针对代码 6-8，编写测试代码，如代码 6-9 所示。

代码 6-9 SVM 预测用户是否值得挽留测试代码

```
package com.tipdm.java.chapter_3_8_6_svm;
import com.tipdm.engine.SparkYarnJob;
import com.tipdm.engine.model.Args;
import com.tipdm.engine.model.SubmitResult;
public class SVM {
    private static String  className="com.tipdm.scala.chapter_3_8_6_svm.SVM";
    private static  String applicationName = "SVM";
    public static void main(String[] args) throws Exception {
        String[] arguments =new String [12];
        arguments[0]="mmconsume_billevent_process";
        arguments[1]="mediamatch_userevent_process";
        arguments[2]="media_index_3m_process";
        arguments[3]="mediamatch_usermsg_process";
        arguments[4]="order_index_process";
        arguments[5]="svm_activate";
        arguments[6]="svm_prediction";
        arguments[7]="10";
```

```
        arguments[8]="1.0";
        arguments[9]="0.01";
        arguments[10]="1.0";
        arguments[11]="user_profile";
        Args innerArgs = Args.getArgs(applicationName,className,arguments);
        SubmitResult submitResult = SparkYarnJob.run(innerArgs);
        SparkYarnJob.monitor(submitResult);
        System.out.println("任务运行成功");
    }
}
```

执行代码 6-9，任务运行完成后，查看 Spark 任务运行情况，其运行结果如图 6-6 所示。

driver-20190222141211-0012	Fri Feb 22 14:12:11 CST 2019	worker-20190222133406-192.168.111.75-7078	FINISHED	1	1024.0 MB	com.tipdm.scala.chapter_3_8_6_svm.SVM			
app-20190222141217-0011	SVM	20	4.0 GB	2019/02/22 14:12:17	spark	FINISHED	7.5 min		

图 6-6　SVM 预测用户是否值得挽留的 Spark 任务运行结果

若任务状态如图 6-6 所示，则表示任务运行成功。任务运行成功后，验证 SVM 运行结果在 Hive 中的存储情况，如代码 6-10 所示。

代码 6-10　验证 SVM 运行结果在 Hive 中的存储情况

```
scala> sqlContext.sql("select * from user_profile.svm_activate").show()
+--------------+------------------+
|param_original|             value|
+--------------+------------------+
|   correctRate| 0.522111712130906|
|  areaUnderROC|0.6657931892455329|
|   areaUnderPR|0.9742817059516753|
+--------------+------------------+
scala>  sqlContext.sql("select  *  from  user_profile.svm_prediction  limit 5").show()
+--------+------------------+------------------+-------------------+-----+
|phone_no|           consume|         join_time|     count_duration|label|
+--------+------------------+------------------+-------------------+-----+
| 1007666|62.333333333333336|10.200475564166666|0.45491304347826084|  1.0|
| 1013290|62.333333333333336|       10.333462235| 0.2507136752136752|  1.0|
| 1016576| 65.66333333333334| 9.983892529166667|0.19679159687325518|  1.0|
| 1017962|138.66666666666666|11.666666666666666|0.32192771084337346|  1.0|
| 1018240|191.99333333333334| 6.471905677500001|0.28752618135376756|  1.0|
+--------+------------------+------------------+-------------------+-----+
```

根据代码 6-10 的结果，可以判断 SVM 预测用户是否值得挽留的代码封装是正确的。至此，对 SVM 预测用户是否值得挽留的参数封装完成。

6.2 用户画像

利用大数据技术进行用户画像。将用户标签化，即以用户为中心，根据用户所有的历史行为构建用户画像，细分出用户的各种特征，有助于企业从整体上深入地了解每一位用户。

6.2.1 用户画像概述

用户画像是指通过各个途径收集到与用户特性和行为有关的属性标签，包括个人基本信息、社会活动信息、操作行为习惯等，通过对这些信息的综合分析，勾勒出该用户的特征与轮廓。

“交互设计之父”艾伦·库伯最早提出了用户画像（Persona）的概念，他认为“用户画像是真实用户的虚拟代表，是建立在一系列真实数据之上的目标用户模型”。通过对用户多方面信息的了解，将多种信息集合在一起形成在一定类型上的特征与气质，即用户的独特的“画像”。用户画像的核心工作是给用户标注标签。标签通常是人为规定的、高度精练的特征标识，如年龄、性别、地域、兴趣等。用户标签计算完成后，将用户的所有标签综合在一起，即可勾勒出该用户的立体画像，如图 6-7 所示。

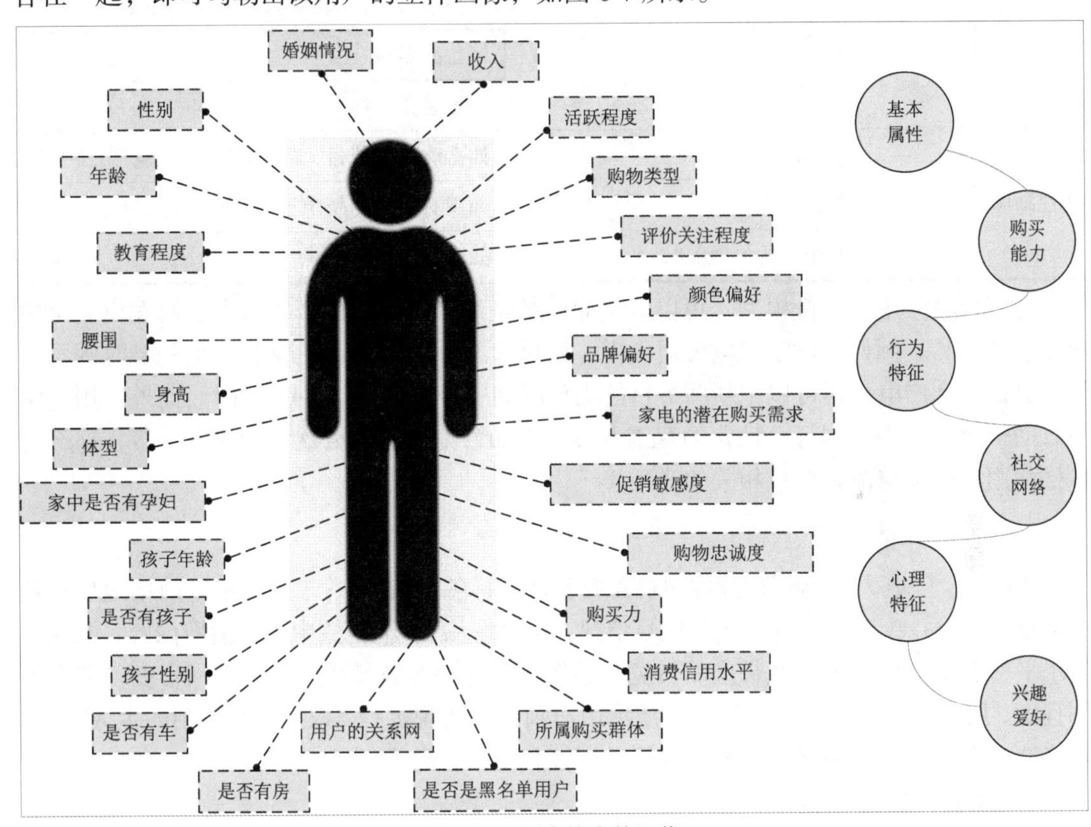

图 6-7 用户的立体画像

一般情况下，用户画像可以围绕显性画像和隐性画像两个方面展开。

显性画像：用户群体的可视化的特征描述，如目标用户的年龄、性别、职业等特征。

隐性画像：用户内在的深层次的特征描述，包含用户使用产品的目的、用户偏好、用户需求、用户使用产品的场景等。

以对广电公司的客户进行用户画像为例，可以围绕显性画像和隐性画像两个方面对用户进行画像，显性画像可包含用户特征（如注册时间、消费水平、支付方式、光机名称、楼盘名称）和业务特征（如业务品牌、Modem 类别、销售品名称、套餐名称）等，隐性画像可包含偏好、收视成员和用户的价值等级等，如表 6-1 所示。

表 6-1　广电用户画像

一级维度	二级维度	特征描述
显性画像	用户特征	注册时间
		消费水平
		支付方式
		光机名称
		楼盘名称
	业务特征	业务品牌
		Modem 类别（单 Modem 或多 Modem）
		销售品名称
		套餐名称
隐性画像	偏好	如喜欢体育节目、新闻节目，不喜欢财经节目
	收视成员	如观看节目的是中年人，而不是儿童
	用户的价值等级	如高价值用户

为用户建立用户画像模型，可以让企业更清楚地了解用户的真实需求，研发出满足用户需求的产品。例如，对于广电公司而言，如果大部分用户喜欢在晚上 7 时—8 时观看新闻节目，那么广电公司可以把新闻节目的播放时间安排在晚上 7 时—8 时。此外，用户画像有助于企业为用户制订个性化的营销方案，例如，如果用户喜欢观看体育节目，那么可以推荐用户购买与体育节目相关的产品。

6.2.2　标签计算

实际上，广电大数据用户画像项目需要实现的标签共有 60 个，为了尽量简单地呈现业务背景知识，并让读者了解真实的项目情况，这里特从 60 个标签中挑选消费内容、电视消费水平、宽带消费水平、宽带产品带宽、销售品名称、业务品牌、电视入网程度、宽带入网程度、用户是否挽留这 9 个标签来演示用户画像的实现过程。根据标签规则的生成方式，可将这 9 个标签分为两部分：阈值计算类标签和字段直贴类标签。

1. 阈值计算类标签

在 3.2.3 小节中，已探索出电视消费水平、宽带消费水平、电视入网程度和宽带入网程度标签的阈值，因此这 4 个阈值计算类标签的计算规则需结合 3.2.3 小节的探索结果来确定。

2. 字段直贴类标签

字段直贴类标签是指可以将数据中某个字段的值直接作为标签值，消费内容、宽带产品带宽、用户是否挽留、销售品名称、业务品牌都属于这类标签。这类标签的计算规则是根据甲方业务人员给出的过滤条件直接通过数据中的某个字段标注标签。

3. 标签的计算规则和实现过程

每个标签的计算规则和实现过程的详细介绍如下。

（1）消费内容

消费内容标签主要通过 mmconsume_billevent_process 中的 fee_code 字段来判断用户的标签。该标签的计算规则如表 6-2 所示。

表 6-2 消费内容标签的计算规则

标签名称	Hive 表	计算规则
消费内容	user_profile.mmconsume_billevent_process	①选择 phone_no、fee_code 字段并去重。 ②根据 fee_code 字段来标注标签。 a. 若 fee_code=0J/0B/OY，则标签为直播。 b. 若 fee_code=0X，则标签为应用。 c. 若 fee_code=0T，则标签为付费频道。 d. 若 fee_code=0W/0L/0Z/0K，则标签为宽带。 e. 若 fee_code=0D，则标签为点播。 f. 若 fee_code=0H，则标签为回看。 g. 若 fee_code=0U，则标签为有线电视收视费

根据表 6-2 的计算规则，在 Spark Shell 中使用 Spark SQL 计算用户的消费内容标签，如代码 6-11 所示。

代码 6-11 计算用户的消费内容标签

```
scala> sqlContext.sql("select distinct phone_no,case when fee_code='0J' or fee_code='0B' or fee_code='0Y' then '直播' when fee_code='0X' then '应用' when fee_code='0T' then '付费频道' when fee_code='0W' or fee_code='0L' or fee_code='0Z' or fee_code='0K' then '宽带' when fee_code='0D' then '点播' when fee_code='0H' then '回看' when fee_code='0U' then '有线电视收视费' end as label,'消费内容' as parent_label from user_profile.mmconsume_billevent_process").show(5)
+--------+------+------------+
|phone_no| label|parent_label|
+--------+------+------------+
| 4864753|   回看|        消费内容|
| 2038755|   直播|        消费内容|
| 4989646|   点播|        消费内容|
| 4316574|   宽带|        消费内容|
| 4982157|   直播|        消费内容|
+--------+------+------------+
only showing top 5 rows
```

（2）电视消费水平

对于电视消费水平的计算，首先需要从 mmconsume_billevent_process 中根据 sm_name 不包含“珠江宽频”的过滤条件，筛选出电视用户的数据，再根据 should_pay 和 favour_fee

字段来进行计算，should_pay 字段指用户需要交纳的金额，favour_fee 指优惠的金额，计算 should_pay 与 favour_fee 的差值，即可得到用户实际需要交纳的金额，进而统计出用户的电视月均消费金额，再根据用户的电视月均消费金额划分用户的电视消费水平。电视消费水平标签的计算规则如表 6-3 所示。

表 6-3　电视消费水平标签的计算规则

标签名称	Hive 表	计算规则
电视消费水平	user_profile.mmconsume_billevent_process	①根据 sm_name 字段不包含“珠江宽频”筛选出电视用户的数据。 ②计算 3 个月的数据中 should_pay−favour_fee 的月平均值 x。 ③标注标签。 a. 若−26.5 < x < 26.5，则标签为电视超低消费。 b. 若 26.5 ≤ x < 26.5+20，则标签为电视低消费。 c. 若 26.5+20 ≤ x < 26.5+40，则标签为电视中等消费。 d. 若 x ≥ 26.5+40，则标签为电视高消费

根据表 6-3 的计算规则，在 Spark Shell 中使用 Spark SQL 计算用户的电视消费水平标签，如代码 6-12 所示。

代码 6-12　计算用户的电视消费水平标签

```
scala> sqlContext.sql("select t2.phone_no,case when fee_per_month>-26.5 and
fee_per_month<26.5 then '电视超低消费' when fee_per_month>=26.5 and fee_per_
month<46.5 then '电视低消费' when fee_per_month>=46.5 and fee_per_month<66.5
then '电视中等消费' when 66.5<=fee_per_month then '电视高消费' end as label,'
电视消费水平' as parent_label from (select t1.phone_no,sum(real_pay)/3 as
fee_per_month from (select phone_no,nvl(should_pay,0)-nvl(favour_fee,0) as
real_pay from user_profile.mmconsume_billevent_process where sm_name like '%
电视%') t1 group by t1.phone_no) t2").show(5)
+--------+----------+------------+
|phone_no|     label|parent_label|
+--------+----------+------------+
| 4124032|     电视高消费|      电视消费水平|
| 3102451|     电视高消费|      电视消费水平|
| 4177269|     电视高消费|      电视消费水平|
| 4262354|     电视高消费|      电视消费水平|
| 4087160|    电视中等消费|      电视消费水平|
+--------+----------+------------+
```

（3）宽带消费水平

宽带消费水平标签的计算与电视消费水平标签的计算类似，不同的是，宽带消费水平标签需要先从 mmconsume_billevent_process 中根据 sm_name 包含“珠江宽频”的过滤条件筛选出宽带用户的数据，再根据 should_pay-favour_fee 计算用户实际需要交纳的金额，进而统计出用户的宽带月均消费金额，再根据用户的宽带月均消费金额划分用户的宽带消费水平。宽带消费水平标签的计算规则如表 6-4 所示。

表 6-4 宽带消费水平标签的计算规则

标签名称	Hive 表	计算规则
宽带消费水平	user_profile.mmconsume_billevent_process	①根据 sm_name 包含“珠江宽频”筛选宽带用户的数据。 ②计算 3 个月的数据 should_pay-favour_fee 的月平均值 Y。 ③标注标签。 a. 若 $Y \leqslant 25$，则标签为宽带低消费。 b. 若 $25 < Y \leqslant 45$，则标签为宽带中消费。 c. 若 $Y > 45$，则标签为宽带高消费

根据表 6-4 的计算规则，在 Spark Shell 中使用 Spark SQL 计算用户的宽带消费水平标签，如代码 6-13 所示。

代码 6-13 计算用户的宽带消费水平标签

```
scala> sqlContext.sql("select t2.phone_no,case when fee_per_month<=25 then '
宽带低消费' when fee_per_month>25 and fee_per_month<=45 then '宽带中消费' when
fee_per_month>45 then '宽带高消费' end as label,'宽带消费水平' as parent_label
from (select t1.phone_no,sum(real_pay)/3 as fee_per_month from (select phone_no,
nvl(should_pay,0)-nvl(favour_fee,0) as real_pay from user_profile.mmconsume_
billevent_process where sm_name='珠江宽频') t1 group by t1.phone_no) t2").
show(5)
+--------+----------+------------+
|phone_no|     label|parent_label|
+--------+----------+------------+
| 3324365|  宽带高消费|   宽带消费水平|
| 4108445|  宽带低消费|   宽带消费水平|
| 3788947|  宽带中消费|   宽带消费水平|
| 4831060|  宽带高消费|   宽带消费水平|
| 4687548|  宽带高消费|   宽带消费水平|
+--------+----------+------------+
only showing top 5 rows
```

（4）宽带产品带宽

宽带产品带宽标签的计算需要先从 order_index_process 中根据 sm_name 包含“珠江宽频”的过滤条件筛选出宽带用户的数据，再根据 phone_no 字段进行分组，取出每个用户中 optdate 字段值最大、effdate 字段值小于或等于当前时间且 expdate 字段值大于或等于当前时间的数据，最后根据 prodname 字段确定标签，prodname 字段的值即用户的宽带产品带宽标签的名称。宽带产品带宽标签的计算规则如表 6-5 所示。

表 6-5　宽带产品带宽标签的计算规则

标签名称	Hive 表	计算规则
宽带产品带宽	user_profile.order_index_process	①根据 sm_name='珠江宽频'来筛选宽带用户。 ②根据字段 phone_no 进行分组，取 optdate 最大且 effdate≤当前时间≤expdate 的数据。 ③根据 prodname 字段确定标签，prodname 字段的值即标签名称

根据表 6-5 的计算规则，在 Spark Shell 中使用 Spark SQL 计算用户的宽带产品带宽标签，如代码 6-14 所示。

代码 6-14　计算用户的宽带产品带宽标签

```
scala> sqlContext.sql("select phone_no,case when prodname=prodname then
prodname end as label,'宽带产品带宽' as parent_label
from(select a.phone_no,a.optdate,a.prodname,a.sm_name,row_number() over (partition
by a.phone_no order by a.optdate desc) rank from (select phone_no,prodname,
expdate,optdate,sm_name from
user_profile.order_index_process where effdate <= from_unixtime(unix_timestamp(),
'yyyy-MM-dd HH:mm:ss') and from_unixtime(unix_timestamp(),'yyyy-MM-dd HH:mm:ss')
<= expdate) a) b where b.rank=1 and b.sm_name='珠江宽频'").show(5)
+--------+------------+------------+
|phone_no|       label|parent_label|
+--------+------------+------------+
| 2000680|宽带空指令产品|  宽带产品带宽|
| 2001120|宽带空指令产品|  宽带产品带宽|
| 2001571|宽带空指令产品|  宽带产品带宽|
| 2003353|宽带空指令产品|  宽带产品带宽|
| 2004244|宽带空指令产品|  宽带产品带宽|
+--------+------------+------------+
only showing top 5 rows
```

（5）销售品名称

销售品名称标签的计算需要先从 order_index_process 中筛选出 cost 字段值大于 0 且 offername 字段不包含“空包”的数据，根据 sm_name 字段区分电视和宽带，sm_name 包含“珠江宽频”的为宽带用户，sm_name 不包含“珠江宽频”的为电视用户。电视用户的销售品分为主销售品和附属销售品两种：同时满足 mode_time 字段为“Y”、offertype 字段为 0、prodstatus 字段为“YY”、effdate 小于或等于当前时间、expdate 大于或等于当前时间、optdate 字段值最大这 6 个条件的数据中的 offername 即电视主销售品；同时满足 mode_time 字段为“Y”、offertype 字段为 1、prodstatus 字段为“YY”、effdate 小于或等于当前时间、expdate 大于或等于当前时间这 5 个条件的数据中的 offername 即电视附属销售品。宽带销售品的判定规则是同时满足 effdate 小于或等于当前时间、expdate 字段大于或等于当前时间且 optdate 字段值最大这 3 个条件，再根据 offername 字段确定标签名称。销售品名称标签的计算规则如表 6-6 所示。

表 6-6 销售品名称标签的计算规则

标签名称	Hive 表	计算规则
销售品名称	user_profile.order_index_process	①过滤 cost 小于或等于 0 且 offername 不包含“空包”的记录。 ②根据 sm_name 区分电视和宽带，sm_name='珠江宽频'的为宽带，sm_name 不包含'珠江宽频'的为电视。 a. 电视主销售品：mode_time='Y'，offertype=0，prodstatus='YY'，effdate≤当前时间≤expdate 且 optdate 最大的数据。 b. 电视附属销售品：mode_time='Y'，offertype=1，prodstatus='YY'且 effdate≤当前时间≤expdate 的数据。 c. 宽带：effdate≤当前时间≤expdate 且 optdate 最大的数据。 ③筛选出来的数据先选择 phone_no、offername 字段并去重，再根据 offername 字段标注标签

根据表 6-6 的计算规则，在 Spark Shell 中使用 Spark SQL 计算用户的销售品名称标签，如代码 6-15 所示。

代码 6-15 计算用户的销售品名称标签

```
scala> sqlContext.sql("select phone_no,case when offername=offername then
offername end as label,'销售品名称' as parent_label from(select phone_no,
offername from (select \
t2.phone_no,t2.optdate,t2.offername,row_number() over (partition by t2.phone_no
order by t2.optdate desc) rank from (select t1.phone_no,t1.offername,
t1.optdate from (select * from user_profile.order_index_process where cost>=0
and offername not like '%空包%') t1 where t1.sm_name like '%电视%' and
t1.mode_time='Y' and t1.offertype=0 and t1.prodstatus='YY' and t1.effdate <=
from_unixtime(unix_timestamp(),'yyyy-MM-dd HH:mm:ss') and from_unixtime
(unix_timestamp(),'yyyy-MM-dd HH:mm:ss') <= t1.expdate) t2) t3 where t3.rank=1
union all select phone_no,offername from (select * from user_profile.
order_index_process where cost>=0 and offername not like '%空包%') t3 where
t3.sm_name like '%电视%' and t3.mode_time='Y' and t3.offertype=1 and t3.
prodstatus='YY' and t3.effdate <= from_unixtime(unix_timestamp(),'yyyy-MM-dd
HH:mm:ss') and from_unixtime(unix_timestamp(),'yyyy-MM-dd HH:mm:ss') <=
t3.expdate union all select phone_no,offername from (select t2.phone_no,
t2.optdate,t2.offername,row_number() over (partition by t2.phone_no order by
t2.optdate desc) rank from (select t1.phone_no,t1.offername,t1.optdate from
(select * from user_profile.order_index_process where cost>=0 and offername
not like '%空包%') t1 where t1.sm_name like '%珠江宽频%' and t1.effdate <=
from_unixtime(unix_timestamp(),'yyyy-MM-dd HH:mm:ss') and from_unixtime(unix_
timestamp(),'yyyy-MM-dd HH:mm:ss') <= t1.expdate) t2 )t3 where t3.rank=1)
tt").show(5)
```

```
+--------+-------------------------------+------------+
|phone_no|                          label| parent_label|
+--------+-------------------------------+------------+
| 2000798| 互动标准包-600 预存-送快乐点 2...|     销售品名称|
| 2001733| 互动标准包-600 预存-送快乐点 2...|     销售品名称|
| 2002011| 租机租卡 - 押 0 元(副卡)(5 元/月...|     销售品名称|
| 2002462| 互动标准包-600 预存-送快乐点 2...|     销售品名称|
| 2003191| 互动标准包-600 预存-送快乐点 2...|     销售品名称|
+--------+-------------------------------+------------+
only showing top 5 rows
```

（6）业务品牌

业务品牌标签的计算是根据 mediamatch_usermsg_process 中的 sm_name 字段确定的，但是需要先删除 sm_name 为“模拟有线电视”或“番通”的数据。业务品牌标签的计算规则如表 6-7 所示。

表 6-7　业务品牌标签的计算规则

标签名称	Hive 表	计算规则
业务品牌	user_profile.mediamatch_usermsg_process	①选择 phone_no、sm_name 字段并去重。 ②删除 sm_name='模拟有线电视'或'番通'的数据。 ③根据 sm_name 字段来标注标签。 a. 若 sm_name='互动电视'，则标签为互动电视。 b. 若 sm_name='数字电视'，则标签为数字电视。 c. 若 sm_name='甜果电视'，则标签为甜果电视。 d. 若 sm_name='珠江宽频'，则标签为珠江宽频

根据表 6-7 的计算规则，在 Spark Shell 中使用 Spark SQL 计算用户的业务品牌标签，如代码 6-16 所示。

代码 6-16　计算用户的业务品牌标签

```
scala> sqlContext.sql("select phone_no,case when sm_name='互动电视' then '互动电视' when sm_name='数字电视' then '数字电视' when sm_name='甜果电视' then '甜果电视' when sm_name='珠江宽频' then '珠江宽频' end as label,'业务品牌' as parent_label from user_profile.mediamatch_usermsg_process where sm_name not like '%模拟有线电视%' or sm_name not like '%番通%'").show(5)
+--------+-------+------------+
|phone_no|  label|parent_label|
+--------+-------+------------+
| 2000248| 互动电视|      业务品牌|
| 2000259| 互动电视|      业务品牌|
| 2000542| 数字电视|      业务品牌|
| 2000589| 互动电视|      业务品牌|
| 2000852| 珠江宽频|      业务品牌|
+--------+-------+------------+
only showing top 5 rows
```

（7）电视入网程度

电视入网程度标签的计算需要先从 mediamatch_usermsg_process 中筛选出 sm_name 字段包含“互动”“甜果”或“数字”的记录，再根据 phone_no 字段进行分组，并计算用户开户时间与当前时间的差值以得到开户时长 T（单位为年）。根据 3.2.3 小节中的入网程度标签阈值可知，若 $T>8$，则标签为老用户，若 $T\leqslant 4$，则标签为新用户，若 $4<T\leqslant 8$，则标签为中等用户。电视入网程度标签的计算规则如表 6-8 所示。

表 6-8 电视入网程度标签的计算规则

标签名称	Hive 表	计算规则
电视入网程度	user_profile.mediamatch_usermsg_process	①筛选 sm_name 包含“互动”“甜果”“数字”的记录。 ②根据 phone_no 字段进行分组，找出 open_time（用户开户时间）字段与当前时间作差得到 T（单位为年）。 ③标注标签。 a. 若 $T>8$，则标签为老用户。 b. 若 $T\leqslant 4$，则标签为新用户。 c. 若 $4<T\leqslant 8$，则标签为中等用户

根据表 6-8 的计算规则，在 Spark Shell 中使用 Spark SQL 计算用户的电视入网程度标签，如代码 6-17 所示。

代码 6-17 计算用户的电视入网程度标签

```
scala> sqlContext.sql("select t1.phone_no,case when T>8 then '老用户' when T>4
and T<=8 then '中等用户' when T<=4 then '新用户' end as label,'电视入网程度' as
parent_label from(select \
phone_no,max(datediff(current_date(),open_time)/365) as T from user_profile.
mediamatch_usermsg_process where sm_name like '%电视%' and open_time is not NULL
group by phone_no) t1").show(5)
+--------+--------+------------+
|phone_no|   label|parent_label|
+--------+--------+------------+
| 2488923| 中等用户|   电视入网程度|
| 2510549|   老用户|   电视入网程度|
| 2583927|   老用户|   电视入网程度|
| 2748771|   老用户|   电视入网程度|
| 2872287|   老用户|   电视入网程度|
+--------+--------+------------+
only showing top 5 rows
```

（8）宽带入网程度

宽带入网程度标签的计算需要先从 mediamatch_usermsg_process 中筛选 sm_name 包含“珠江宽频”、force 为“宽带生效”、sm_code 为 b0 的数据，再根据 phone_no 字段进行分组，计算每个用户的开户时间与当前时间的差值 T。根据 3.2.3 小节中的入网程度标签阈值可知，若 $T>6$，则标签为老用户，若 $T\leqslant 2$，则标签为新用户，若 $2<T\leqslant 6$，则标签为中等用户。宽带入网程度标签的计算规则如表 6-9 所示。

表 6-9　宽带入网程度标签的计算规则

标签名称	Hive 表	规则
宽带入网程度	user_profile.mediamatch_usermsg_process	①选择 sm_name 包含“珠江宽频”、force 为“宽带生效”、sm_code=b0 的数据。 ②根据 phone_no 字段进行分组，找出 open_time 字段（用户的开户时间）与当前时间作差得到 T。 ③标注标签。 a. 若 $T>6$，则标签为老用户。 b. 若 $T\leqslant 2$，则标签为新用户。 c. 若 $2<T\leqslant 6$，则标签为中等用户

根据表 6-9 的计算规则，在 Spark Shell 中使用 Spark SQL 计算用户的宽带入网程度标签，如代码 6-18 所示。

代码 6-18　计算用户的宽带入网程度标签

```
scala> sqlContext.sql("select t1.phone_no,case when T>6 then '老用户' when T>2
and T<=6 then '中等用户' when T<=2 then '新用户'  end as label,'宽带入网程度' as
parent_label from (select phone_no,max(datediff(current_date(),open_time)/365)
as T from user_profile.mediamatch_usermsg_process where sm_name='珠江宽频' and
force like '%宽带生效%' and sm_code='b0' group by phone_no) t1").show(5)
+--------+--------+------------+
|phone_no|   label|parent_label|
+--------+--------+------------+
| 4081643|  中等用户|   宽带入网程度|
| 4092173|  中等用户|   宽带入网程度|
| 4234076|  中等用户|   宽带入网程度|
| 4286510|  中等用户|   宽带入网程度|
| 4509978|  中等用户|   宽带入网程度|
+--------+--------+------------+
only showing top 5 rows
```

（9）用户是否挽留

在 6.1 小节中使用了 SVM 算法预测用户的挽留状态，并将预测结果保存到 Hive 的 user_profile 库的 svm_prediction 中，因此，用户是否挽留标签可根据 svm_prediction 的 label 字段进行计算，label 等于 1 的为挽留用户，label 等于 0 的为非挽留用户。用户是否挽留标签的计算规则如表 6-10 所示。

表 6-10 用户是否挽留标签的计算规则

标签名称	Hive 表	计算规则
用户是否挽留	user_profile.svm_prediction	①选择 phone_no、label 两列数据并去重。 ②标注标签。 a. 若 label=1，则标签为挽留用户。 b. 若 label=0，则标签为非挽留用户

根据表 6-10 的计算规则，在 Spark Shell 中使用 Spark SQL 计算用户是否挽留标签，如代码 6-19 所示。

代码 6-19 计算用户是否挽留标签

```
scala> sqlContext.sql("select phone_no,case when label=1 then '挽留用户' when
label=0 then '非挽留用户'  end as label,'用户是否挽留' as parent_label from
user_profile.svm_prediction").show(5)
+--------+---------+------------+
|phone_no|    label|parent_label|
+--------+---------+------------+
| 2028358| 非挽留用户|   用户是否挽留|
| 2203435| 非挽留用户|   用户是否挽留|
| 2146628| 非挽留用户|   用户是否挽留|
| 2000121| 非挽留用户|   用户是否挽留|
| 2073340| 非挽留用户|   用户是否挽留|
+--------+---------+------------+
only showing top 5 rows
```

本小节介绍的标签计算都是在 Spark Shell 中实现的。在实际项目中，通常会对代码进行封装，并把计算的结果保存到数据库中，以供其他系统使用。下一小节将介绍用户画像工程的实现。

6.2.3 用户画像工程实现

本小节主要介绍的是在工程中实现用户画像，用户画像工程实现的主要步骤如下。

（1）由 6.2.2 小节的标签计算可知，所有标签的计算都可以通过 Spark SQL 执行一条 SQL 语句来实现，因此，为了代码重用，可以实现一个 SQLEngine 类，该类的功能是执行一条 SQL 语句并把执行的结果保存到数据库中。

（2）提供一个 JSON 文件，其中包含所有需要用到的标签规则。

（3）编写一个 Java 实现类 LabelSQL，该类的功能是读取标签规则的 JSON 文件。

（4）编写最终的实现类 SQLResolve，该类的功能是解析 SQLEngine 类所需的参数值，并提交运行。

下面对用户画像工程每一个步骤的实现过程及代码逐一进行介绍。首先介绍 SQLEngine 类的代码，如代码 6-20 所示。

代码 6-20 SQLEngine 类

```
package com.tipdm.scala.chapter_3_9_3_user_profile
import java.util.Properties
import org.apache.spark.sql.SaveMode
import org.apache.spark.sql.hive.HiveContext
import org.apache.spark.{SparkConf, SparkContext}
object SQLEngine {
  def main(args: Array[String]): Unit = {
    val appName = args(0)
    val sql = args(1)
    val outputTable = args(2)
    val conf = new SparkConf()
    val sc = new SparkContext(conf).setJars(Seq("/opt/cloudera/parcels/
CDH-5.7.3-1.cdh5.7.3.p0.5/lib/hive/lib/mysql-connector-java-5.1.7-bin.jar"))
    val sqlContext = new HiveContext(sc)
    val data = sqlContext.sql(sql)
    val dbType = args(3)
    val saveMode = args(4)
    dbType match {
      case "hive" => data.write.mode(saveMode).saveAsTable(outputTable)
      case "rdbms" => {
        val url = args(5)
        val connectionProperties = new Properties()
        connectionProperties.setProperty("user", args(6))
        connectionProperties.setProperty("password", args(7))
        data.write.mode(saveMode).jdbc(url, outputTable, connectionProperties)
      }
    }
    sc.stop()
  }
}
```

在代码 6-20 中，设置了 8 个参数，每个参数的说明如下。

（1）appName：程序的名称。

（2）sql：要执行的 SQL 语句，如计算业务品牌标签，则需要将代码 6-16 中的 SQL 语句作为参数。

（3）outputTable：最后保存的表名，若保存到 MySQL 的 user_label 表中，则“user_label”是这个参数的参数值。

（4）dbType：保存的数据库类型（Hive 或传统数据库），如果结果保存到 MySQL 中，那么其参数值为“rdbms”；如果保存到 Hive 中，那么其参数值为“hive”。

（5）saveMode：保存的模式，若覆盖原来表中的内容，则其参数值是“overwrite”；若追加到原来的表中，则参数值是“append”。

（6）url：目标数据库的 URL，若保存到 Hive 中，则不需要这个参数，参数值置空即可，若保存到 MySQL 中，则其参数值类似于“jdbc:mysql://192.168.111.75:3306/user_profile”，实际中只需要修改对应的节点 IP 地址和数据库名称即可。

（7）user：数据库的用户名，若数据库的用户名是“root”，则其参数值为“root”。

（8）password：数据库的密码，若数据库的密码是“root”，则其参数值为“root”。

针对代码 6-20 编写测试代码，代码实现的功能是查询 mediamatch_usermsg_process 中 phone_no、run_time、sm_name 和 run_name 字段的前 10 条记录，并把查询到的结果保存到 IP 地址为 192.168.111.75 的节点的 MySQL 的 user_profile 库的 sqlEngine_test 中，如代码 6-21 所示。

代码 6-21 SQLEngine 类的测试代码

```
package com.tipdm.java.chapter_3_9_3_user_profile;
import com.tipdm.engine.SparkYarnJob;
import com.tipdm.engine.model.Args;
import com.tipdm.engine.model.SubmitResult;
public class SQLResolveTest {
   private static String className=" com.tipdm.scala.chapter_3_9_3_user_profile.
SQLEngine";
   private static  String applicationName = "SQLEngine";
   public static void main(String[] args) throws Exception {
      String[] arguments = new String[8];
      arguments[0] = "SQLEngine";
      arguments[1] = "select phone_no,run_time,sm_name,run_name from mediamatch_
usermsg_process limit 10";
      arguments[2] = "sqlEngine_test";
      arguments[3] = "rdbms";
      arguments[4] = "overwrite";
      arguments[5] = "jdbc:mysql://192.168.111.75:3306/user_profile";
      arguments[6] = "root";
      arguments[7] = "root";
      Args innerArgs = Args.getArgs(applicationName,className,arguments);
      SubmitResult submitResult = SparkYarnJob.run(innerArgs);
      SparkYarnJob.monitor(submitResult);
      System.out.println("任务运行成功");
   }
}
```

执行代码 6-21，任务运行完成后，查看 Spark 任务运行情况，其运行结果如图 6-8 所示。

driver-20190226163522-0088	Tue Feb 26 16:35:22 CST 2019	worker-20190226134040-192.168.111.78-7078	FINISHED	1	1024.0 MB	com.tipdm.scala.chapter_3_9_3_user_profile.SQLEngine		
app-20190226163528-0113	SQLEngine 1551170105736	12	8.0 GB	2019/02/26 16:35:28	spark	FINISHED	36 s	

图 6-8 Spark 任务的运行结果

若任务状态如图 6-8 所示，则表示任务运行成功。任务运行成功后，查看 MySQL 中的数据存储情况，如图 6-9 所示。

phone_no	run_time	sm_name	run_name
2000080	2013-07-27 17:17:32	互动电视	正常
2000298	2013-07-28 18:05:23	数字电视	正常
2000325	2013-08-04 14:33:51	互动电视	正常
2000761	2013-08-07 16:06:10	数字电视	正常
2000947	2013-08-16 17:32:58	珠江宽频	正常
2001160	2013-08-17 18:22:20	珠江宽频	正常
2001484	2017-07-17 13:40:03	互动电视	正常
2002302	2018-02-28 09:14:30	互动电视	主动暂停
2003105	2013-09-26 18:09:32	珠江宽频	正常
2003160	2014-11-30 23:30:36	珠江宽频	主动暂停

图 6-9 查看 MySQL 中的数据存储情况

根据图 6-9 所示的结果，可以判断 SQLEngine 类的代码是正确的，因此，在工程中可以利用 SQLEngine 类的代码来实现用户画像。

广电大数据用户画像项目中需要实现的标签个数较多，不同标签的计算规则不一样。在实际项目中，业务任务可以在标签管理界面中添加标签及标签的计算规则，在后台数据库中生成对应的标签及标签的计算规则，这个过程的实现较为复杂。为简化标签规则的生成过程，这里提供了一个包含目前使用的所有标签规则的 JSON 文件，后续直接从该 JSON 文件中读取标签规则即可。JSON 文件的部分内容如代码 6-22 所示。

代码 6-22　JSON 文件的部分内容

```
{
  "消费内容": [
    {
      "label": "直播",
      "rules": "fee_code=\"0J\" or fee_code=\"0B\" or fee_code=\"0Y\""
    },
    {
      "label":"应用",
      "rules":"fee_code=\"0X\""
    },
    {
      "label":"付费频道",
      "rules":"fee_code=\"0T\""
    },
    {
      "label":"宽带",
      "rules":"fee_code=\"0W\" or fee_code=\"0L\" or fee_code=\"0Z\" or fee_
code=\"0K\""
    },
    {
      "label":"点播",
      "rules":"fee_code=\"0D\""
    },
    {
      "label":"回看",
      "rules":"fee_code=\"0H\""
    },
    {
      "label":"有线电视收视费",
      "rules":"fee_code=\"0U\""
    }
  ],
……
  "宽带入网程度":[
    {
      "label":"老用户",
      "rules":"T>6",
      "comment":"注释，T 为当前时间减去开户时间（单位：年）"
```

```
    },
    {
      "label":"中等用户",
      "rules":"T>2 and T<=6"
    },
    {
      "label":"新用户",
      "rules":"T<=2"
    }
  ]
}
```

注意：代码 6-22 所示的标签规则中省略了部分内容，具体内容可参考工程文件 user_profile_project\src\main\resources\rules\consume_content.json。

因为 SQLEngine 类需要一条完整的 SQL 语句作为参数，所以需要将标签规则的 JSON 格式根据标签名称解析成 SQL 语句。读取代码 6-22 所示的标签规则并将其解析为 SQL 语句，如代码 6-23 所示。

代码 6-23　读取标签规则并将其解析成 SQL 语句

```
package com.tipdm.java.label;
import com.alibaba.fastjson.JSONArray;
import com.alibaba.fastjson.JSONObject;
import java.io.InputStream;
import java.io.StringWriter;
import org.apache.commons.io.IOUtils;
import java.io.IOException;
import java.util.regex.Matcher;
import java.util.regex.Pattern;
public class LabelSQL {
    /**
     * @param label: 标签名称
     * @param table: 数据源的表名称
     * @return 根据标签得到 SQL 语句
     * @throws IOException
     */
    public static String getLabel(String label, String table)throws IOException {
        String sql = null;
        String condition = getSQL(label);
        if (label.equals("消费内容")) {
            sql = "select distinct phone_no,case " + condition + " from " + table;
        } else if (label.equals("电视消费水平")) {
            sql = "select t2.phone_no,case " + condition + " from\n" +
                  "(select t1.phone_no,sum(real_pay)/3 as fee_per_month  from
(select phone_no,nvl(should_pay,0)-nvl(favour_fee,0) as real_pay from " +
table + " where sm_name like '%电视%') t1 group by t1.phone_no) t2";
        } else if (label.equals("宽带消费水平")) {
            sql = "select t2.phone_no,case " + condition + " from\n" +
```

```
                    "(select t1.phone_no,sum(real_pay)/3 as fee_per_month  from
(select phone_no,nvl(should_pay,0)-nvl(favour_fee,0) as real_pay from " +
table + " where sm_name='珠江宽频') t1 group by t1.phone_no) t2";
        } else if (label.equals("宽带产品带宽")) {
            sql = "select b.phone_no, case " + condition + " from(select a.
phone_no,a.optdate,a.prodname,a.sm_name,row_number() over (partition by a.
phone_no order by a.optdate desc) rank from (select phone_no,prodname,expdate,
optdate,sm_name from " + table + " where effdate <= from_unixtime(unix_timestamp(),
'yyyy-MM-dd HH:mm:ss') and from_unixtime(unix_timestamp(),'yyyy-MM-dd HH:
mm:ss') <= expdate) a) b where b.rank=1 and b.sm_name='珠江宽频'";
        } else if(label.equals("用户是否挽留")){
            sql="select phone_no,case "+condition+" from "+table;
        }else if (label.equals("销售品名称")) {
            sql = "select phone_no,case " + condition + " from(\n" +
                    "select phone_no,offername from \n" +
                    "(select t2.phone_no,t2.optdate,t2.offername,row_number()
over (partition by t2.phone_no order by t2.optdate desc) rank from\n" +
                    "(select t1.phone_no,t1.offername,t1.optdate from \n" +
                    "(select * from " + table + " where cost>=0 and offername not
like '%空包%') t1 where t1.sm_name like '%电视%' and t1.mode_time='Y' and t1.
offertype=0 and t1.prodstatus='YY' and t1.effdate <= from_unixtime(unix_timestamp(),
'yyyy-MM-dd HH:mm:ss') and from_unixtime(unix_timestamp(),'yyyy-MM-dd HH:mm:
ss') <= t1.expdate) t2) t3 where t3.rank=1 union all select  phone_no,offername
from \n" +
                    "(select * from " + table + " where cost>=0 and offername not
like '%空包%') t3 where t3.sm_name like '%电视%' and t3.mode_time='Y' and t3.
offertype=1 and t3.prodstatus='YY' and t3.effdate <= from_unixtime(unix_timestamp(),
'yyyy-MM-dd HH:mm:ss') and from_unixtime(unix_timestamp(),'yyyy-MM-dd HH:mm:
ss') <= t3.expdate\n" +
                    "union all select phone_no,offername from \n" +
                    "(select t2.phone_no,t2.optdate,t2.offername,row_number()
over (partition by t2.phone_no order by t2.optdate desc) rank from\n" +
                    "(select t1.phone_no,t1.offername,t1.optdate from \n" +
                    "(select * from " + table + " where cost>=0 and offername not
like '%空包%') t1 where t1.sm_name like '%珠江宽频%' and t1.effdate <= from_
unixtime(unix_timestamp(),'yyyy-MM-dd HH:mm:ss') and from_unixtime(unix_
timestamp(),'yyyy-MM-dd HH:mm:ss') <= t1.expdate) t2 )t3 where t3.rank=1) tt";
        } else if (label.equals("业务品牌")) {
            sql = "select phone_no,case " + condition + " from " + table + " where sm_
name not like '%模拟有线电视%' or sm_name not like '%番通%'";
        } else if (label.equals("电视入网程度")) {
            sql = "select t1.phone_no,case " + condition + " from(\n" +
                    "select
phone_no,max(datediff(current_date(),open_time)/365)
as T from " + table + " where sm_name like '%电视%' and open_time is not NULL
group by phone_no) t1";
        } else if (label.equals("宽带入网程度")) {
            sql = "select t1.phone_no,case " + condition + " from\n" +
```

```
                "(select phone_no,max(datediff(current_date(),open_time)/365)
as T from " + table + " where sm_name='珠江宽频' and force like '%宽带生效%'
and sm_code='b0' group by phone_no) t1";
        } else {
            System.out.printf("标签名称没有找到");
            System.exit(1);
        }
        return sql;
    }
    /**
     *
     * @param label: 二级标签名称
     * @return: 拼接标签的 when 语句
     * @throws IOException
     */
    public static String getSQL(String label) throws IOException {
        InputStream is = LabelSQL.class.getClassLoader().getResourceAsStream
("rules/consume_content.json");
        StringWriter writer = new StringWriter();
        IOUtils.copy(is, writer, "UTF-8");
        String json = writer.toString();
        JSONObject jsonObject = JSONObject.parseObject(json);
        JSONArray jsonArray = jsonObject.getJSONArray(label);
        StringBuilder sql = new StringBuilder();
        if (jsonArray.size() > 0) {
            for (int i = 0; i < jsonArray.size(); i++) {
                JSONObject obj = jsonArray.getJSONObject(i);
                String rules = (String) obj.get("rules");
                sql.append("when").append(" ").append(rules).append(" ");
                String la = (String) obj.get("label");
                Pattern pattern = Pattern.compile("[\\u4e00-\\u9fa5]");
                Matcher matcher = pattern.matcher(la.charAt(0) + "");
                if (matcher.matches()) {
                    sql.append("then ").append("\"" + la + "\"" + " ");
                } else {
                    sql.append("then ").append(la + " ");
                }
            }
            sql.append(" end as label,\"" + label + "\" as parent_label");
        } else {
            System.out.println("rules/consume_content.json 文件中没有此标签");
            System.exit(1);
        }
        return sql.toString();
    }
}
```

针对代码 6-23 编写测试代码，通过代码 6-20 和代码 6-23 计算每个标签并把计算结果保存到 MySQL 的 user_profile 库的 user_label 表中，如代码 6-24 所示。

代码 6-24　用户画像测试代码

```
package com.tipdm.java.chapter_3_9_3_user_profile;
import com.tipdm.engine.SparkYarnJob;
import com.tipdm.engine.engine.type.EngineType;
import com.tipdm.engine.model.Args;
import com.tipdm.engine.model.SubmitResult;
import com.tipdm.java.label.LabelSQL;
import java.util.HashMap;
import java.util.Map;
public class SQLResolve {
   private static String className = "com.tipdm.scala.chapter_3_9_3_user_
profile.SQLEngine";
   private static String applicationName = "SQLResolve";
   public static void main(String[] args) throws Exception {
      Map<String, String> labelTable = new HashMap<>();
      labelTable.put("消费内容", "user_profile.mmconsume_billevent_process");
      labelTable.put("电视消费水平", " user_profile.mmconsume_billevent_process");
      labelTable.put("宽带消费水平", " user_profile.mmconsume_billevent_process");
      labelTable.put("宽带产品带宽", " user_profile.order_index_process");
      labelTable.put("用户是否挽留"," user_profile.svm_prediction");
      labelTable.put("销售品名称", " user_profile.order_index_process");
      labelTable.put("业务品牌", " user_profile.mediamatch_usermsg_process");
      labelTable.put("电视入网程度", " user_profile.mediamatch_usermsg_process");
      labelTable.put("宽带入网程度", " user_profile.mediamatch_usermsg_process");
      int flag = 0;
      String[] arguments = new String[8];
      String sql = null;
      for (Map.Entry<String, String> entry : labelTable.entrySet()) {
         sql = LabelSQL.getLabel(entry.getKey(), entry.getValue());
         System.out.println("SQL: " + sql);
         if (flag == 0) {
            arguments[4] = "overwrite";
         } else {
            arguments[4] = "append";
         }
         arguments[0] = "LabelName: " + entry.getKey();
         arguments[1] = sql;
         arguments[2] = "user_label";
         arguments[3] = "rdbms";
         arguments[5] = "jdbc:mysql://192.168.111.75:3306/user_profile";
         arguments[6] = "root";
         arguments[7] = "root";
         Args innerArgs = Args.getArgs(applicationName,className,arguments,
EngineType.SPARK);
         SubmitResult submitResult = SparkYarnJob.run(innerArgs);
         SparkYarnJob.monitor(submitResult);
         flag++;
         // 间隔10s发起下一个任务
         Thread.sleep(1000 * 10L);
      }
```

```
            System.out.println("任务运行成功");
        }
}
```

执行代码 6-24，任务运行完成后，查看 Spark 任务运行情况，用户画像测试结果如图 6-10 所示。

Completed Applications

Application ID	Name	Cores	Memory per Node	Submitted Time	User	State	Duration
app-20211206174728-0008	SQLResolve 1638784043367	20	5.0 GB	2021/12/06 17:47:28	spark	FINISHED	1.1 min
app-20211206174612-0007	SQLResolve 1638783967378	20	5.0 GB	2021/12/06 17:46:12	spark	FINISHED	1.0 min
app-20211206174305-0006	SQLResolve 1638783780537	20	5.0 GB	2021/12/06 17:43:05	spark	FINISHED	2.6 min
app-20211206174232-0005	SQLResolve 1638783747095	20	5.0 GB	2021/12/06 17:42:32	spark	FINISHED	10 s
app-20211206173903-0004	SQLResolve 1638783538604	20	5.0 GB	2021/12/06 17:39:03	spark	FINISHED	2.8 min
app-20211206173323-0003	SQLResolve 1638783198038	20	5.0 GB	2021/12/06 17:33:23	spark	FINISHED	5.2 min
app-20211206172703-0002	SQLResolve 1638782818823	20	5.0 GB	2021/12/06 17:27:03	spark	FINISHED	5.9 min
app-20211206172455-0001	SQLResolve 1638782690796	20	5.0 GB	2021/12/06 17:24:55	spark	FINISHED	1.6 min
app-20211206172049-0000	SQLResolve 1638782443027	20	5.0 GB	2021/12/06 17:20:49	spark	FINISHED	3.8 min

Completed Drivers

Submission ID	Submitted Time	Worker	State	Cores	Memory	Main Class
driver-20211206174726-0008	Mon Dec 06 17:47:26 CST 2021	worker-20211206171934-192.168.1.204-7078	FINISHED	2	2.0 GB	com.tipdm.scala.chapter_3_9_3_user_profile.SQLEngine
driver-20211206174610-0007	Mon Dec 06 17:46:10 CST 2021	worker-20211206171934-192.168.1.204-7078	FINISHED	2	2.0 GB	com.tipdm.scala.chapter_3_9_3_user_profile.SQLEngine
driver-20211206174303-0006	Mon Dec 06 17:43:03 CST 2021	worker-20211206171934-192.168.1.204-7078	FINISHED	2	2.0 GB	com.tipdm.scala.chapter_3_9_3_user_profile.SQLEngine
driver-20211206174229-0005	Mon Dec 06 17:42:29 CST 2021	worker-20211206171934-192.168.1.204-7078	FINISHED	2	2.0 GB	com.tipdm.scala.chapter_3_9_3_user_profile.SQLEngine
driver-20211206173901-0004	Mon Dec 06 17:39:01 CST 2021	worker-20211206171934-192.168.1.204-7078	FINISHED	2	2.0 GB	com.tipdm.scala.chapter_3_9_3_user_profile.SQLEngine
driver-20211206173320-0003	Mon Dec 06 17:33:20 CST 2021	worker-20211206171934-192.168.1.204-7078	FINISHED	2	2.0 GB	com.tipdm.scala.chapter_3_9_3_user_profile.SQLEngine
driver-20211206172701-0002	Mon Dec 06 17:27:01 CST 2021	worker-20211206171934-192.168.1.204-7078	FINISHED	2	2.0 GB	com.tipdm.scala.chapter_3_9_3_user_profile.SQLEngine
driver-20211206172453-0001	Mon Dec 06 17:24:53 CST 2021	worker-20211206171934-192.168.1.204-7078	FINISHED	2	2.0 GB	com.tipdm.scala.chapter_3_9_3_user_profile.SQLEngine
driver-20211206172046-0000	Mon Dec 06 17:20:46 CST 2021	worker-20211206171934-192.168.1.204-7078	FINISHED	2	2.0 GB	com.tipdm.scala.chapter_3_9_3_user_profile.SQLEngine

图 6-10　用户画像测试结果

任务运行成功后，查看存储在 MySQL 中的标签结果，如图 6-11 所示。

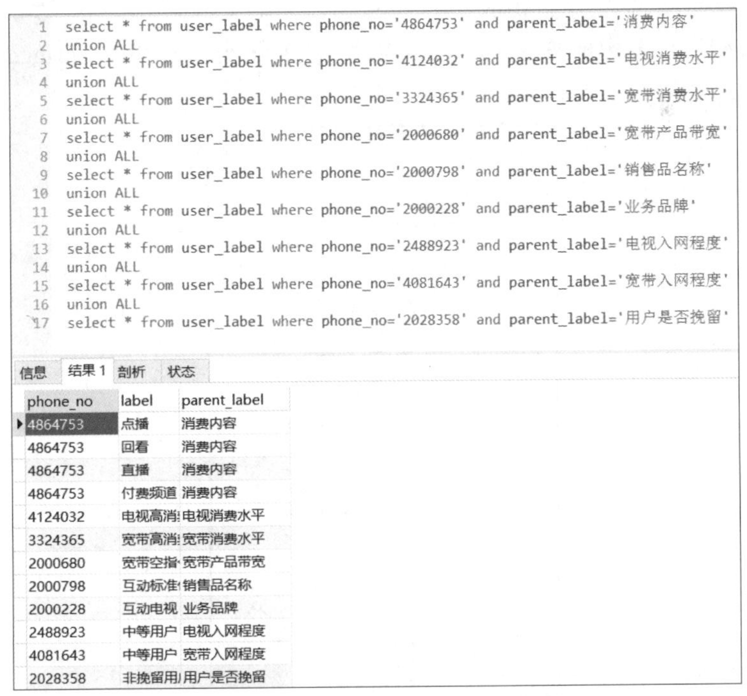

```
1  select * from user_label where phone_no='4864753' and parent_label='消费内容'
2  union ALL
3  select * from user_label where phone_no='4124032' and parent_label='电视消费水平'
4  union ALL
5  select * from user_label where phone_no='3324365' and parent_label='宽带消费水平'
6  union ALL
7  select * from user_label where phone_no='2000680' and parent_label='宽带产品带宽'
8  union ALL
9  select * from user_label where phone_no='2000798' and parent_label='销售品名称'
10 union ALL
11 select * from user_label where phone_no='2000228' and parent_label='业务品牌'
12 union ALL
13 select * from user_label where phone_no='2488923' and parent_label='电视入网程度'
14 union ALL
15 select * from user_label where phone_no='4081643' and parent_label='宽带入网程度'
16 union ALL
17 select * from user_label where phone_no='2028358' and parent_label='用户是否挽留'
```

信息　结果 1　剖析　状态

phone_no	label	parent_label
4864753	点播	消费内容
4864753	回看	消费内容
4864753	直播	消费内容
4864753	付费频道	消费内容
4124032	电视高消	电视消费水平
3324365	宽带高消	宽带消费水平
2000680	宽带空指	宽带产品带宽
2000798	互动标准	销售品名称
2000228	互动电视	业务品牌
2488923	中等用户	电视入网程度
4081643	中等用户	宽带入网程度
2028358	非挽留用	用户是否挽留

图 6-11　存储在 MySQL 中的标签结果

图 6-11 查询的是 6.2.2 小节中通过每个标签的计算结果得到的第一个 phone_no，查询到的结果与 6.2.2 小节的标签计算结果一致，说明用户画像工程的实现是正确的。

6.3 用户画像可视化

为了让用户画像系统的管理者能够更加方便地观察用户画像数据的维度、指标，可将用户标签数据以图表的方式直观迅速地展现出来，以方便系统管理者了解目标用户群体的行为习惯、消费偏好、消费水平等特征。

6.3.1 用户画像可视化简介

可视化是利用计算机图形学和图像处理技术，将数据转换成图形或图像在屏幕上显示出来，并进行交互处理的理论、方法和技术。用户画像的可视化是指通过可视化的相关技术使用户的标签数据以直观的图表形式在界面中展示出来。

在实际的项目中，用户画像可视化的需求较为复杂，因此只选取了用户画像的 9 个标签来展现，具体的需求如下。

（1）根据输入的用户编号来展示用户的标签结果，展示的方式要有表格和图形两种方式。

（2）要对输入的用户编号进行基本的验证。

（3）当用户标签结果为空时要有提示。

考虑到项目的需求及开发的成本，用户画像可视化工程主要使用 Spring Boot 框架进行管理。同时，为了减少工作量，前端主要使用 Vue.js 和 Element UI 框架实现用户标签的展示。

6.3.2 可视化工程实现

因为用户标签结果存储在 MySQL 中，所以可视化工程的后端主要负责从 MySQL 中读取并构造数据，返回给前端界面展示。

根据 6.2 节中用户画像标签结果的表结构，结合 Java 分层思想及 Spring Boot 开发的规约，构建图 6-12 所示的用户画像可视化工程结构。

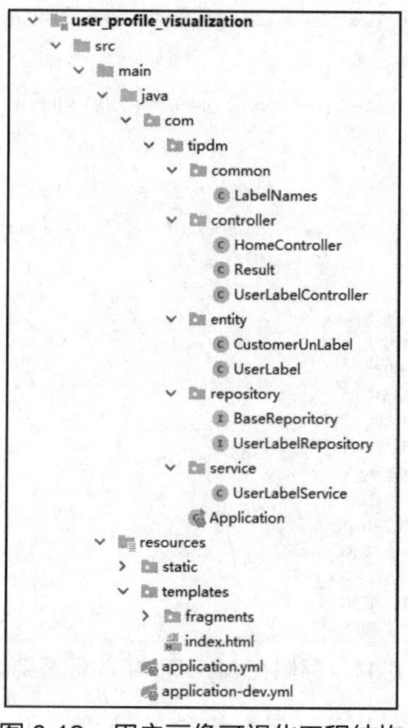

图 6-12　用户画像可视化工程结构

用户标签结果的表结构中只有phone_no、label、parent_label这3个字段，缺少主键，而根据Spring Boot JPA的规范，在操作无主键的表时会报错，为了避免这个问题发生，在创建Java实体类时，使用用户标签表中的3个字段作为联合主键，如代码6-25所示。

代码6-25 创建实体类CustomerUnLabel

```
package com.tipdm.entity;
import javax.persistence.EmbeddedId;
import javax.persistence.Entity;
import javax.persistence.Table;
import javax.persistence.UniqueConstraint;
import java.io.Serializable;
@Entity
@Table(name = "user_label",
      uniqueConstraints = {@UniqueConstraint(columnNames = {"phone_no", "label",
"parent_label"})})
public class CustomerUnLabel<PK extends Serializable> implements Serializable {
   private static final long serialVersionUID = 5260200516679644119L;
   @EmbeddedId
   private UserLabel customerPortrayalId;
   public UserLabel getCustomerPortrayalId() {
      return customerPortrayalId;
   }
   public void setCustomerPortrayalId(UserLabel customerPortrayalId) {
      this.customerPortrayalId = customerPortrayalId;
   }
}
```

根据标签结果展示的数据格式要求，需要在业务层（Service）中构建合适的数据结构，如代码6-26所示。

代码6-26 在业务层构建数据结构

```
package com.tipdm.service;
import com.tipdm.common.LabelNames;
import com.tipdm.entity.CustomerUnLabel;
import com.tipdm.entity.UserLabel;
import com.tipdm.repository.UserLabelRepository;
import org.springframework.beans.factory.annotation.Autowired;
import org.springframework.beans.factory.annotation.Value;
import org.springframework.context.annotation.PropertySource;
import org.springframework.stereotype.Service;
import org.springframework.transaction.annotation.Transactional;
import java.util.*;
@Service
@Transactional
@PropertySource("classpath:application.yml")
public class UserLabelService {
   @Autowired
   private UserLabelRepository userLabelRepository;
   @Autowired
```

```
    private LabelNames labelNames;
    @Value("${number}")
    private int number;
    public List getLabelsByPhoneNo(Long phoneNo) {
        List<CustomerUnLabel> userLabels = userLabelRepository.findByPhoneNo
(phoneNo);
        List<UserLabel> useLabelLists = new ArrayList<>();
        for (CustomerUnLabel userLabel : userLabels) {
            UserLabel label = userLabel.getCustomerPortrayalId();
            useLabelLists.add(label);
        }
        return useLabelLists;
    }
    public Map<String, Object> findLabel(Long phoneNo) {
        List<String> parentLabels = labelNames.getLabelNames();
        List<CustomerUnLabel> userLabels = userLabelRepository.findByPhoneNo
(phoneNo);
        int size = userLabels.size();
        Map<String, Object> map = new HashMap<>();
        List<String> parent = new ArrayList<>();
        for (int i = 0; i < number; i++) {
            parent.add(parentLabels.get(i));
            List<String> child = new ArrayList<>();
            for (CustomerUnLabel userLabel : userLabels) {
                UserLabel label = userLabel.getCustomerPortrayalId();
                if (parentLabels.get(i).equals(label.getParent_label())) {
                    child.add(label.getLabel());
                }
            }
            map.put(parentLabels.get(i), child);
        }
        map.put("isNull", size == 0);
        map.put("parentName", parent);
        return map;
    }
}
```

代码 6-26 中有两个方法，其中，getLabelsByPhoneNo()方法返回的是 List 类型，用于用户标签的表格展示；findLabel()方法返回的是 Map 类型，用于用户标签的图形化展示。

用户画像可视化工程 user_profile_visualization 中的 index.html 是根据项目需求编写的标签展示界面，index.html 的主要部分如代码 6-27 所示。

代码 6-27　index.html 的主要部分

```
<!--用户标签图形展示-->
<div class="upDiv" v-show="isShow">
    <div class="userPortrayal">
        <div class="upSor" v-for="name in cusporData.map.parentName">
            <ul>
                <li v-for="child in cusporData.map[name]">{{child}}</li>
```

```
            </ul>
            <div class="tit">{{name}}</div>
        </div>
    </div>
</div>
<!--标签详情表格显示-->
<el-table :data="labelDetails" style="width: 100%">
    <el-table-column prop="phone_no" label="用户编号" width="180">
    </el-table-column>
    <el-table-column prop="parent_label" label="父标签" width="180">
    </el-table-column>
    <el-table-column prop="label" label="标签">
    </el-table-column>
</el-table>
<script >
  new Vue({
        el: "#app",
        data() {
            return {
                input_id: '',
                innerVisible:false,
                outerVisible:false,
                dialogLoading:false,
                labelDetails:[],
                cusporData:{
                    map:{
                        parentName:[]
                    }
                },
                isShow:false,
            }
        },
        methods:{
            labelDetail(){
                var _self = this;
                var id = this.input_id;
                if(isNaN(id) || !id.trim().length>0){
                    alert("用户编号不能为空并且是数字类型")
                }else{
                    $.ajax({
                        url: 'labels/user/'+id,
                        method: "get",
                        contentType: "application/json;charset=UTF-8",
                        success: function (data) {
                            _self.labelDetails = data;
                            _self.innerVisible = true;
                        },
                        error: function () {
                        }
                    });
```

```
            }
        },
        open() {
            var id = this.input_id;
            if(isNaN(id) || !id.trim().length>0){
                alert("用户编号不能为空并且是数字类型")
            }else {
                this.outerVisible = true;
                this.dialogLoading = true;
                var that = this;
                $.ajax({
                    url: 'labels/users/'+id+'/all/',
                    method: "get",
                    contentType: "application/json;charset=UTF-8",
                    success: function (res) {
                        that.cusporData = res.data;
                        that.cusporData.map=res.data;
                        if(that.cusporData.map["isNull"]){
                            that.isShow=false;
                        }else {
                            that.isShow=true;
                        }
                        that.formatData(that.cusporData);
                        // 若二级标签下无三级标签，则添加“无”标签
                        that.dialogLoading = false;
                    },
                });
            }
        },
    }
  });
</script>
```

6.3.3 结果展示

将可视化工程打包成 user_profile_visualization-0.0.1-SNAPSHOT.jar 的 JAR 包，并将其上传到 server1（IP 地址为 192.168.111.73）上，启动用户画像可视化工程，如代码 6-28 所示。

代码 6-28 启动用户画像可视化工程

```
[root@server1 ~]#java -jar user_profile_visualization-0.0.1-SNAPSHOT.jar
SLF4J: The requested version 1.7.16 by your slf4j binding is not compatible
with [1.6]
SLF4J:  See  http://www.slf4j.org/codes.html#version_mismatch  for  further
details.
  .   ____          _            __ _ _
 /\\ / ___'_ __ _ _(_)_ __  __ _ \ \ \ \
( ( )\___ | '_ | '_| | '_ \/ _` | \ \ \ \
 \\/  ___)| |_)| | | | | || (_| |  ) ) ) )
  '  |____| .__|_| |_|_| |_\__, | / / / /
```

```
=========|_|===============|___/=/_/_/_/
 :: Spring Boot ::        (v1.5.10.RELEASE)
2019-02-22 15:25:29.339  INFO 30356 --- [           main] com.tipdm.Application
......
2019-02-22  15:25:39.492   INFO  30356  ---  [                main]  s.b.c.e.t.
TomcatEmbeddedServletContainer  :  Tomcat  started  on  port(s):  8001  (http)
2019-02-22  15:25:39.498   INFO  30356  ---  [                main]  com.tipdm.
Application                  : Started Application in 11.44 seconds (JVM running
for 13.111)
```

工程启动成功后，通过浏览器访问 192.168.111.73:8001/user_profile。工程访问的端口和路径可以在用户画像可视化工程 user_profile_visualization 的 application-dev.yml 配置文件中修改。广电大数据用户画像展示平台如图 6-13 所示。

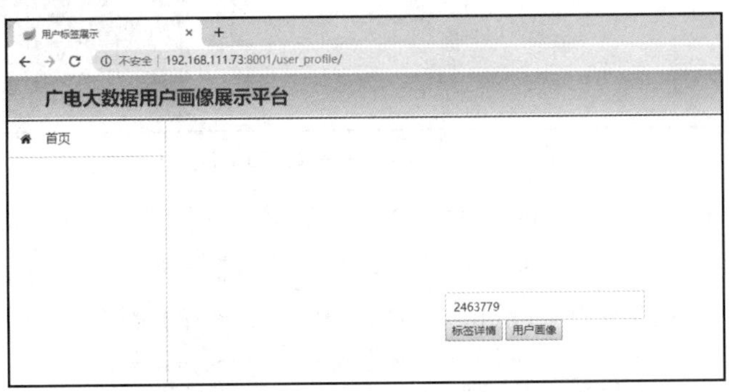

图 6-13　广电大数据用户画像展示平台

在图 6-13 中，有一个默认的用户 2463779，单击“标签详情”按钮，可以看到该用户的所有标签结果，如图 6-14 所示。

用户标签详情

用户编号	父标签	标签
2463779	电视入网程度	老用户
2463779	消费内容	直播
2463779	消费内容	点播
2463779	消费内容	付费频道
2463779	消费内容	回看
2463779	业务品牌	互动电视
2463779	电视消费水平	电视高消费
2463779	用户是否挽留	挽留用户
2463779	销售品名称	优惠购机—(499元)(26.5元/月)
2463779	销售品名称	精彩点-包月-20元
2463779	销售品名称	直播108-包月-10元
2463779	销售品名称	院线大片-包月-49元
2463779	销售品名称	少儿乐园-包月-25元
2463779	销售品名称	全时移回看-包月-20元

图 6-14　用户 2463779 的所有标签结果

单击图 6-13 中的“用户画像”按钮，可以看到用户 2463779 的画像，如图 6-15 所示。

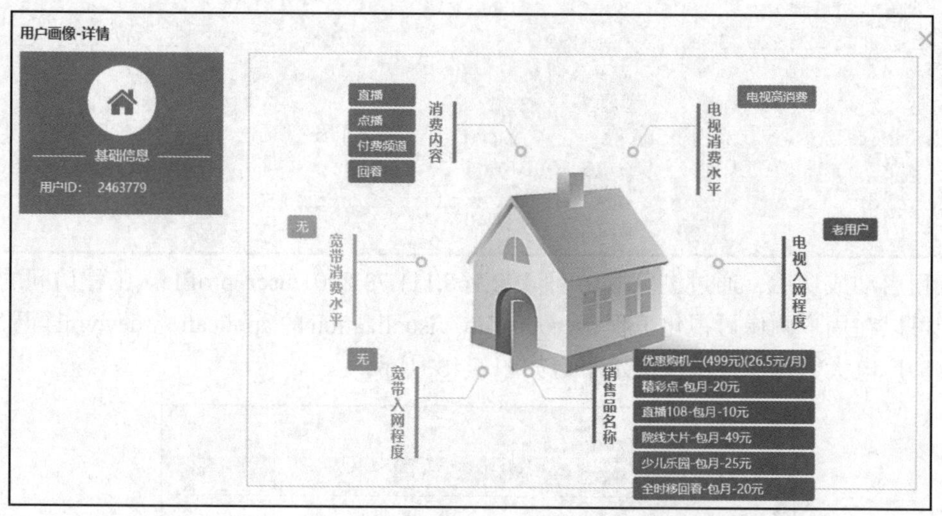

图 6-15　用户 2463779 的画像

图 6-15 中只显示了用户 2463779 的 6 个标签结果，这是因为展示画布的大小有限，且对全部标签结果进行展示会影响界面美观。至于展示标签的数量及顺序，可以通过修改工程的 application.yml 配置文件中的 number 值和 labelNames 的排序来实现，如图 6-16 所示。

```
application.yml
2   spring:
3     profiles:
4         active: dev
5
6     #用户画像页面二级标签显示个数
7     number: 6
8
9     #用户画像显示二级标签名称
10  labelName:
11    labelNames:
12      - "消费内容"
13      - "电视消费水平"
14      - "宽带消费水平"
15      - "电视入网程度"
16      - "宽带入网程度"
17      - "销售品名称"
18      - "业务品牌"
19      - "宽带产品带宽"
20      - "用户是否挽留"
```

图 6-16　application.yml 配置文件

至此，简化版的用户画像可视化工程已基本实现，满足用户画像可视化的基本需求。在实际项目中，可能还需要有用户基本信息的展示，同时，用户标签展示可以根据标签名称（如业务品牌）来显示相应的标签等。

小结

用户画像是广电大数据用户画像项目的核心，本章虽然只挑选了部分标签进行计算，但是尽量多地展现了整个用户画像的重点过程。对广电用户进行用户画像的过程包括标签计算、工程封装和用户画像可视化。标签计算部分提供了阈值计算和字段直贴类的标签计算方式和对应的计算过程；工程封装部分介绍了 SVM 模型构建与预测过程封装和用户画像计算封装，并对封装的工程进行了测试；用户画像可视化部分通过 Spring Boot 开发了一个可视化界面，以图表的形式对用户画像进行展示。

第 7 章 广电大数据用户画像——任务调度实现

当项目开发到一定程度或者项目开发完成后，就需要进行项目上线和运维。其中，项目上线是指 IT 项目研发完成、上线准备工作完成后，开始测试、试运行、联网运行。本章将部署 XXL-JOB 分布式调度平台，并在 XXL-JOB 上部署广电大数据用户画像项目的模拟生产数据、数据传输、数据预处理、SVM 预测用户是否挽留、用户画像和实时统计订单信息等任务，模拟广电大数据用户画像系统线上运营。

学习目标

（1）熟悉广电大数据用户画像项目中相关任务的调度策略的设计。

（2）熟悉在 XXL-JOB 分布式任务调度平台上进行调度系统配置的过程。

（3）掌握广电大数据用户画像系统的调度任务的配置及执行的方法。

大多数企业在项目实施时会使用任务调度功能，任务调度是指基于给定时间点、给定时间间隔或给定执行次数自动执行任务。在执行任务调度前，需要先设计任务调度策略，根据不同的任务需求，可对任务设置不同的定时调度。

7.1 调度策略

广电大数据用户画像项目主要包含数据采集、数据传输、大数据平台计算、用户画像可视化这 4 个模块。

（1）数据采集主要是模拟产生数据到 Elasticsearch 集群中。

（2）数据传输指将 Elasticsearch 数据传输到 Hive 中。

（3）大数据平台计算包括数据预处理、SVM 预测用户是否挽留、用户画像和 Kafka 结合 Spark Streaming 实时统计订单信息。

（4）用户画像可视化指将用户画像的结果在可视化界面中以图表的形式进行展示。

广电大数据用户画像系统需要定时运行模拟产生数据、数据传输、数据预处理、SVM 预测用户是否挽留、用户画像、实时统计订单信息这 6 个任务，对这 6 个任务的运行时间的设置说明如下。

1. 模拟产生数据

模拟产生数据包括模拟产生账单数据、订单数据和用户收视行为数据。因为账单数据是每月 1 日生成的，并且白天会有较多人不定时地使用集群，所以，为了确保集群有足够的资源运行任务，将模拟产生账单数据任务的启动时间设置在每月 1 日的 20:10:00。而订单数据和用户收视行为数据是每天都有记录产生的，因此，需要每天模拟产生新的数据。模拟产生账单数据的任务大概需要 40min，为确保集群有足够的资源可用，模拟产生订单数据任务可以在模拟产生账单数据任务完成之后启动，因此，将模拟产生订单数据任务的启动时间设置在每天 21:10:00。而模拟产生订单数据的任务一般在 1h 之内可以完成，因此，将模拟产生用户收视行为数据任务的启动时间设置在每天 22:10:00。

2. 数据传输

数据传输是指分别从 Elasticsearch 集群中读取 mediamatch_usermsg、mediamatch_userevent、order_index、mmconsume_billevents 和 media_index 数据到 Hive 中，这些都是后续进行用户画像需要使用的数据。因为用户画像标签只需要一个月更新一次，所以数据传输的任务也可一个月更新一次。考虑到数据量比较大，数据传输任务运行比较缓慢，并且为了保证运行数据传输任务时模拟产生数据的任务已经完成，数据传输任务在模拟产生用户收视行为数据任务完成（1h 内完成）之后启动。因为数据传输任务完成时间可能需要 1～2h，所以每个数据传输任务的启动时间最好间隔 2h。因此，将从 Elasticsearch 集群读取 mediamatch_usermsg 数据传输到 Hive 中的任务的启动时间设置为每月 1 日 23:10:00；将读取 mediamatch_userevent 数据传输到 Hive 中的任务的启动时间设置为每月 2 日 1:10:00；将读取 order_index 数据传输到 Hive 中的任务的启动时间设置为每月 2 日 3:10:00；将读取 mmconsume_billevents 数据传输到 Hive 中的任务的启动时间设置为每月 2 日 5:10:00；因为白天有较多人不定时使用集群，所以为了错开集群使用高峰期，将读取 media_index 数据传输到 Hive 中的任务的启动时间设置为每月 2 日 20:10:00。

3. 数据预处理

数据预处理任务必须在数据传输任务完成之后执行，执行数据传输最后一个任务，即读取 media_index 数据传输到 Hive 中的任务，大概需要 5h，因此，数据预处理任务的启动时间可设置为每月 3 日 2:10:00。

4. SVM 预测用户是否挽留

SVM 预测用户是否挽留使用的输入数据是经过预处理的数据，因此 SVM 预测用户是否挽留任务需要在数据预处理任务完成之后启动。数据预处理任务大概需要 0.5h，故将 SVM 预测用户是否挽留任务的启动时间设置为每月 3 日 2:40:00。

5. 用户画像

用户画像使用的输入数据是经过预处理的数据，因此用户画像任务的启动设置在数据预处理任务完成之后。此外，由于 SVM 预测用户是否挽留任务需要使用较多的集群资源，为避免集群资源不足，用户画像任务与 SVM 预测用户是否挽留任务的启动时间应该错开，SVM 预测用户是否挽留任务大概需要 16min，因此，将用户画像任务的启动时间设置为每月 3 日 3:10:00。

6. 实时统计订单信息

模拟产生订单数据的定时任务总会有结束的时候，因此，实时统计订单信息任务可以定时运行。依据用于模拟订单实时流的数据量和模拟速率，每月运行一次实时统计订单信息的任务比较合适。考虑到集群的错峰使用，将实时统计订单信息任务的启动时间设置为每月 4 日 1:10:00。

根据以上 6 个任务的定时设置，整个广电大数据用户画像系统的在线任务定时策略如表 7-1 所示。

表 7-1　整个广电大数据用户画像系统的在线任务定时策略

任务名称	定时策略
模拟产生账单数据	每月 1 日 20:10:00
模拟产生订单数据	每天 21:10:00
模拟产生用户收视行为数据	每天 22:10:00
mediamatch_usermsg 数据传输到 Hive 中	每月 1 日 23:10:00
mediamatch_userevent 数据传输到 Hive 中	每月 2 日 1:10:00
order_index 数据传输到 Hive 中	每月 2 日 3:10:00
mmconsume_billevents 数据传输到 Hive 中	每月 2 日 5:10:00
media_index 数据传输到 Hive 中	每月 2 日 20:10:00
数据预处理	每月 3 日 2:10:00
SVM 预测用户是否挽留	每月 3 日 2:40:00
用户画像	每月 3 日 3:10:00
实时统计订单信息	每月 4 日 1:10:00

根据第 3～6 章中对表 7-1 的每个任务的实现结果，总结出所有任务的输入/输出参数设置，如表 7-2 所示。

表 7-2　所有任务的输入/输出参数设置

任务名称	输入	输出
模拟产生账单数据	Hive 的 user_profile 库中的 mmconsume_billevents	Elasticsearch 集群中的 mmconsume_billevents/doc
模拟产生订单数据	Hive 的 user_profile 库中的 order_index_v3	Elasticsearch 集群中的 order_index_v3/doc
模拟产生用户收视行为数据	Hive 的 user_profile 库中的 media_index	Elasticsearch 集群中的 media_indexyyyyww/doc
mediamatch_usermsg 数据传输到 Hive 中	Elasticsearch 集群中的 mediamatch_usermsg/doc	Hive 的 user_profile 库中的 mediamatch_usermsg
mediamatch_userevent 数据传输到 Hive 中	Elasticsearch 集群中的 mediamatch_userevent/doc	Hive 的 user_profile 库中的 mediamatch_userevent

续表

任务名称	输入	输出
order_index 数据传输到 Hive 中	Elasticsearch 集群中的 order_index_v3/doc	Hive 的 user_profile 库中的 order_index_v3
mmconsume_billevents 数据传输到 Hive 中	Elasticsearch 集群中的 mmconsume_billevents/doc	Hive 的 user_profile 库中的 mmconsume_billevents
media_index 数据传输到 Hive 中	Elasticsearch 集群中的 media_indexyyyyww/doc	Hive 的 user_profile 库中的 media_index
数据预处理	Hive 的 user_profile 库中的 mediamatch_usermsg、mediamatch_userevent、order_index_v3、mmconsume_billevents、media_index	Hive 的 user_profile 库中的 mediamatch_usermsg_process、mediamatch_userevent_process、order_index_process、mmconsume_billevent_process、media_index_process
SVM 预测用户是否挽留	Hive 的 user_profile 库中的 mediamatch_usermsg_process、mediamatch_userevent_process、order_index_process、mmconsume_billevent_process、media_index_process	Hive 的 user_profile 库中的 svm_activate、svm_prediction
用户画像	Hive 的 user_profile 库中的 mediamatch_usermsg_process、mediamatch_userevent_process、order_index_process、mmconsume_billevent_process、media_index_process、svm_prediction	MySQL 的 user_profile 库中的 user_label
实时统计订单信息	node4 节点的/data/order.csv 文件	Redis 的 db0 中的各键值对

7.2 调度实现

有了明确的任务调度策略即可进行任务调度的实现。在符合需求的调度系统中，配置广电大数据用户画像系统的调度任务，包括配置模拟产生数据、数据传输、数据预处理、SVM 预测用户是否挽留、用户画像和实时统计订单信息的调度任务。

1. 调度系统配置及二次开发

XXL-JOB 是一个轻量级分布式任务调度平台，该平台的核心设计目标是开发迅速、学习简单、轻量级、易扩展、开箱即用。XXL-JOB 分布式任务调度平台已在 GitHub 上开源。

关于 XXL-JOB 分布式任务调度平台的相关介绍及使用，读者可以参考 XXL 开源社区中 XXL-JOB 分布式任务调度平台的快速入门、任务详解、任务管理的相关内容，快速构建实战环境。XXL-JOB 分布式任务调度平台的后台有多个模块，本节主要简单介绍和广电大数据用户画像项目相关的几个核心模块（详细设计请参考 XXL 开源社区中的相关内容）。在本项目的工程代码中，已将 XXL-JOB 分布式任务调度平台加入工程，如图 7-1 所示。

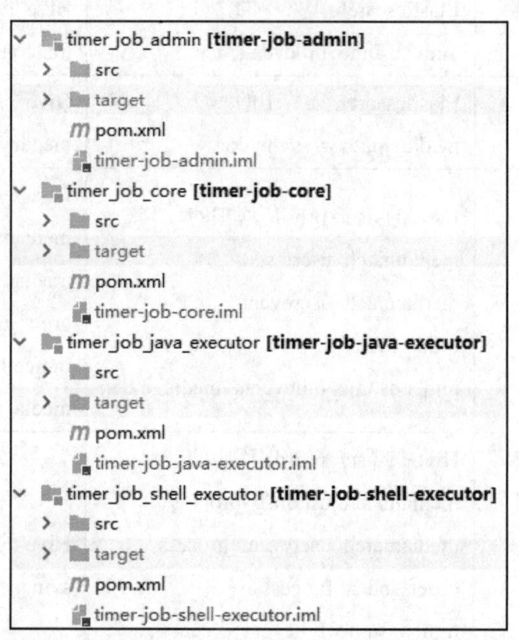

图 7-1　将 XXL-JOB 分布式任务调度平台加入工程

在图 7-1 中，分布式任务调度平台分为 3 个模块：任务调度中心（timer-job-admin）模块、任务核心（timer-job-core，主要是任务处理器，即 JobHandler）模块和执行器模块。其中，执行器模块包含 timer-job-java-executor（Java 执行器）和 timer-job-shell- executor（Shell 执行器）。

任务调度中心是一个 Web 工程，提供图形化用户界面，用户可以在任务调度中心的界面中进行任务创建、配置、执行、定时调度、日志查看、监控等操作。

任务核心模块可以理解为一个中间件或基础模块，供其他模块调用，不需要部署（任务核心模块的部分代码已在原 GitHub 基础上经过调整，以适应本项目）。

执行器模块作为客户端执行的具体任务，是一个 Spring Boot 工程，可直接使用 JDK 运行。例如，在任务调度中心中创建一个 Shell 任务，并把此任务分配给 Shell 执行器，当调度程序发起任务时，Shell 执行器会在任务部署的节点上执行该 Shell 脚本（底层会在该节点上创建一个临时脚本文件并执行。此外，如果有参数，则将参数传入执行）。任务执行完成后，会反馈该任务的执行状态和日志到任务调度中心。

在 Linux 中部署 XXL-JOB，需要先在 GitHub 上下载源码，部署的工程有 timer-job-admin（WAR 包方式）、timer-job-shell-executor（JAR 包方式）、timer-job-java-executor（JAR 包方式）。

针对 timer-job-admin 工程，修改 src/main/resources/目录下的 xxl-job-admin.properties 文件和 log4j.xml 文件的内容。xxl-job-admin.properties 文件修改的内容如代码 7-1 所示。

代码 7-1　xxl-job-admin.properties 文件修改的内容

```
xxl.job.db.driverClass=com.mysql.jdbc.Driver
xxl.job.db.url=jdbc:mysql://192.168.111.75:3306/xxl-job-user-profile?useUnicode=
true&characterEncoding=UTF-8
xxl.job.db.user=root
xxl.job.db.password=root
```

在 node1 节点上的 MySQL 中运行 resources/db/xxl-job-user-profile.sql 文件，进行初始化。

log4j.xml 文件修改的内容如代码 7-2 所示。

代码 7-2　log4j.xml 文件修改的内容

```
<appender name="FILE" class="org.apache.log4j.DailyRollingFileAppender">
<param name="file" value="/tmp/applogs/xxl-job-user-profile/xxl-job-admin.
log"/>
```

xxl-job-admin.properties 文件和 log4j.xml 文件修改完成后，在 IntelliJ IDEA 中进行编译，将 timer-job-admin 工程编译成 WAR 包，如图 7-2 所示。

编译完成后，在输出路径 target 下，可以发现编译好的 timer-job-admin WAR 包，如图 7-3 所示。

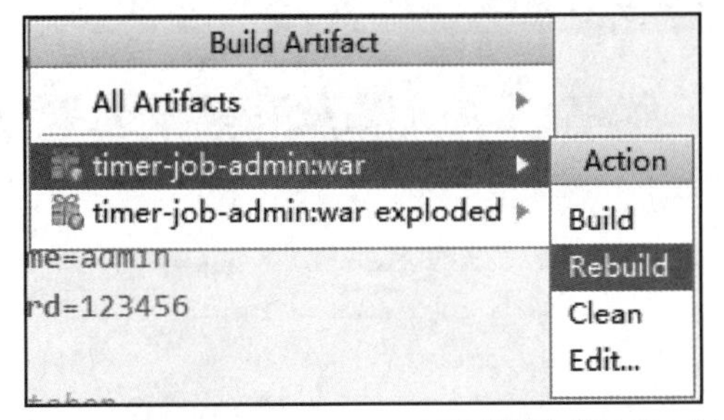

图 7-2　将 timer-job-admin 工程编译成 WAR 包

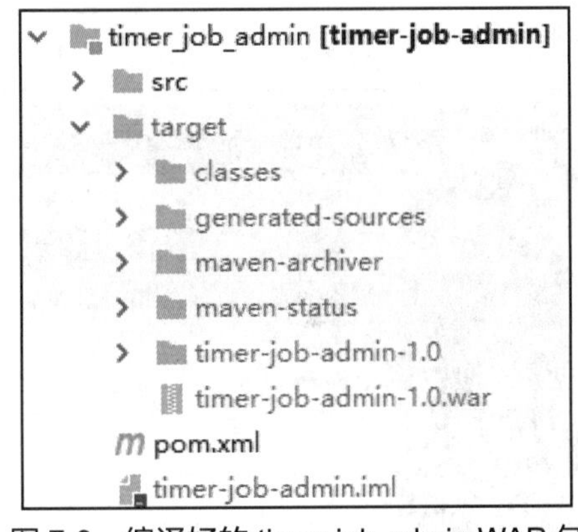

图 7-3　编译好的 timer-job-admin WAR 包

将 timer-job-admin 部署在 server2 的/root/xxl-job-user-profile/apache-tomcat-8.0.32 上（注意，Tomcat 下载并解压后，只需将其复制到/root/xxl-job-user-profile 目录下即可）。把 timer-job-admin 的 WAR 包部署在 Tomcat 上，只需要简单地将 timer-job-admin 的 WAR 包复制到 Tomcat 的 webapps 目录下即可。为了方便访问，把该 WAR 包重命名为 admin.war。

timer-job-admin 部署完成后，在 Tomcat 的 bin 目录下，执行命令“./startup.sh”，并启动 Tomcat，如图 7-4 所示。

```
[root@server2 bin]# ./startup.sh
Using CATALINA_BASE:   /root/xxl-job-user-profile/apache-tomcat-8.0.32
Using CATALINA_HOME:   /root/xxl-job-user-profile/apache-tomcat-8.0.32
Using CATALINA_TMPDIR: /root/xxl-job-user-profile/apache-tomcat-8.0.32/temp
Using JRE_HOME:        /usr/java/jdk
Using CLASSPATH:       /root/xxl-job-user-profile/apache-tomcat-8.0.32/bin/bootstrap.jar:/root/xxl-job-user-profile/ap
ache-tomcat-8.0.32/bin/tomcat-juli.jar
Tomcat started.
```

图 7-4　启动 timer-job-admin 的 Tomcat

通过浏览器访问“http://192.168.111.74:8083/admin”，即可进入 timer-job-admin Tomcat 登录界面，如图 7-5 所示。

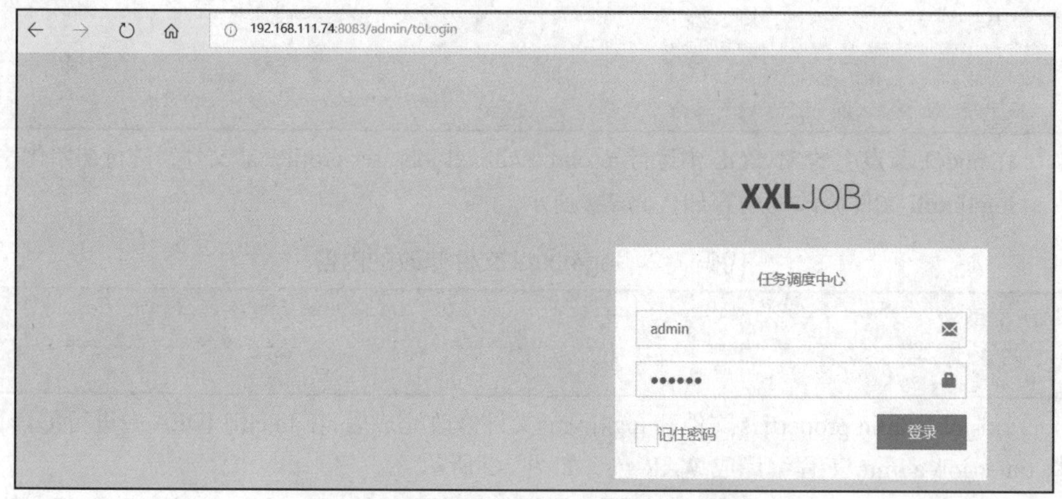

图 7-5　timer-job-admin Tomcat 登录界面

输入用户名和密码，单击“登录”按钮，即可进入 timer-job-admin 任务调度中心界面，如图 7-6 所示。

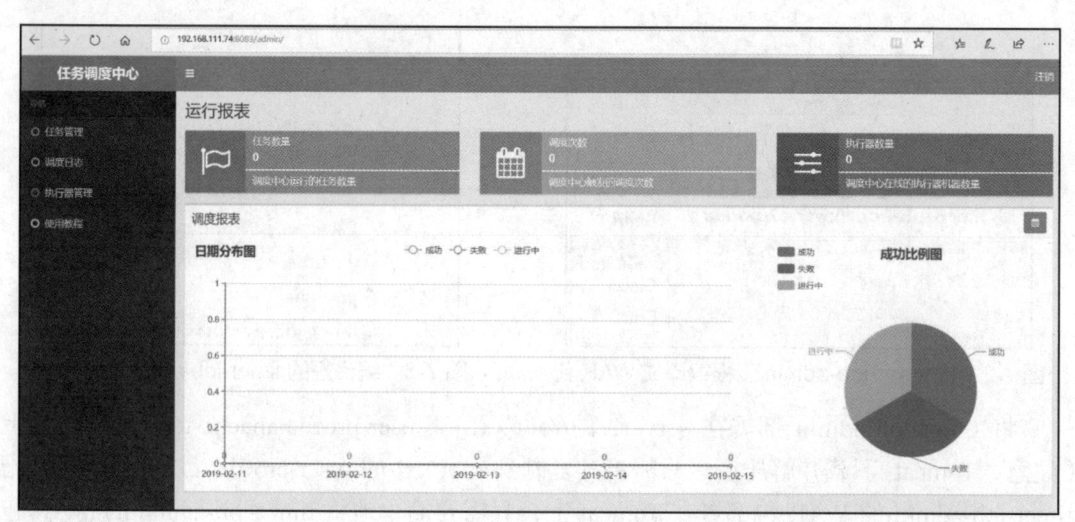

图 7-6　timer-job-admin 任务调度中心界面

在图 7-6 中，可以看到目前的任务数量、调度次数、执行器数量都是 0。对调度系统进行二次开发，部署一个 Shell 执行器和一个 Java 执行器，用于对广电大数据用户画像项目的调度。

经查询发现，node4 节点的资源比较充足，因此选择在 node4 节点上部署 Shell 执行器和 Java 执行器。

对 Shell 执行器进行配置，需要先修改 timer-job-shell-executor 工程 src/main/resources 目录下的 application.properties 文件和 logback.xml 文件。application.properties 文件修改的内容如代码 7-3 所示。

代码 7-3 application.properties 文件修改的内容

```
xxl.job.admin.addresses=http://192.168.111.74:8083/admin
xxl.job.executor.appname=xxl-job-executor-shell
xxl.job.executor.ip=192.168.111.78
xxl.job.executor.port=9999
xxl.job.executor.logpath=/tmp/applogs/xxl-job/jobhandler
```

在代码 7-3 中，需要修改任务调度监控的部署路径（部署的路径为浏览器访问路径 http://192.168.111.74:8083/admin），appname 设置的是执行器名称，executor.ip 设置的是运行 executor 的节点（运行 executor 的节点为 node4 节点）的 IP 地址，executor.port 设置的是 Shell 执行器服务的端口号，logpath 设置的是日志路径。

对于 logback.xml 文件，需要修改其中的日志路径，如代码 7-4 所示。

代码 7-4 logback.xml 文件修改的内容

```
<property name="log.path" value="/tmp/applogs/xxl-job-user-profile/xxl-job-executor-sample-springboot.log"/>
```

application.properties 文件和 logback.xml 文件修改完成后，重新编译 timer-job-shell-executor 工程，编译成功之后，在 timer-job-shell-executor 工程的 target 目录下找到打包后的 JAR 包，将 JAR 包复制到 node4 节点的/root/xxl-job-user-profile/目录下，如图 7-7 所示。

```
[root@node4 xxl-job-user-profile]# pwd
/root/xxl-job-user-profile
[root@node4 xxl-job-user-profile]# ls
killprocess.sh  start_shell.sh  stop.sh  timer-job-shell-executor-1.0.jar
```

图 7-7 复制 timer-job-shell-executor 工程的 JAR 包到 node4 节点上

JAR 包复制完成后，启动 timer-job-shell-executor 工程，如代码 7-5 所示。

代码 7-5 启动 timer-job-shell-executor 工程

```
[root@node4 xxl-job-user-profile]# java -jar timer-job-shell-executor-1.0.jar
  .   ____          _            __ _ _
 /\\ / ___'_ __ _ _(_)_ __  __ _ \ \ \ \
( ( )\___ | '_ | '_| | '_ \/ _` | \ \ \ \
 \\/  ___)| |_)| | | | | || (_| |  ) ) ) )
  '  |____| .__|_| |_|_| |_\__, | / / / /
 =========|_|==============|___/=/_/_/_/
 :: Spring Boot ::       (v1.5.10.RELEASE)
……
10:18:36.785 logback [Thread-2] INFO  o.e.jetty.server.ServerConnector - Started ServerConnector@6d581981{HTTP/1.1}{192.168.111.78:9999}
10:18:36.787 logback [Thread-2] INFO  org.eclipse.jetty.server.Server - Started @9070ms
10:18:36.788 logback [Thread-2] INFO c.x.j.c.r.n.jetty.server.JettyServer - >>>>>>>>>>> xxl-job jetty server start success at port:9999.
……
10:18:41.624 logback [Thread-6] INFO  c.x.j.c.t.ExecutorRegistryThread - >>>>>>>>>>> xxl-job registry success, registryParam:RegistryParam{registGroup='EXECUTOR', registryKey='xxl-job-executor-shell', registryValue='192.168.111.
```

```
78:9999'}, registryResult:ReturnT [code=200, msg=null, content=null]
……
10:18:38.225 logback [main] INFO  com.xxl.job.executor.Application - Started
Application in 9.165 seconds (JVM running for 10.508)
```

Timer-job-shell-executor 工程启动成功后，在任务调度中心中新增 Shell 执行器，如图 7-8 所示。

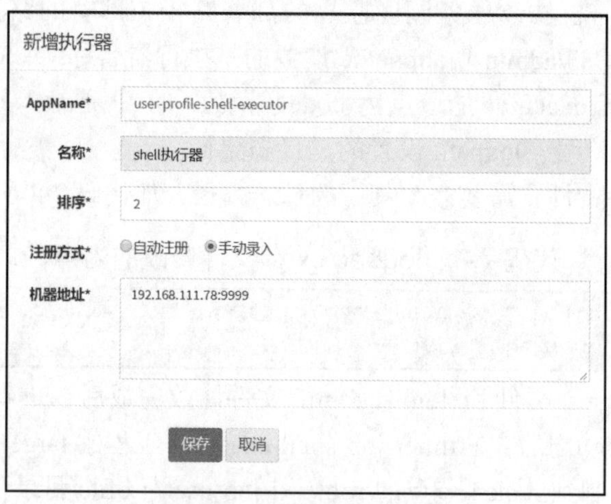

图 7-8　新增 Shell 执行器

添加 Shell 执行器后，在图 7-6 所示的任务调度中心界面中可以看到添加的 Shell 执行器，如图 7-9 所示。

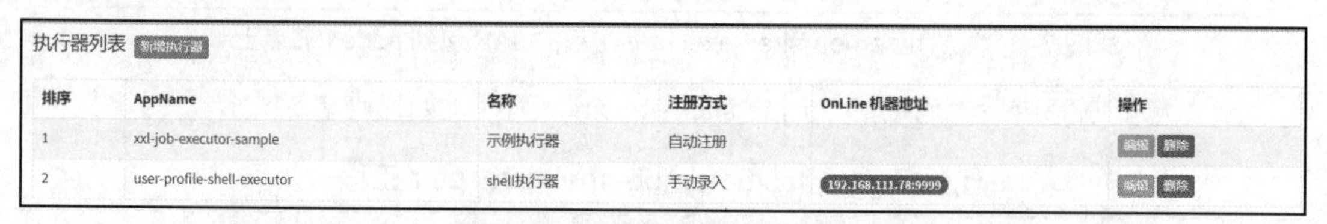

执行器列表 新增执行器

排序	AppName	名称	注册方式	OnLine 机器地址	操作
1	xxl-job-executor-sample	示例执行器	自动注册		编辑 删除
2	user-profile-shell-executor	shell执行器	手动录入	192.168.111.78:9999	编辑 删除

图 7-9　添加的 Shell 执行器

在任务调度中心界面中选择“任务管理”选项，进入任务管理界面，添加一个任务，并将该任务分配给 Shell 执行器，如图 7-10 所示。

图 7-10　将任务分配给 Shell 执行器

新增任务后，在任务管理界面中即可看到添加的任务，如图 7-11 所示。

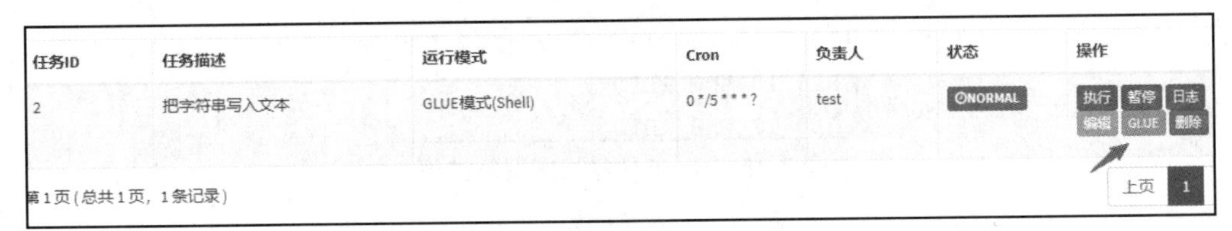

任务ID
任务描述
运行模式
Cron
负责人
状态
操作
2
把字符串写入文本
GLUE模式(Shell)
0 */5 * * * ?
test
NORMAL
执行
暂停
日志
编辑
GLUE
删除
第 1 页 (总共 1 页，1 条记录)
上页
1

图 7-11　任务管理界面中添加的任务

单击图 7-11 所示的“GLUE”按钮，编辑 Shell 脚本，如图 7-12 所示。

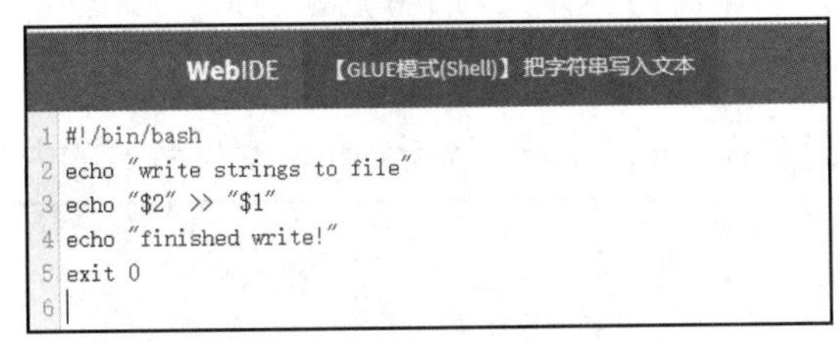

图 7-12　编辑 Shell 脚本

在图 7-12 所示的脚本中，会将任务参数中的第二个参数的内容（即字符串）写入第一个参数对应的文件（即写入第一个参数提供的路径下）。在图 7-11 所示的界面中，单击“执行”按钮，会立即执行任务（在创建好任务后，任务已经是定时触发状态，单击“执行”按钮会立即触发任务并执行一次）。任务执行多次后（一次立即触发和多次定时触发），查看调度日志，如图 7-13 所示。

调度日志

执行器 全部　任务 全部　状态 全部　调度时间 2019-01-22 00:00:00 - 2019-01-22 23:59:59　搜索　清理

每页 10 条记录

任务ID	调度时间	调度结果	调度备注	执行时间	执行结果	执行备注	操作
2	2019-01-22 11:15:00	成功	查看	2019-01-22 11:15:00	成功	无	执行日志
2	2019-01-22 11:13:33	成功	查看	2019-01-22 11:13:34	成功	无	执行日志

第 1 页 (总共 1 页，2 条记录)　上页 1 下页

图 7-13　查看调度日志

查看该 Shell 任务的运行报表，如图 7-14 所示。

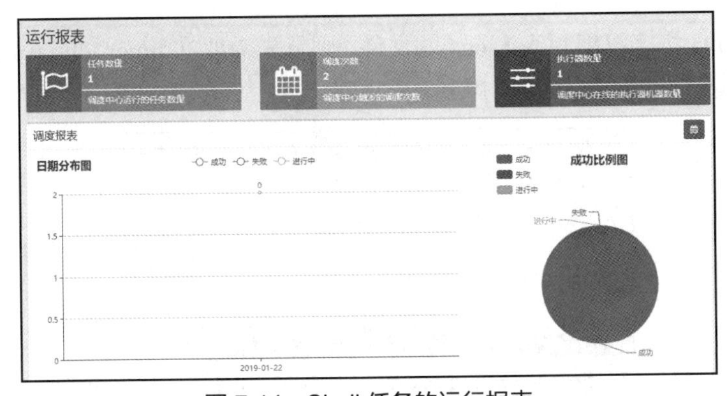

图 7-14　Shell 任务的运行报表

查看 Shell 任务的输出，Shell 任务最终运行结果如图 7-15 所示。

```
[root@node4 ~]# cat /tmp/a.txt
this is the info
this is the info
```

图 7-15　Shell 任务最终运行结果

部署 Java 执行器的方法与部署 Shell 执行器的方法类似，先修改 timer-job-java-executor 工程 src/main/resources 目录下的 application.properties 文件和 logback.xml 文件。application.properties 文件修改的内容如代码 7-6 所示。logback.xml 文件修改的内容参照代码 7-4 即可。

代码 7-6　application.properties 文件修改的内容

```
xxl.job.admin.addresses=http://192.168.111.74:8083/admin
xxl.job.executor.appname=xxl-job-executor-shell
xxl.job.executor.ip=192.168.111.78
xxl.job.executor.port=997
xxl.job.executor.logpath=/tmp/applogs/xxl-job/jobhandler
```

application.properties 文件和 logback.xml 文件修改完成后，重新编译 timer-job-java-executor 工程，编译成功后，在 timer-job-java-executor 工程的 target 目录下找到打包后的 JAR 包，并将 JAR 包复制到 node4 节点的/root/xxl-job-user-profile/目录下。参考代码 7-5 所示的内容启动 timer-job-java-executor 工程，并参考图 7-8 新增一个 Java 执行器，Java 执行器添加成功后，单击任务调度中心界面的“执行器管理”选项，可查看执行器列表，如图 7-16 所示。

执行器列表 新增执行器

排序	AppName	名称	注册方式	OnLine 机器地址	操作
1	xxl-job-executor-sample	示例执行器	自动注册		编辑 删除
2	user-profile-shell-executor	shell执行器	手动录入	192.168.111.78:9999	编辑 删除
3	user-profile-java-executor	java执行器	手动录入	192.168.111.78:997	编辑 删除

图 7-16　任务调度中心的执行器列表

通过 Java 执行器调度 Java 任务，需要编写 Java 类及实现 JavaJobInterface 接口，并在任务调度中心中进行相关参数设置。

如要通过 Java 将字符串写入 Linux 的文件，则首先需要在 timer-job-java-executor 模块中添加 Java 执行代码的路径，如图 7-17 所示，路径设置为/src/main/java/com/tipdm.java/jobs。

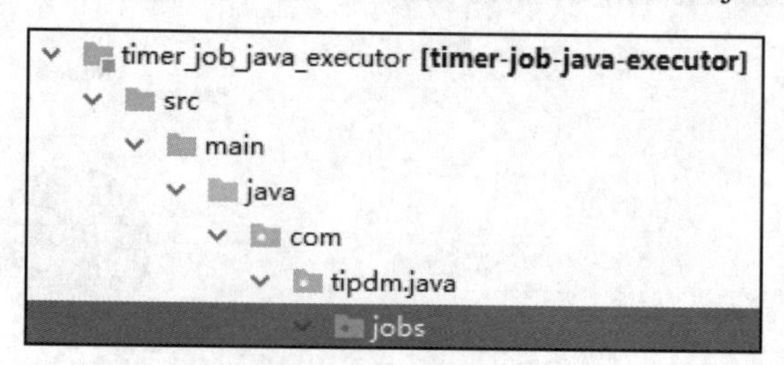

图 7-17　timer-job-java-executor 中添加 Java 执行代码的路径

在图 7-17 所示的 jobs 包下新建 DemoTest 类，如代码 7-7 所示。

代码 7-7　新建 DemoTest 类

```
package com.tipdm.java.jobs;
import com.tipdm.java.JavaJobInterface;
import com.xxl.job.core.biz.model.ReturnT;
import java.io.BufferedWriter;
import java.io.File;
import java.io.FileWriter;
import java.util.Map;
/**
 * 示例 Java Executor 任务
*/
public class DemoTest implements JavaJobInterface {
    @Override
    public ReturnT<String> execute(String className, Map<String, String> args)
throws Exception {
        String input = args.get("--input");
        String info = args.get("--info");
        File file =new File(input);
        //若文件不存在，则创建文件
        if(!file.exists()){
            file.createNewFile();
        }
        //若文件存在，则将 info 追加至文件中
        FileWriter fileWriter = new FileWriter(file,true);
        BufferedWriter bufferWriter = new BufferedWriter(fileWriter);
        bufferWriter.write(info);
        bufferWriter.close();
        return ReturnT.SUCCESS;
    }
}
```

重新编译、部署 timer-job-java-executor 工程，成功启动 timer-job-java-executor 工程后，为图 7-16 中的 Java 执行器添加一个示例任务，如图 7-18 所示。

执行器*	java执行器	任务描述*	把字符串写入文本
路由策略*	第一个	Cron*	0 */5 * * * ?
运行模式*	BEAN模式	JobHandler*	javaJobHandler
任务参数*	com.tipdm.java.jobs.DemoTest --input /tn	子任务ID*	请输入子任务的任务ID,如存在多个则逗
阻塞处理策略*	com.tipdm.java.jobs.DemoTest --input /tmp/b.txt --info hello		失败告警
负责人*	test	报警邮件*	请输入报警邮件，多个邮件地址则逗号

图 7-18　timer-job-java-executor 中添加示例任务

添加示例任务后，在任务管理列表中即可看到该任务，如图 7-19 所示。单击“执行”按钮，即可通过 Java 执行器执行 Java 任务。

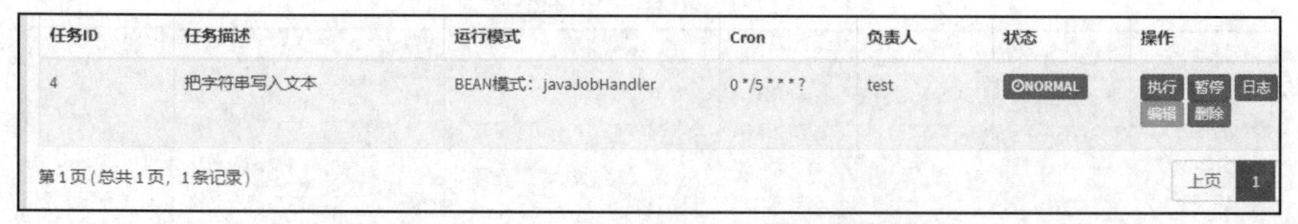

任务ID	任务描述	运行模式	Cron	负责人	状态	操作
4	把字符串写入文本	BEAN模式：javaJobHandler	0 */5 * * * ?	test	⊙NORMAL	执行 暂停 日志 编辑 删除

第 1 页 (总共 1 页，1 条记录)　上页 1

图 7-19　timer-job-java-executor 任务管理列表

2. 定时任务配置

按照表 7-1 所设计的任务调度策略，在调度系统中配置广电大数据用户画像系统的调度任务。首先，分别单击图 7-11 和图 7-16 所示界面中的“删除”按钮，删除调度系统中的任务、执行器，将调度系统还原成图 7-6 所示的界面。下面开始配置模拟生产数据、数据传输、数据预处理、SVM 预测用户是否挽留、用户画像和实时统计订单信息等定时任务。调度任务若涉及 Shell 脚本，则均使用 node4 节点上的 Shell 执行器，若涉及 Java 程序，则均使用 node4 节点上的 Java 执行器。

（1）模拟产生数据

由 4.1.2 小节可知模拟产生数据是由一个 Shell 脚本实现的，因此模拟产生数据的任务可由 Shell 执行器实现。模拟产生数据需要模拟产生账单数据、订单数据和用户收视行为数据，分别用 3 个 Shell 脚本实现，因此需要 3 个调度任务，具体配置如下。

① 配置模拟产生账单数据任务。在任务调度中心的任务管理模块中新增模拟产生账单数据任务，如图 7-20 所示。“执行器”设置为“shell 执行器；路由策略选择“第一个”，表示将选择执行器注册的地址的第一台机器执行，如果第一台机器出现故障，则任务失败；“运行模式”设置为“GLUE 模式(Shell)”；在“任务参数”后输入“2018-07-01 192.168.111.75 server3 /root/qwm/user_profile_project-1.0.jar 9200 year_month mmconsume_billevents doc”，每个参数之间用空格隔开；“负责人”可根据实际需求设置，如这里设置为“user_profile”；“任务描述”中输入的内容主要为描述任务的功能，如“每月 1 号的 20:10:00 模拟产生账单数据”；在“Cron”后输入任务运行的时间表达式，如“0 10 20 1 * ? *”；失败处理策略选择“失败告警”。

执行器*	shell执行器	任务描述*	每月1号的20:10:00模拟产生账单数据
路由策略*	第一个	Cron*	0 10 20 1 * ? *
运行模式*	GLUE模式(Shell)	JobHandler*	请输入JobHandler
任务参数*	2018-07-01 192.168.111.75 server3 /root/q	子任务ID*	请输入子任务的任务ID,如存在多个则逗
阻塞处理策略*	单机串行	失败处理策略*	失败告警
负责人*	user_profile	报警邮件*	请输入报警邮件，多个邮件地址则逗号

保存　取消

图 7-20　新增模拟产生账单数据任务

图 7-20 所示的信息设置完成后单击“保存”按钮，即可完成新增任务，在任务管理列表中可查看新增的任务，如图 7-21 所示。

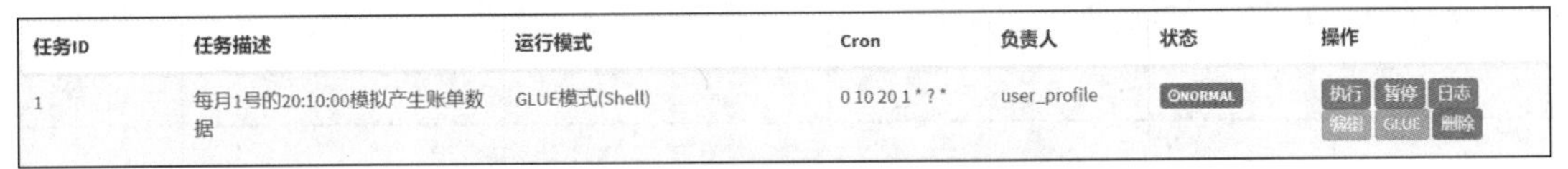

任务ID	任务描述	运行模式	Cron	负责人	状态	操作
1	每月1号的20:10:00模拟产生账单数据	GLUE模式(Shell)	0 10 20 1 * ? *	user_profile	NORMAL	执行 暂停 日志 编辑 GLUE 删除

图 7-21　任务管理列表

单击图 7-21 中的“GLUE”按钮，编辑任务的 Shell 脚本，脚本内容如代码 7-8 所示。

代码 7-8　模拟产生账单数据的 Shell 脚本

```
#!/bin/bash
#Shell 脚本开始运行的时间
start_time=$(date +%Y-%m-%d-%H:%M:%S)
log_file=/root/qwm/mmconsume/log.out
#Hive 脚本
hive_script=/root/qwm/mmconsume/create_mmconsume_billevents_1d.hql
echo -e "Shell 脚本开始运行时间 : $start_time" >> $log_file
#计算当前时间与 2018-07-01 相差的月数
datatime=$1
curr_ymd=$(date +%Y-%m-%d)
curr_time2=$(($(date +%s -d $curr_ymd) - $(date +%s -d $datatime)))
month_delta=$((curr_time2/(60*60*24*30)))
echo "当前时间与 2018-07-01 相差的月数: $month_delta" >> $log_file
#在 Hive 中创建表 mmconsume_billevents_1d
echo -e "drop table if exists mmconsume_billevents_1d;\ncreate table
mmconsume_billevents_1d as select terminal_no,phone_no,fee_code,concat
(add_months(year_month,${month_delta}),' ','00:00:00')as year_month,owner_
name,owner_code,sm_name,should_pay,favour_fee from mmconsume_billevents;
">${hive_script}
hive -f /root/qwm/mmconsume/create_mmconsume_billevents_1d.hql
#将 mmconsume_billevents_1d 表的数据以 CSV 格式导出
#hive -e 'SET hive.cli.print.header=false; SELECT * FROM mmconsume_billevents_
1d' | sed -e 's/\t/;/g' > /root/qwm/data/mmconsume_billevents_1d.csv
#删除 Elasticsearch 集群中的 mmconsume_billevents_test 数据
curl -XDELETE http://$2:9200/$7
#将 mmconsume_billevents_1d 表的数据导入 Elasticsearch 集群
spark-submit --class com.tipdm.scala.datasource.Hive2Elasticsearch--executor-
memory 10g --total-executor-cores 20 --master spark://$3: 7077--jars /opt/
cloudera/parcels/CDH-5.7.3-1.cdh5.7.3.p0.5/lib/spark/lib/elasticsearch-spark-
13_2.10-6.3.2.jar $4 mmconsume_billevents_1d $2 $5 $6 $7 $8
#将 mmconsume_billevents_1d.csv 数据导入 Elasticsearch 集群
#nohup /data/logstash-6.3.2/bin/logstash -f /root/qwm/mmconsume/mmconsume_
billevents_1d.conf --path.data=/root/qwm/mmconsume/mmconsume_billevents_1d &
end_time=$(date +%Y-%m-%d-%H:%M:%S)
echo "任务结束运行时间: $end_time" >> $log_file
```

为测试定时任务能否按时执行，将执行时间改为每天 20:10:00，并将参数改为“2018-07-01 192.168.111.75 server3 /root/qwm/user_profile_project-1.0.jar 9200 year_month mmconsume_billevents_test doc”，测试模拟产生账单数据保存到 Elasticsearch 集群中的 mmconsume_billevents_test/doc。测试任务运行完成后，在调度日志界面中可查看该定时任务的执行结果，如图 7-22 所示。

1	2019-02-21 20:10:00	成功	查看	2019-02-21 21:46:09	成功	无	执行日志

图 7-22　测试模拟产生账单数据任务的执行结果

根据模拟产生 mmconsume_billevents 的规则可知，测试任务运行的时间（2019-02-21 20:10:00）减去原始账单数据中的最大时间（2018-07-01）得到的月份差 month_delta 等于 7，将 mmconsume-billevents-test 数据中的 year_month 字段都加上月份差 7 即可得到新模拟得到的数据。因此，模拟生成的数据应包含 year_month 字段为任务运行时间的当前年月的数据，如测试任务运行时间是“2019-02-21　20:10:00”，则生成的数据中应包含“2019-02-01”的数据。查看 Elasticsearch 集群的 mmconsume_billevents_test/doc 中的数据，其结果如图 7-23 所示。从图 7-23 中可以看到，数据中有 year_month 字段包含“2019-02-01 00:00:00”的数据，说明任务调度成功。

搜索 mmconsume_billevents_test (1227087 个文档) 的文档，查询条件：
must　doc.year_month.keyword　wildcard　*2019-02*
搜索　返回格式：Table　显示数量：10　显示查询语句
查询 5 个分片中用的 5 个. 203483 命中. 耗时 0.408 秒

_index	_type	_id	_score ▲	terminal_no	phone_no	fee_code	year_month	owner_name	owner_code	sm_name	should_pay	favour_fee
mmconsume_billevents_test	doc	-W1VJ2kBHxwflOJLl_oR	1	2400184270	4991634	0T	2019-02-01 00:00:00	HC级	00	甜果电视	2.5	0.0
mmconsume_billevents_test	doc	D21VJ2kBHxwflOJLl_sR	1	2400184344	4991756	0K	2019-02-01 00:00:00	HC级	00	珠江宽频	70.0	70.0
mmconsume_billevents_test	doc	ym5VJ2kBHxwflOJLlwIU	1	1300007025	1924009	0H	2019-02-01 00:00:00	HE级	00	互动电视	19.5	0.0
mmconsume_billevents_test	doc	_GdVJ2kBkPPgJQzSl8kC	1	1900151861	2268999	0T	2019-02-01 00:00:00	HC级	00	数字电视	7.5	8.0
mmconsume_billevents_test	doc	0KtVJ2kB3DxxIGYwtndY	1	1500100279	3364128	0B	2019-02-01 00:00:00	HC级	00	甜果电视	26.5	0.0
mmconsume_billevents_test	doc	vm1VJ2kBHxwflOJLl_4T	1	1100170652	2602075	0T	2019-02-01 00:00:00	HC级	00	互动电视	2.5	0.0
mmconsume_billevents_test	doc	wW1VJ2kBHxwflOJLl_4T	1	1100027375	2610824	0B	2019-02-01 00:00:00	HC级	00	甜果电视	26.5	0.0
mmconsume_billevents_test	doc	hG1VJ2kBHxwflOJLl_sS	1	2000165194	5056837	0Y	2019-02-01 00:00:00	HC级	00	数字电视	5.0	0.0
mmconsume_billevents_test	doc	iWdVJ2kBkPPgJQzSl9EX	1	1200133348	1453089	0K	2019-02-01 00:00:00	HC级	00	珠江宽频	58.34	0.0
mmconsume_billevents_test	doc	kWdVJ2kBkPPgJQzSl9EX	1	1200065221	1448730	0H	2019-02-01 00:00:00	HC级	00	数字电视	7.5	8.0

图 7-23　mmconsume_billevents_test/doc 的数据

② 配置模拟产生订单数据任务。参考配置模拟产生账单数据任务的过程，配置模拟产生订单数据任务，在任务调度中心的任务管理模块中新增模拟产生订单数据任务，如图 7-24 所示。将“任务参数”改为“2018-10-10 create_order_index_1d.hql 2018-01-01 server3/root/qwm /user_profile_project-1.0.jar 192.168.111.75 9200 optdate order_index_v3 doc”，每个参数之间用空格隔开；将“任务描述”改为“每天 21:10:00 模拟产生订单数据”；在“Cron”后输入任务运行的时间表达式，如“0 10 21 * * ? *”。

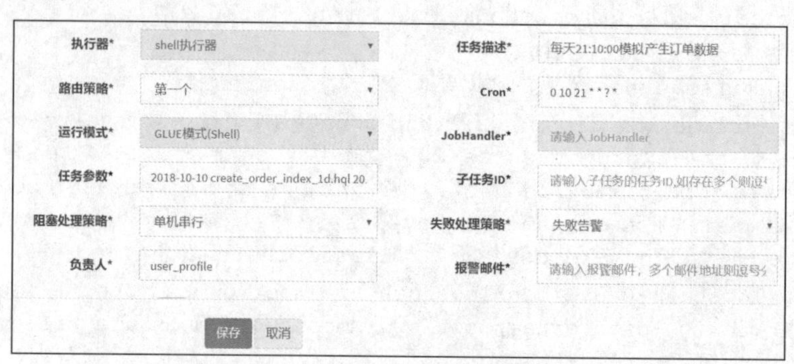

图 7-24　新增模拟产生订单数据任务

图 7-24 所示的信息设置完成后单击“保存”按钮，即可在任务管理列表中看到新增的任务，如图 7-25 所示。

任务ID	任务描述	运行模式	Cron	负责人	状态	操作
2	每天21:10:00模拟产生订单数据	GLUE模式(Shell)	0 10 21 * * ? *	user_profile	NORMAL	执行 暂停 日志 编辑 GLUE 删除

图 7-25 任务管理列表

单击图 7-25 中的“GLUE”按钮，编辑任务的 Shell 脚本，脚本内容如代码 7-9 所示。

代码 7-9 模拟产生订单数据的 Shell 脚本

```
#!/bin/bash
#任务启动的时间
task_start_time=$1
#Hive 脚本
hive_script=/root/qwm/order_index/$2
#Shell 脚本开始运行的时间
start_time=$(date +%Y-%m-%d-%H:%M:%S)
#日志保存的路径
log_file=/root/qwm/order_index/log.out
echo -e "\nstart time : $start_time" >> $log_file
echo "program run day: $task_start_time" >> $log_file
#任务启动的时间与数据的开始时间（2018-01-01）间隔的天数
datatime=$3
time1=$(($(date +%s -d $task_start_time) - $(date +%s -d $datatime)))
delta=$((time1/(60*60*24)))
echo "任务启动的时间与数据的开始时间（2018-01-01）间隔的天数:$delta" >> $log_file
#当前时间与任务启动的时间间隔的天数
curr_ymd=$(date +%Y-%m-%d)
curr_time2=$(($(date +%s -d $curr_ymd) - $(date +%s -d $task_start_time)))
curr_delta=$((curr_time2/(60*60*24)))
#当前时间与数据开始时间间隔的天数
curr_time3=$(($(date +%s -d $curr_ymd) - $(date +%s -d $datatime)))
curr_delta1=$((curr_time3/(60*60*24)))
echo "当前时间与任务启动的时间${task_start_time}间隔的天数:$curr_delta" >> $log_file
echo "当前时间与数据开始时间 2018-01-01 间隔的天数:$curr_delta1" >> $log_file
#在 Hive 中创建表 order_index_1d,若任务启动的时间(task_start_time)是 2018-10-09,则 order_
#index_1d 中的数据是原始数据中 optdate 字段为 2018-01-01 的数据,且计算当前时间与 optdate
#字段的时间差 delta, orderdate、expdate、effdate 字段分别加上时间差 delta
echo -e "drop table if exists order_index_1d;\ncreate table order_index_
1d as select phone_no,owner_name,concat(CURRENT_DATE,' ',from_unixtime(unix_
timestamp(optdate),\"HH:mm:ss\")) as optdate,prodname,sm_name,offerid,offername,
business_name,owner_code,prodprcid,prodprcname,concat(date_add(effdate,datediff
(CURRENT_DATE,optdate)),' ',from_unixtime(unix_timestamp(effdate),\"HH:mm:
ss\")) as effdate,concat(date_add(expdate,datediff(CURRENT_DATE,optdate)),
' ',from_unixtime(unix_timestamp(expdate),\"HH:mm:ss\")) as expdate,concat
(date_add(orderdate,datediff(CURRENT_DATE,optdate)),' ',from_unixtime(unix_
timestamp(orderdate),\"HH:mm:ss\")) as orderdate,cost,mode_time,prodstatus,
run_name,orderno,offertype from order_index_v3 where date_add(optdate, ${delta})
= date_sub(CURRENT_DATE,${curr_delta});">${hive_script}
hive -f /root/qwm/order_index/${hive_script}
#将 order_index_1d 表的数据导入 Elasticsearch 集群
```

```
spark-submit --class com.tipdm.scala.datasource.Hive2Elasticsearch --executor-memory 10g --total-executor-cores 20 --master spark://$4:7077 --jars /opt/cloudera/parcels/CDH-5.7.3-1.cdh5.7.3.p0.5/lib/spark/lib/elasticsearch-spark-13_2.10-6.3.2.jar $5 order_index_1d $6 $7 $8 $9 ${10}
end_time=$(date +%Y-%m-%d-%H:%M:%S)
echo "end time : $end_time" >> $log_file
```

为测试模拟产生订单数据任务能否按时执行，将执行时间改为每天 21:10:00，并将参数改为“2018-10-10 create_order_index_1d.hql 2018-01-01 server3 /root/qwm/user_profile_project-1.0.jar 192.168.111.75 9200 optdate order_test4 doc”，测试模拟产生订单数据保存到 Elasticsearch 集群的 order_test4/doc 中。测试任务运行完成后，在调度日志界面中查看该定时任务的执行结果，如图 7-26 所示。

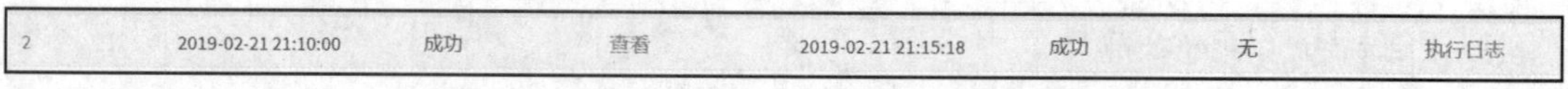

2	2019-02-21 21:10:00	成功	查看	2019-02-21 21:15:18	成功	无	执行日志

图 7-26　测试模拟产生订单数据任务的执行结果

查看 Elasticsearch 集群的 order_test4/doc 中的数据，其结果如图 7-27 所示。从图 7-27 中可以看到 optdate 字段的时间为“2019-02-21”开头，与图 7-26 的调度时间是一致的，说明任务调度成功。

搜索 order_test4 (86879 个文档) 的文档，查询条件：
must　match_all　+　-
搜索　返回格式：Table　显示数量：10　显示查询语句
查询 5 个分片中用的 5 个. 86879 命中. 耗时 0.279 秒

_index	_type	_id	_score ▲	phone_no	owner_name	optdate	prodname	sm_name	offerid	offername	business_name	owner_code
order_test4	doc	feMwEGkBUCupFqzlobIR	1	1715805	HC级	2019-02-21 00:41:52	标清直播基本包_广州	互动电视	GZ122757	互动+联合宽带-59元包-月享	到期暂停状态	00
order_test4	doc	g-MwEGkBUCupFqzlobIR	1	1609614	HC级	2019-02-21 00:35:33	联合宽带30M	珠江宽频	GZ123464	宽带30M提速_	到期暂停状态	00
order_test4	doc	n-MwEGkBUCupFqzlobIS	1	1311522	HE级	2019-02-21 00:29:44	互动空指令产品	甜果电视	GZ123387	月享套餐前3月优惠	到期暂停状态	00
order_test4	doc	o-MwEGkBUCupFqzlobIS	1	5388480	HC级	2019-02-21 00:47:55	基本包	数字电视	GZ000004	数字基本包-配机类	到期暂停状态	00
order_test4	doc	peMwEGkBUCupFqzlobIS	1	1333187	HC级	2019-02-21 00:38:27	甜果TV(精彩点_点播)	互动电视	00133381	快乐点-2年-600元	到期暂停状态	00
order_test4	doc	rOMwEGkBUCupFqzlobIS	1	4995679	HC级	2019-02-21 00:34:08	标清直播基本包_广州	甜果电视	GZ123142	甜果+联合宽带-49包-月享	到期暂停状态	00
order_test4	doc	reMwEGkBUCupFqzlobIS	1	4984346	HC级	2019-02-21 00:32:37	带宽12M	珠江宽频	GZ123411	互动+宽带-12M-3月-月享优惠	到期暂停状态	00
order_test4	doc	sOMwEGkBUCupFqzlobIS	1	4093469	EE级	2019-02-21 00:41:31	联合宽带8M	珠江宽频	GZ100065	宽带-8M-包月	到期暂停状态	01
order_test4	doc	yuMwEGkBUCupFqzlobIS	1	2489412	HC级	2019-02-21 00:33:25	高尔夫.网球频道	互动电视	GZ123042	赠送两套付费频道-1年-0元	到期暂停状态	00
order_test4	doc	zOMwEGkBUCupFqzlobIS	1	4982324	HC级	2019-02-21 00:18:08	联合宽带30M	珠江宽频	GZ123720	互动+宽带-30M-3月-89月享优惠	到期暂停状态	00

图 7-27　order_index4/doc 数据结果

③ 配置模拟产生用户收视行为数据任务。参考配置模拟产生账单数据任务的过程，配置模拟产生用户收视行为数据任务，在任务调度中心的任务管理模块中新增模拟产生用户收视行为数据任务，如图 7-28 所示。将“任务参数”改为“2018-10-10 create_media_1d.hql 2018-05-02 server3 /root/qwm/user_profile_project-1.0.jar 192.168.111.85 9200 origin_time media_index media”，每个参数之间用空格隔开；将“任务描述”改为“每天 22:10:00 模拟产生用户收视行为数据”；在“Cron”后输入任务运行的时间表达式，如“0 10 22 * * ? *”。

执行器*	shell执行器	任务描述*	每天22:10:00模拟产生用户收视行为数据
路由策略*	第一个	Cron*	0 10 22 * * ? *
运行模式*	GLUE模式(Shell)	JobHandler*	请输入JobHandler
任务参数*	2018-10-10 create_media_1d.hql 2018-05-	子任务ID*	请输入子任务的任务ID,如存在多个则逗
阻塞处理策略*	单机串行	失败处理策略*	失败告警
负责人*	user_profile	报警邮件*	请输入报警邮件，多个邮件地址则逗号分

保存　取消

图 7-28　新增模拟产生用户收视行为数据任务

图 7-28 所示的信息设置完成后单击“保存”按钮，即可在任务管理列表中看到新增的任务，如图 7-29 所示。

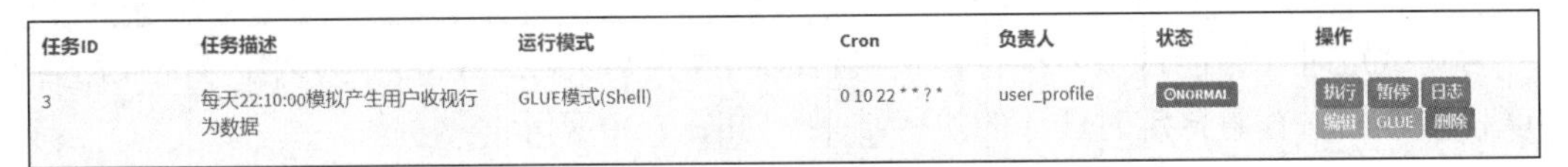

任务ID	任务描述	运行模式	Cron	负责人	状态	操作
3	每天22:10:00模拟产生用户收视行为数据	GLUE模式(Shell)	0 10 22 * * ? *	user_profile	NORMAL	执行 暂停 日志 编辑 GLUE 删除

图 7-29　任务管理列表

单击图 7-29 中的“GLUE”按钮，编辑任务的 Shell 脚本，脚本内容如代码 7-10 所示。

代码 7-10　模拟产生用户收视行为数据的 Shell 脚本

```
#!/bin/bash
#设置任务启动的时间
task_start_time=$1
#Hive 脚本
hive_script=$2
#Shell 脚本开始运行的时间
start_time=$(date +%Y-%m-%d-%H:%M:%S)
log_file=/root/qwm/log.out
echo -e "\nstart time : $start_time" >> $log_file
echo "program run day: $task_start_time" >> $log_file
#任务启动的时间与数据的开始时间（2018-05-02）间隔的天数
datatime=$3
time1=$(($(date +%s -d $task_start_time) - $(date +%s -d $datatime)))
delta=$((time1/(60*60*24)))
echo "任务启动的时间与数据的开始时间（2018-05-02）间隔的天数:$delta" >> $log_file
#当前时间与任务启动的时间间隔的天数
curr_ymd=$(date +%Y-%m-%d)
curr_time2=$(($(date +%s -d $curr_ymd) - $(date +%s -d $task_start_time)))
curr_delta=$((curr_time2/(60*60*24)))
echo "当前时间与任务启动的时间${task_start_time}间隔的天数:$curr_delta" >> $log_file
#在 Hive 中创建表 media_1d，若 task_start_time 是 2018-09-18，则 media_1d 中的数据是
#选择原始数据中 2018-05-02 的数据，且其中 origin_time 和 end_time 的时间都改为 2018-
#09-18，到了 2018-09-19，则 media_1d 中的数据是 2018-05-03 的数据，且其中 origin_time
#和 end_time 的时间都改为 2018-09-19
echo -e "drop table if exists media_1d;\ncreate table media_1d as select
terminal_no,phone_no,duration,station_name,concat(CURRENT_DATE,' ',from_
unixtime(unix_timestamp(origin_time),\"HH:mm:ss\")) as origin_time,concat
(CURRENT_DATE,' ',from_unixtime(unix_timestamp(end_time),\"HH:mm:ss\")) as
end_time,owner_code,owner_name,vod_cat_tags,resolution,audio_lang,region,
res_name,res_type,vod_title,category_name,program_title,sm_name,first_show_
time from media_index_3m where date_add(origin_time,${delta}) = date_sub
(CURRENT_DATE,${curr_delta});">${hive_script}
hive -f /root/qwm/${hive_script}
#将 media_1d 表的数据导入 Elasticsearch 集群
spark-submit --class com.tipdm.scala.datasource.Hive2Elasticsearch
--executor-memory 10g --total-executor-cores 20 --master spark://$4:7077
--jars /opt/cloudera/parcels/CDH-5.7.3-1.cdh5.7.3.p0.5/lib/spark/lib/
elasticsearch-spark-13_2.10-6.3.2.jar $5 media_1d $6 $7 $8 $9 ${10}
```

```
#Shell 脚本结束运行的时间
end_time=$(date +%Y-%m-%d-%H:%M:%S)
echo "end time : $end_time" >> $log_file
```

为测试定时任务能否按时执行，把执行时间改为每天 22:10:00。测试任务完成后，在调度日志界面中查看该定时任务的执行结果，如图 7-30 所示。

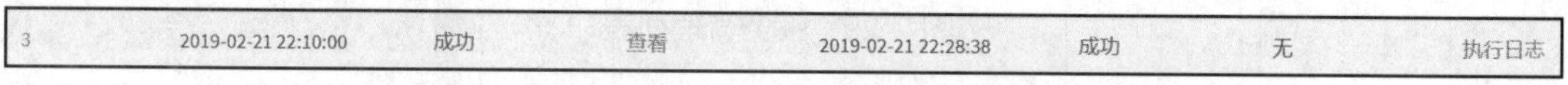

3	2019-02-21 22:10:00	成功	查看	2019-02-21 22:28:38	成功	无	执行日志

图 7-30　测试模拟产生用户收视行为数据的执行结果

因为任务执行的时间是“2019-02-21”，这个时间处于 2019 年的第 8 周，所以产生的数据应该保存在 Elasticsearch 集群的 media_index20198/doc 中。查看 Elasticsearch 集群中的 media_index20198/doc 数据，其结果如图 7-31 所示。从图 7-31 中可以看到“origin_time”字段和“end_time”字段中包含“2019-02-21”开头的数据，说明任务调度成功。

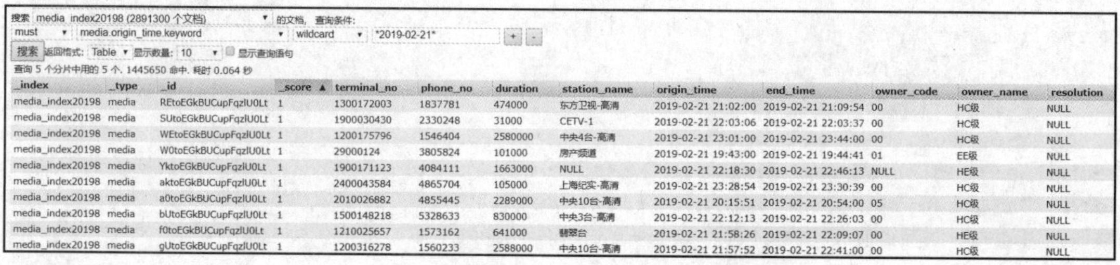

搜索 media_index20198 (2891300 个文档) 的文档，查询条件：
must media.origin_time.keyword wildcard *2019-02-21* + -
搜索 返回格式：Table 显示数量：10 显示查询语句
查询 5 个分片中用的 5 个. 1445650 命中. 耗时 0.064 秒

_index	_type	_id	_score ▲	terminal_no	phone_no	duration	station_name	origin_time	end_time	owner_code	owner_name	resolution
media_index20198	media	REtoEGkBUCupFqzlU0Lt	1	1300172003	1837781	474000	东方卫视-高清	2019-02-21 21:02:00	2019-02-21 21:09:54	00	HC级	NULL
media_index20198	media	SUtoEGkBUCupFqzlU0Lt	1	1900030430	2330248	31000	CETV-1	2019-02-21 22:03:06	2019-02-21 22:03:37	00	HC级	NULL
media_index20198	media	WEtoEGkBUCupFqzlU0Lt	1	1200175796	1546404	2580000	中央4台-高清	2019-02-21 23:01:00	2019-02-21 23:44:00	00	HC级	NULL
media_index20198	media	W0toEGkBUCupFqzlU0Lt	1	29000124	3805824	101000	房产频道	2019-02-21 19:43:00	2019-02-21 19:44:41	01	EE级	NULL
media_index20198	media	YktoEGkBUCupFqzlU0Lt	1	1900171123	4084111	1663000	NULL	2019-02-21 22:18:30	2019-02-21 22:46:13	NULL	HE级	NULL
media_index20198	media	aktoEGkBUCupFqzlU0Lt	1	2400043584	4865704	105000	上海纪实-高清	2019-02-21 23:28:54	2019-02-21 23:30:39	00	HC级	NULL
media_index20198	media	a0toEGkBUCupFqzlU0Lt	1	2010026882	4855445	2289000	中央10台-高清	2019-02-21 20:15:51	2019-02-21 20:54:00	05	HC级	NULL
media_index20198	media	bUtoEGkBUCupFqzlU0Lt	1	1500148218	5328633	830000	中央3台-高清	2019-02-21 22:12:13	2019-02-21 22:26:03	00	HC级	NULL
media_index20198	media	f0toEGkBUCupFqzlU0Lt	1	1210025657	1573162	641000	翡翠台	2019-02-21 21:58:26	2019-02-21 22:09:07	00	HE级	NULL
media_index20198	media	gUtoEGkBUCupFqzlU0Lt	1	1200316278	1560233	2588000	中央10台-高清	2019-02-21 21:57:52	2019-02-21 22:41:00	00	HC级	NULL

图 7-31　media_index20198/doc 数据结果

（2）数据传输

数据传输指从 Elasticsearch 集群读取数据到 Hive 中，包括读取 media_index 数据、mediamatch_usermsg 数据、mediamatch_userevent 数据、order_index 数据和 mmconsume_billevents 数据。由 4.2.1 小节可知，可通过 Java 实现类来运行数据从 Elasticsearch 集群传输到 Hive 中的任务，因此数据传输的任务可以通过 Java 执行器来运行。

下面以将 Elasticsearch 集群中的 mediamatch_usermsg 数据传输到 Hive 中为例，介绍数据传输任务的配置。

在 timer-job-java-executor 工程的 src/main/java/com/tipdm/java/jobs/ 目录下新建目录 user_profile/datasource，并在 user_profile/datasource 目录下新建 Usermsg2Hive 类，该类的代码如代码 7-11 所示。因为实际项目中 mediamatch_usermsg 传输到 Hive 中时是根据 run_time 字段选取当前时间之前 50 年的数据，所以 Usermsg2Hive 类的代码中的第 7 个参数应该设置为当前时间，而不是与代码 4-31 中一样设置为“2018-08-01”。

代码 7-11　新建 Usermsg2Hive 类

```
package com.tipdm.java.jobs.user_profile.datasource;
import com.tipdm.engine.SparkYarnJob;
import com.tipdm.java.JavaJobInterface;
import com.xxl.job.core.biz.model.ReturnT;
import org.slf4j.Logger;
import org.slf4j.LoggerFactory;
import java.text.SimpleDateFormat;
```

```
import java.util.Date;
import java.util.Map;
public class Usermsg2Hive implements JavaJobInterface {
    private static Logger log = LoggerFactory.getLogger(Usermsg2Hive.class);
    private static String sparkClassName = "com.tipdm.scala.chapter_3_5_1_
datasource.Elasticsearch2Hive";
    private static String applicationName = "Usermsg2Hive";
    public static void main(String[] args) throws Exception {
        String className = "com.tipdm.java.jobs.user_profile.datasource.
Usermsg2Hive";
        Usermsg2Hive usermsg2Hive = new Usermsg2Hive();
        usermsg2Hive.execute(className, null);
    }
    @Override
    public ReturnT<String> execute(String jobClassName, Map<String, String>
args) throws Exception {
        SimpleDateFormat df = new SimpleDateFormat("yyyy-MM-dd HH:mm:ss");
        String[] arguments =new String [9];
        arguments[0]="mediamatch_usermsg";
arguments[1]="terminal_no,phone_no,sm_name,run_name,sm_code,owner_name,owner_
code,run_time,addressoj,estate_name,open_time,force";
        arguments[2]="run_time";
        arguments[3]="yyyy-MM-dd HH:mm:ss";
        arguments[4]="50";
        arguments[5]="Y";
        arguments[6]="mediamatch_usermsg/doc";
        arguments[7]=df.format(new Date());
        arguments[8]="user_profile";
        SparkYarnJob.runAndMonitor(applicationName,sparkClassName,arguments);
        log.info("任务运行成功");
        return ReturnT.SUCCESS;
    }
}
```

重新编译打包 timer-job-java-executor 工程，将新生成的 JAR 包上传到 node4 节点的/root/xxl-job-user-profile 目录下，并参考代码 7-5 启动 timer-job-java-executor 工程。

工程启动后，在任务调度中心首页的任务管理模块中新增数据传输任务，如图 7-32 所示。其中，“执行器”设置为“java 执行器”；“运行模式”设置为“BEAN 模式”；“任务参数”设置为“com.tipdm. java.jobs.user_profile.datasource.Usermsg2Hive”；“负责人”设置为“user_profile”；“任务描述”设置为“每个月 1 号晚上 23:10:00 从 es 读取 mediamatch_usermsg 数据”；“Cron”设置为“0 10 23 1 * ? *”；“JobHandler”设置为“javaJobHandler”。

图 7-32 所示的信息设置完成后，即可在任务管理列表中看到新增的任务，如图 7-33 所示。

为测试定时任务能否按时执行，将执行时间改为每天 18:50:00，为了使任务尽快运行完成，将代码 7-11 中的 arguments[4]的值改为“10”，并重新编译打包 timer-job-java-executor 工程，将新生成的 JAR 包上传到 node4 节点的/root/xxl-job-user-profile 目录下并重启工程。

每天 18:50:00 之后，在调度日志界面中查看该定时任务的执行结果，如图 7-34 所示。

执行器*	java执行器	任务描述*	每个月1号晚上23:10:00从es读取mediam
路由策略*	第一个	Cron*	0 10 23 1 * ? *
运行模式*	BEAN模式	JobHandler*	javaJobHandler
任务参数*	com.tipdm.java.jobs.user_profile.datasou	子任务ID*	请输入子任务的任务ID,如存在多个则逗
阻塞处理策略*	单机串行	失败处理策略*	失败告警
负责人*	user_profile	报警邮件*	请输入报警邮件，多个邮件地址则逗号

保存 取消

图 7-32　新增数据传输任务

任务ID	任务描述	运行模式	Cron	负责人	状态	操作
4	每个月1号晚上23:10:00从es读取mediamatch_usermsg数据	BEAN模式：javaJobHandler	0 10 23 1 * ? *	user_profile	NORMAL	执行 暂停 日志 编辑 删除

图 7-33　任务管理列表

4	2019-02-21 18:50:00	成功	查看	2019-02-21 19:30:03	成功	无	执行日志

图 7-34　mediamatch_usermsg 数据传输到 Hive 中的定时任务的执行结果

测试的时候，读取 Elasticsearch 集群中的 mediamatch_usermsg/doc 传输到 Hive 的 user_profile 库的 mediamatch_usermsg 中，任务运行结束后，在 HDFS 中查看 mediamatch_usermsg 的存储情况，如图 7-35 所示。从图 7-35 中可以看出，mediamatch_usermsg 的内容的修改时间大概在“2019-02-21 19:07”到“2019-02-21 19:30”之间，这一时间正好在任务的调度时间（2019-02-21 18:50:00）与任务的执行时间（2019-02-21 19:30:03）之间。图 7-34 和图 7-35 的结果说明了将 Elasticsearch 集群中的 mediamatch_usermsg 数据传输到 Hive 中的任务是按时执行并且执行成功的。

Contents of directory /user/hive/warehouse/user_profile.db/mediamatch_usermsg

Goto : /user/hive/warehouse/user go

Go to parent directory

Name	Type	Size	Replication	Block Size	Modification Time	Permission	Owner	Group
.hive-staging_hive_2019-02-21_19-07-50_819_4915531139604898473-1	dir				2019-02-21 19:07	rwxrwxrwt	spark	hive
part-00000	file	0 B	3	128 MB	2019-02-21 19:07	rwxrwxrwt	spark	hive
part-00001	file	115.25 MB	3	128 MB	2019-02-21 19:13	rwxrwxrwt	spark	hive
part-00002	file	0 B	3	128 MB	2019-02-21 19:07	rwxrwxrwt	spark	hive
part-00003	file	0 B	3	128 MB	2019-02-21 19:07	rwxrwxrwt	spark	hive
part-00004	file	0 B	3	128 MB	2019-02-21 19:07	rwxrwxrwt	spark	hive
part-00005	file	57.55 MB	3	128 MB	2019-02-21 19:29	rwxrwxrwt	spark	hive
part-00006	file	0 B	3	128 MB	2019-02-21 19:07	rwxrwxrwt	spark	hive
part-00007	file	0 B	3	128 MB	2019-02-21 19:07	rwxrwxrwt	spark	hive
part-00008	file	0 B	3	128 MB	2019-02-21 19:07	rwxrwxrwt	spark	hive
part-00009	file	0 B	3	128 MB	2019-02-21 19:07	rwxrwxrwt	spark	hive
part-00010	file	57.69 MB	3	128 MB	2019-02-21 19:28	rwxrwxrwt	spark	hive
part-00011	file	0 B	3	128 MB	2019-02-21 19:07	rwxrwxrwt	spark	hive
part-00012	file	57.74 MB	3	128 MB	2019-02-21 19:26	rwxrwxrwt	spark	hive
part-00013	file	0 B	3	128 MB	2019-02-21 19:07	rwxrwxrwt	spark	hive
part-00014	file	0 B	3	128 MB	2019-02-21 19:07	rwxrwxrwt	spark	hive
part-00015	file	57.75 MB	3	128 MB	2019-02-21 19:29	rwxrwxrwt	spark	hive

图 7-35　在 HDFS 中查看 mediamatch_usermsg 的存储情况

其他的数据传输任务配置可参考 mediamatch_usermsg 数据的传输，所有数据传输任务配置完成后，在任务管理列表中可看新增的所有数据传输任务，如图 7-36 所示。（具体的代码实现详见教材配套的代码，可参考 user_profile/datasource 目录下的 MediaIndex2Hive 类、Order_index2Hive 类、Userevent2Hive 类、Billevents2Hive 类。）

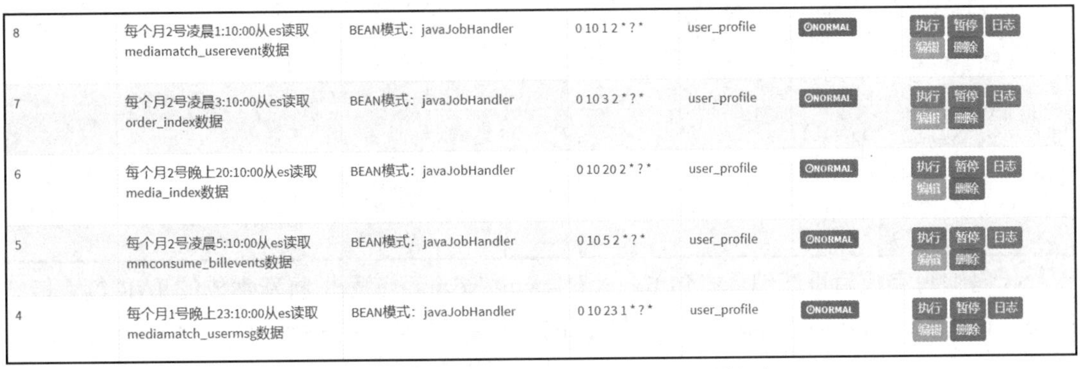

8	每个月2号凌晨1:10:00从es读取 mediamatch_userevent数据	BEAN模式：javaJobHandler	0 10 1 2 * ? *	user_profile	⊙NORMAL	执行 暂停 日志 编辑 删除
7	每个月2号凌晨3:10:00从es读取 order_index数据	BEAN模式：javaJobHandler	0 10 3 2 * ? *	user_profile	⊙NORMAL	执行 暂停 日志 编辑 删除
6	每个月2号晚上20:10:00从es读取 media_index数据	BEAN模式：javaJobHandler	0 10 20 2 * ? *	user_profile	⊙NORMAL	执行 暂停 日志 编辑 删除
5	每个月2号凌晨5:10:00从es读取 mmconsume_billevents数据	BEAN模式：javaJobHandler	0 10 5 2 * ? *	user_profile	⊙NORMAL	执行 暂停 日志 编辑 删除
4	每个月1号晚上23:10:00从es读取 mediamatch_usermsg数据	BEAN模式：javaJobHandler	0 10 23 1 * ? *	user_profile	⊙NORMAL	执行 暂停 日志 编辑 删除

图 7-36　任务管理列表中所有的数据传输任务

（3）数据预处理

数据预处理任务也是通过 Java 执行器运行的，因此需要在 timer-job-java-executor 工程中编写数据预处理的实现类。在工程 timer_job_java_executor/src/main/java/com/tipdm/java/jobs/user_profile 目录下新建目录 dataprocess，并在 dataprocess 目录下新建 DataProcess 类，其代码如代码 7-12 所示。

代码 7-12　新建 DataProcess 类

```
package com.tipdm.java.jobs.user_profile.dataprocess;
import com.tipdm.engine.SparkYarnJob;
import com.tipdm.java.JavaJobInterface;
import com.xxl.job.core.biz.model.ReturnT;
import org.slf4j.Logger;
import org.slf4j.LoggerFactory;
import java.util.Map;
public class DataProcess implements JavaJobInterface {
    private static Logger log = LoggerFactory.getLogger(DataProcess.class);
    private static String sparkClassName = " com.tipdm.scala.chapter_3_6_processing.
DataProcess";
    private static String applicationName = "DataProcess";
    public static void main(String[] args) throws Exception {
        String className = "com.tipdm.java.jobs.DataProcess";
        DataProcess dataProcess = new DataProcess();
        dataProcess.execute(className, null);
    }
    @Override
    public ReturnT<String> execute(String jobClassName, Map<String, String>
args) throws Exception {
        String[] arguments = new String[10];
        arguments[0] = "user_profile.media_index_3m";
        arguments[1] = "user_profile.media_index_3m_process";
```

```
        arguments[2] = "user_profile.mediamatch_userevent";
        arguments[3] = "user_profile.mediamatch_userevent_process";
        arguments[4] = "user_profile.mediamatch_usermsg";
        arguments[5] = "user_profile.mediamatch_usermsg_process";
        arguments[6] = "user_profile.mmconsume_billevents";
        arguments[7] = "user_profile.mmconsume_billevent_process";
        arguments[8] = "user_profile.order_index_v3";
        arguments[9] = "user_profile.order_index_process";
        SparkYarnJob.runAndMonitor(applicationName,sparkClassName,arguments);
        log.info("任务运行成功");
        return ReturnT.SUCCESS;
    }
}
```

代码编写完成后重新编译打包 timer-job-java-executor 工程，将新生成的 JAR 包上传到 node4 节点的/root/xxl-job-user-profile 目录下，并参考代码 7-5 启动 timer-job-java-executor 工程。

工程启动后，在任务调度中心首页的任务管理模块中新增数据预处理任务，如图 7-37 所示。其中，“执行器”设置为“java 执行器”；“运行模式”设置为“BEAN 模式”；“任务参数”设置为“com.tipdm. java.jobs.user_profile.dataprocess.DataProcess”；“负责人”设置为“user_profile”；“任务描述”设置为“每个月 3 号凌晨 2:10:00 进行数据预处理”；“Cron”设置为“0 10 2 3 * ? *”；“JobHandler”设置为“javaJobHandler”。

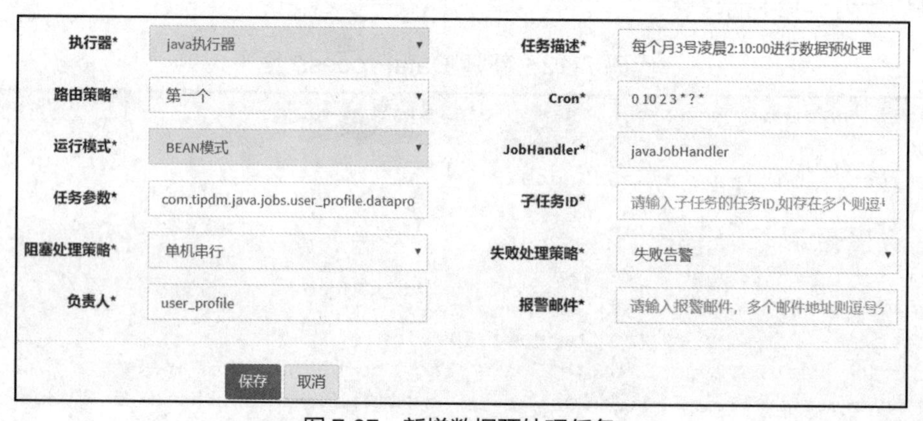

图 7-37　新增数据预处理任务

图 7-37 所示的信息设置完成后，即可在任务管理列表中看到新增的任务，如图 7-38 所示。

任务ID	任务描述	运行模式	Cron	负责人	状态	操作
9	每个月3号凌晨2:10:00进行数据预处理	BEAN模式：javaJobHandler	0 10 2 3 * ? *	user_profile	NORMAL	执行 暂停 日志 编辑 删除

图 7-38　任务管理列表中新增的任务

为测试任务能否按时执行，将任务的执行时间改为每天 9:20:00，并在每天 9:20:00 之后，在调度日志界面中查看任务是否按时启动并执行成功，如图 7-39 所示。在测试这个任务的时候，集群中有较多任务在运行，因此运行效率比较低，需要近 1h 才能执行完成。

9	2019-02-22 09:20:00	成功	查看	2019-02-22 10:15:20	成功	查看	执行日志

图 7-39　数据预处理任务的执行结果

数据预处理任务执行成功后会修改 Hive 中的 user_profile 库中的 mediamatch_usermsg_process、mediamatch_userevent_process、mmconsume_billevent_process、order_index_process 和 media_index_3m_process。因此，任务运行成功后可以在 HDFS 中查看这些表的修改时间。以查看 media_index_3m_process 为例，查看该表在 HDFS 中的存储情况，如图 7-40 所示。该表的修改时间是在任务的调度时间“2019-02-22 09:20:00”之后，说明任务是按时执行且执行成功的。

Contents of directory /user/hive/warehouse/user_profile.db/media_index_3m_process

Goto : /user/hive/warehouse/user go

Go to parent directory

Name	Type	Size	Replication	Block Size	Modification Time	Permission	Owner	Group
_SUCCESS	file	0 B	3	128 MB	2019-02-22 09:46	rw-r--r--	spark	hive
_common_metadata	file	2.31 KB	3	128 MB	2019-02-22 09:47	rw-r--r--	spark	hive
_metadata	file	679.02 KB	3	128 MB	2019-02-22 09:47	rw-r--r--	spark	hive
part-r-00000-1e738140-d1da-4fc8-a737-b3d73bbe155c.gz.parquet	file	13.29 MB	3	128 MB	2019-02-22 09:39	rw-r--r--	spark	hive
part-r-00001-1e738140-d1da-4fc8-a737-b3d73bbe155c.gz.parquet	file	13.32 MB	3	128 MB	2019-02-22 09:39	rw-r--r--	spark	hive
part-r-00002-1e738140-d1da-4fc8-a737-b3d73bbe155c.gz.parquet	file	13.36 MB	3	128 MB	2019-02-22 09:39	rw-r--r--	spark	hive
part-r-00003-1e738140-d1da-4fc8-a737-b3d73bbe155c.gz.parquet	file	13.23 MB	3	128 MB	2019-02-22 09:39	rw-r--r--	spark	hive
part-r-00004-1e738140-d1da-4fc8-a737-b3d73bbe155c.gz.parquet	file	13.33 MB	3	128 MB	2019-02-22 09:39	rw-r--r--	spark	hive
part-r-00005-1e738140-d1da-4fc8-a737-b3d73bbe155c.gz.parquet	file	13.28 MB	3	128 MB	2019-02-22 09:39	rw-r--r--	spark	hive

图 7-40　查看 media_index_3m_process 在 HDFS 中的存储情况

（4）SVM 预测用户是否挽留

SVM 预测用户是否挽留任务也可通过 Java 执行器实现，因此需要在 timer-job-java-executor 工程的 src/main/java/com/tipdm/java/jobs/user_profile 目录下新建目录 svm，并在 svm 目录下新建 SVM 类，其代码如代码 7-13 所示。

代码 7-13　新建 SVM 类

```
package com.tipdm.java.jobs.user_profile.svm;
import com.tipdm.engine.SparkYarnJob;
import com.tipdm.java.JavaJobInterface;
import com.xxl.job.core.biz.model.ReturnT;
import org.slf4j.Logger;
import org.slf4j.LoggerFactory;
import java.util.Map;
public class SVM implements JavaJobInterface {
   private static Logger log = LoggerFactory.getLogger(SVM.class);
   private static String sparkClassName = "com.tipdm.scala.chapter_3_8_6_svm.
SVM";
   private static String applicationName = "SVM";
   public static void main(String[] args) throws Exception {
      String className = "com.tipdm.java.jobs.user_profile.svm.SVM";
      SVM svm = new SVM();
      svm.execute(className, null);
   }
   @Override
   public ReturnT<String> execute(String jobClassName, Map<String, String>
```

```
args) throws Exception {
        String[] arguments = new String[12];
        arguments[0] = "mmconsume_billevent_process";
        arguments[1] = "mediamatch_userevent_process";
        arguments[2] = "media_index_3m_process";
        arguments[3] = "mediamatch_usermsg_process";
        arguments[4] = "order_index_process";
        arguments[5] = "svm_activate";
        arguments[6] = "svm_prediction";
        arguments[7]="10";
        arguments[8]="1.0";
        arguments[9]="0.01";
        arguments[10]="1.0";
        arguments[11]="user_profile";
        SparkYarnJob.runAndMonitor(applicationName,sparkClassName,arguments);
        log.info("任务运行成功");
        return ReturnT.SUCCESS;
    }
}
```

代码编写完成后重新编译打包 timer-job-java-executor 工程，将新生成的 JAR 包上传到 node4 节点的/root/xxl-job-user-profile 目录下，并参考代码 7-5 启动 timer-job-java-executor 工程。

工程启动之后，在任务调度中心首页的任务管理模块中新增 SVM 预测用户是否挽留任务，如图 7-41 所示。其中，“执行器”设置为“java 执行器”；“运行模式”设置为“BEAN 模式”；“任务参数”设置为“com.tipdm.java.jobs.user_profile.svm.SVM”；“负责人”设置为“user_profile”；“任务描述”设置为“每个月 3 号凌晨 2:40:00 运行 SVM 预测用户是否挽留任务”；“Cron”设置为“0 40 2 3 * ? *”；“JobHandler”设置为“javaJobHandler”。

图 7-41　新增 SVM 预测用户是否挽留任务

图 7-41 所示的信息设置完成后，即可在任务管理列表中看到新增的任务，如图 7-42 所示。

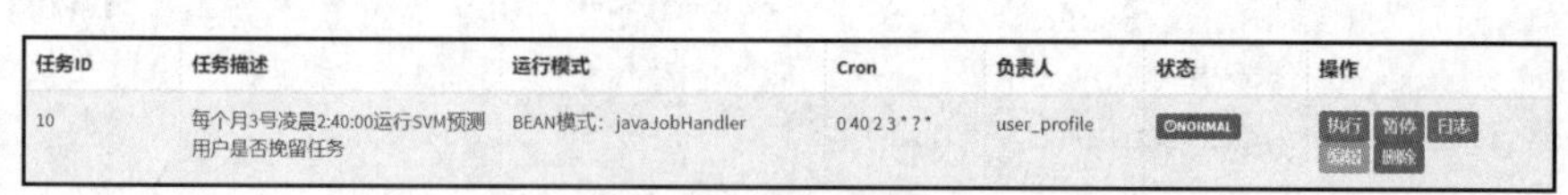

任务ID	任务描述	运行模式	Cron	负责人	状态	操作
10	每个月3号凌晨2:40:00运行SVM预测用户是否挽留任务	BEAN模式：javaJobHandler	0 40 2 3 * ? *	user_profile	NORMAL	执行 暂停 日志 编辑 删除

图 7-42　任务管理列表

为测试定时任务能否按时执行，将执行时间改为每天 16:42:00，并在每天 16:42:00 之后，在调度日志界面中查看任务是否按时启动并执行成功，如图 7-43 所示。

10	2019-02-22 16:42:00	成功	查看	2019-02-22 16:49:26	成功	无	执行日志

图 7-43　SVM 预测用户是否挽留任务的执行结果

每次运行 SVM 预测用户是否挽留任务都会更新 Hive 的 user_profile 库中的 svm_activate 和 svm_prediction，因此，任务运行成功后，可以在 HDFS 中查看这两个表是否按时更新，其结果如图 7-44 所示。从图 7-44 中可以看出，svm_activate 的更新时间是“2019-02-22 16:47”，svm_prediction 的更新时间是“2019-02-22 16:49”，这两个时间都在任务开始执行的时间（2019-02-22 16:42:00）与任务运行结束的时间（2019-02-22 16:49:26）之间，说明 SVM 预测用户是否挽留任务是按时执行并且执行成功的。

svm_activate	dir				2019-02-22 16:47	rwxrwxrwt	spark	hive
svm_prediction	dir				2019-02-22 16:49	rwxrwxrwt	spark	hive

图 7-44　svm_activate 和 svm_prediction 在 HDFS 中的更新结果

（5）用户画像

用户画像任务通过 Java 执行器实现，在 timer-job-java-executor 工程的 src/main/java/com/tipdm/java/jobs/user_profile 目录下新建目录 label，将 user_profile_project 工程下的 src/main/java/com/tipdm/java/label/LabelSQL.java 复制到 label 目录下，并在 label 目录下新建 SQLResolve 类，其代码如代码 7-14 所示。此外，在 timer-job-java-executor 工程的 src/main/resources/目录下创建 rules 目录，并把 user_profile_project 工程中 src/main/resources/rules/目录下的 consume_content.json 文件复制到 rules 目录下。

代码 7-14　新建 SQLResolve 类

```
package com.tipdm.java.chapter_3_9_3_user_profile;
import com.tipdm.engine.SparkYarnJob;
import com.tipdm.engine.engine.type.EngineType;
import com.tipdm.engine.model.Args;
import com.tipdm.engine.model.SubmitResult;
import com.tipdm.java.label.LabelSQL;
import java.util.HashMap;
import java.util.Map;
public class SQLResolve {
    private static String className = "com.tipdm.scala.chapter_3_9_3_user_
profile.SQLEngine";
    private static String applicationName = "SQLResolve";
    public static void main(String[] args) throws Exception {
        Map<String, String> labelTable = new HashMap<>();
        labelTable.put("消费内容", "user_profile.mmconsume_billevent_process");
        labelTable.put("电视消费水平", "user_profile.mmconsume_billevent_process");
        labelTable.put("宽带消费水平", "user_profile.mmconsume_billevent_process");
        labelTable.put("宽带产品带宽", "user_profile.order_index_process");
        labelTable.put("用户是否挽留","user_profile.svm_prediction");
        labelTable.put("销售品名称", "order_index_process");
        labelTable.put("业务品牌", "user_profile.mediamatch_usermsg_process");
        labelTable.put("电视入网程度", "user_profile.mediamatch_usermsg_process");
```

```
        labelTable.put("宽带入网程度", "user_profile.mediamatch_usermsg_process");
        int flag = 0;
        String[] arguments = new String[8];
        String sql = null;
        for (Map.Entry<String, String> entry : labelTable.entrySet()) {
            sql = LabelSQL.getLabel(entry.getKey(), entry.getValue());
            System.out.println("SQL: " + sql);
            if (flag == 0) {
                arguments[4] = "overwrite";
            } else {
                arguments[4] = "append";
            }
            arguments[0] = "LabelName: " + entry.getKey();
            arguments[1] = sql;
            arguments[2] = "user_label";
            arguments[3] = "rdbms";
            arguments[5] = "jdbc:mysql://192.168.111.75:3306/user_profile";
            arguments[6] = "root";
            arguments[7] = "root";
            Args innerArgs = Args.getArgs(applicationName,className,arguments,
EngineType.SPARK);
            SubmitResult submitResult = SparkYarnJob.run(innerArgs);
            SparkYarnJob.monitor(submitResult);
            flag++;
            // 间隔 10s 发起下一个任务
            Thread.sleep(1000 * 10L);
        }
        System.out.println("任务运行成功");
    }
}
```

重新编译打包 timer-job-java-executor 工程，将新生成的 JAR 包上传到 node4 节点的/root/xxl-job-user-profile 目录下，并参考代码 7-5 启动 timer-job-java-executor 工程。

工程启动后，在任务调度中心首页的任务管理模块中新增用户画像任务，如图 7-45 所示。其中，“执行器”设置为“java 执行器”；“运行模式”设置为“BEAN 模式”；“任务参数”设置为“com.tipdm. java.jobs.user_profile.label.SQLResolve”；“负责人”设置为“user_profile”；“任务描述”设置为“每个月 3 号凌晨 3:10:00 运行用户画像任务”；“Cron”设置为“0 10 3 3 * ? *”；“JobHandler”设置“javaJobHandler”。

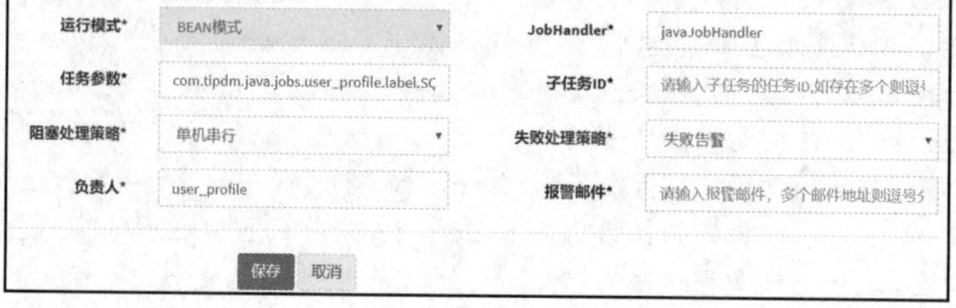

图 7-45　新增用户画像任务

图 7-45 所示的信息设置完成后，即可在任务管理列表中看到新增的任务，如图 7-46 所示。

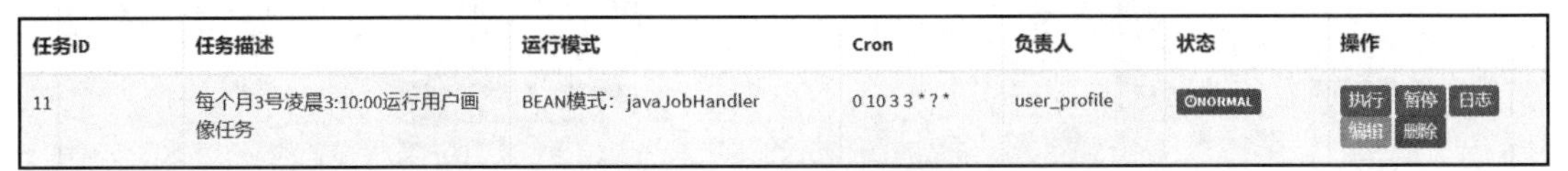

任务ID	任务描述	运行模式	Cron	负责人	状态	操作
11	每个月3号凌晨3:10:00运行用户画像任务	BEAN模式：javaJobHandler	0 10 3 3 * ? *	user_profile	NORMAL	执行 暂停 日志 编辑 删除

图 7-46 任务管理列表中新增的任务

为测试定时任务能否按时执行，将任务的执行时间改为每天 11:05:00，并在每天的 11:05:00 之后，在调度日志界面中查看任务是否按时启动并执行成功，如图 7-47 所示。

11	2019-02-22 11:05:00	成功	查看	2019-02-22 11:48:36	成功	无	执行日志

图 7-47 用户画像任务的执行结果

任务运行成功后会在 MySQL 的 user_profile 库中创建 user_label，并把用户画像标签结果保存在该表中，查看该表的对象信息，如图 7-48 所示。从图 7-48 中可以看到，表的创建时间为“2019-02-22 11:05:35”，最后的修改时间为“2019-02-22 11:47:15”，这个时间正好在任务的调度时间（2019-02-22 11:05:00）之后，说明任务是按时执行并执行成功的。

常规 DDL

名	值
名	user_label
数据库	user_profile
组名	
行	12620420
表类型	MyISAM
自动递增值	
行格式	Dynamic
修改日期	2019-02-22 11:47:15
创建日期	2019-02-22 11:05:35
检查时间	
索引长度	1.00 KB (1,024)
数据长度	572.70 MB (600,519,956)
最大数据长度	256.00 TB (281,474,976,710,655)
数据可用空间	0 bytes (0)
排序规则	utf8_general_ci
创建选项	
注释	

图 7-48 user_label 的对象信息

（6）实时统计订单信息

实时统计订单信息的定时任务分为启动 Kafka 生产者和启动 Spark Streaming 程序。其中，使用 Shell 执行器启动 Kafka 生产者，在任务调度中心首页的任务管理模块中新增产生实时订单实时流数据任务，如图 7-49 所示。其中，“执行器”设置为“shell 执行器”；“运行模式”设置为“GLUE 模式(Shell)”；“任务参数”设置为“/root/user_profile_project-1.0.jar com.tipdm.java.streaming.Kafka Producer /data/order.csv”；“负责人”设置为“user_profile”；“任务描述”设置为“每个月 4 号凌晨 1:10:00 运行产生实时订单实时流数据任务”；“Cron”设置为“0 10 1 4 * ? *”。

图 7-49　产生实时订单实时流数据任务

图 7-49 所示的信息设置完成后单击“保存”按钮，即可在任务管理列表中看到新增的任务，如图 7-50 所示。

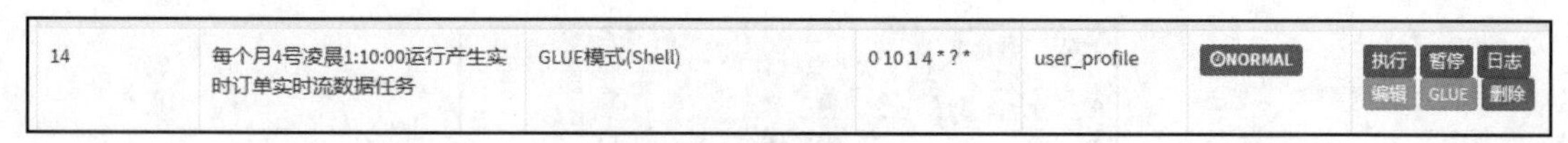

图 7-50　任务管理列表

单击图 7-50 中的“GLUE”按钮，编辑任务脚本，如代码 7-15 所示。

代码 7-15　产生实时订单实时流数据任务脚本

```
1  #!/bin/bash
2  #JAR 包的路径
3  jar_path=$1
4  #主类名称
5  MainClass=$2
6  #订单数据文件
7  file=$3
8  #Shell 脚本开始运行的时间
9  start_time=$(date +%Y-%m-%d-%H:%M:%S)
10 log_file=/root/qwm/log.out
11 echo -e "\nstart time : $start_time" >> $log_file
12 echo "program run day: $task_start_time">> $log_file
13 java -cp.:$jar_path -Djava.ext.dirs=/opt/cloudera/parcels/CDH-5.7.3-1.
   cdh5.7.3.p0.5/jars/ $MainClass $file
14 #Shell 脚本结束运行的时间
15 end_time=$(date +%Y-%m-%d-%H:%M:%S)
16 echo "end time:$end_time">> $log_file
17 exit 0
```

使用 Shell 执行器配置 Spark Streaming 实时统计订单信息任务，在任务调度中心首页的任务管理模块中新增 Spark Streaming 实时统计订单信息任务，如图 7-51 所示。其中，将“执行器”设置为“shell 执行器”；“运行模式”设置为“GLUE 模式(Shell)”；任务参数为“/root/user_ profile_ project-1.0.jar server3"；"负责人"设置为"user_profile"；"任务描述"设置为"每个月 4 号凌晨 1:10:00 运行实时统计订单信息任务"；"Cron"设置为"0 10 1 4 * ? *"。

更新任务

执行器*	shell执行器	任务描述*	每个月4号凌晨1:10:00运行实时统计订单
路由策略*	第一个	Cron*	0 10 1 4 * ? *
运行模式*	GLUE模式(Shell)	JobHandler*	请输入JobHandler
任务参数*	/root/user_profile_project-1.0.jar server3	子任务ID*	请输入子任务的任务ID,如存在多个则逗
阻塞处理策略*	单机串行	失败处理策略*	失败告警
负责人*	user_profile	报警邮件*	请输入报警邮件，多个邮件地址则逗号

保存　取消

图 7-51　Spark Streaming 实时统计订单信息任务

图 7-51 所示的信息设置完成后单击"保存"按钮，即可在任务管理列表中看到新增的任务，如图 7-52 所示。

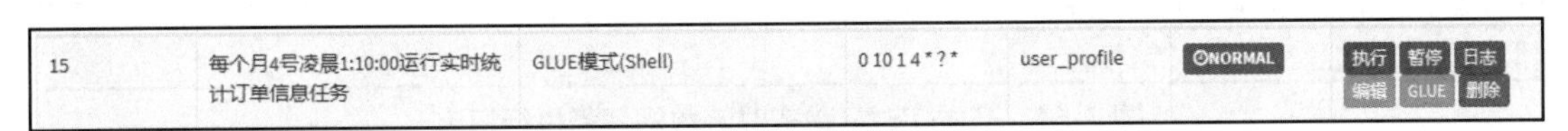

图 7-52　任务管理列表

单击图 7-52 中的"GLUE"按钮，编辑任务脚本，脚本内容如代码 7-16 所示。

代码 7-16　Spark Streaming 实时统计订单信息任务脚本

```
#!/bin/bash
#Shell 脚本开始运行的时间
start_time=$(date +%Y-%m-%d-%H:%M:%S)
#日志保存的路径
log_file=/root/qwm/log.out
echo -e "\nstart time : $start_time" >> $log_file
#将 order_index_1d 表的数据导入 Elasticsearch 集群
spark-submit  --class  com.tipdm.scala.chapter_3_7_streaming.KafakStream
--executor-memory 10g --total-executor-cores 20 --master spark://$2:7077
--jars /opt/cloudera/parcels/CDH-5.7.3-1.cdh5.7.3.p0.5/lib/spark/lib/* $1
#Shell 脚本结束运行的时间
end_time=$(date +%Y-%m-%d-%H:%M:%S)
echo "end time : $end_time" >> $log_file
```

为测试定时任务配置是否正确，将产生实时订单实时流数据任务的执行时间临时改为 2019-02-22 19:39:03，将 Spark Streaming 实时统计订单信息任务的执行时间临时改为 2019-02-22 19:39:07，2019-02-23 早上观察这两个任务的执行情况，查看实时统计订单信息任务日志，如图 7-53 所示。

任务ID	调度时间	调度结果	调度备注	执行时间	执行结果	执行备注	操作
15	2019-02-22 19:39:07	成功	查看			无	执行日志 终止任务
14	2019-02-22 19:39:03	成功	查看			无	执行日志 终止任务

图 7-53　实时统计订单信息任务日志

从图 7-53 中可以看到任务还在执行中，这是因为订单数据文件比较大，模拟读取该数据文件的速率较小，所以该任务执行时间比较长。可以查看执行日志的情况，单击产生实时订单实时流数据任务（任务 ID 为 14）的“执行日志”链接，该任务执行日志如图 7-54 所示。

```
50365413 T1 ku.Logging$class.warn WARN Produce request with correlation id 20076 failed due to [order,2]: org.apache.kafka.common.errors.NotLeaderForPartitionException
50365418 T1 ku.Logging$class.info Back off for 100 ms before retrying send. Remaining retries = 3
50365520 T1 ku.Logging$class.info Fetching metadata from broker BrokerEndPoint(0,192.168.111.76,9092) with correlation id 20077 for 1 topic(s) Set(order)
50365523 T1 ku.Logging$class.info Connected to 192.168.111.76:9092 for producing
50365530 T1 ku.Logging$class.info Disconnecting from 192.168.111.76:9092
50365533 T1 ku.Logging$class.info Disconnecting from node4:9092
50365534 T1 ku.Logging$class.info Disconnecting from node3:9092
50365543 T1 ku.Logging$class.info Connected to node4:9092 for producing
50373419 T1 ku.Logging$class.info Connected to node3:9092 for producing
index:10000,content:1687466;HE级;2015-07-17 13:25:45;整转赠送-CHC;数字电视;BO002327;整转包(月租26.5元);正常状态;NULL;NULL;NULL;2008-06-19 00:00:00;2008-07-17 18:39:20;2008-06-16 18:39:20;NULL;N;BY;正常
50665868 T1 ku.Logging$class.info Back off for 100 ms before retrying send. Remaining retries = 3
50665969 T1 ku.Logging$class.info Fetching metadata from broker BrokerEndPoint(2,192.168.111.78,9092) with correlation id 20212 for 1 topic(s) Set(order)
50665973 T1 ku.Logging$class.info Connected to 192.168.111.78:9092 for producing
50665979 T1 ku.Logging$class.info Disconnecting from 192.168.111.78:9092
50665981 T1 ku.Logging$class.info Disconnecting from node4:9092
50665981 T1 ku.Logging$class.info Disconnecting from node2:9092
50665982 T1 ku.Logging$class.info Disconnecting from node3:9092
50666015 T1 ku.Logging$class.info Connected to node2:9092 for producing
50675612 T1 ku.Logging$class.info Connected to node4:9092 for producing
50675661 T1 ku.Logging$class.info Connected to node3:9092 for producing
50708858 T1 ku.Logging$class.info Fetching metadata from broker BrokerEndPoint(1,192.168.111.77,9092) with correlation id 20237 for 1 topic(s) Set(order)
50708861 T1 ku.Logging$class.info Connected to 192.168.111.77:9092 for producing
50708863 T1 ku.Logging$class.info Disconnecting from 192.168.111.77:9092
50708864 T1 ku.Logging$class.info Disconnecting from node4:9092
50708866 T1 ku.Logging$class.info Disconnecting from node2:9092
50708866 T1 ku.Logging$class.info Disconnecting from node3:9092
50708868 T1 ku.Logging$class.info Connected to node2:9092 for producing
50716676 T1 ku.Logging$class.info Connected to node4:9092 for producing
50717027 T1 ku.Logging$class.info Connected to node3:9092 for producing
```

图 7-54　产生实时订单实时流数据任务执行日志

单击 Spark Streaming 实时统计订单信息任务（任务 ID 为 15）的“执行日志”链接，该任务执行日志如图 7-55 所示。

```
19/02/23 09:30:00 INFO scheduler.TaskSetManager: Starting task 0.0 in stage 27.0 (TID 81, node3, partition 0,NODE_LOCAL, 2076 bytes)
19/02/23 09:30:02 INFO storage.BlockManagerInfo: Added broadcast_27_piece0 in memory on node3:50329 (size: 1841.0 B, free: 530.0 MB)
19/02/23 09:30:03 INFO scheduler.TaskSetManager: Finished task 0.0 in stage 27.0 (TID 81) in 2563 ms on node3 (1/3)
19/02/23 09:30:04 INFO scheduler.TaskSetManager: Starting task 1.0 in stage 27.0 (TID 82, node1, partition 1,ANY, 2076 bytes)
19/02/23 09:30:04 INFO scheduler.TaskSetManager: Starting task 2.0 in stage 27.0 (TID 83, node1, partition 2,ANY, 2076 bytes)
19/02/23 09:30:04 INFO storage.BlockManagerInfo: Added broadcast_27_piece0 in memory on node1:56857 (size: 1841.0 B, free: 530.0 MB)
19/02/23 09:30:04 INFO storage.BlockManagerInfo: Added broadcast_27_piece0 in memory on node1:35624 (size: 1841.0 B, free: 530.0 MB)
19/02/23 09:30:05 INFO scheduler.TaskSetManager: Finished task 2.0 in stage 27.0 (TID 83) in 1681 ms on node1 (2/3)
19/02/23 09:30:05 INFO scheduler.TaskSetManager: Finished task 1.0 in stage 27.0 (TID 82) in 1685 ms on node1 (3/3)
19/02/23 09:30:05 INFO scheduler.TaskSchedulerImpl: Removed TaskSet 27.0, whose tasks have all completed, from pool
19/02/23 09:30:05 INFO scheduler.DAGScheduler: ResultStage 27 (foreach at KafkaStream.scala:93) finished in 5.175 s
19/02/23 09:30:05 INFO scheduler.DAGScheduler: Job 27 finished: foreach at KafkaStream.scala:93, took 5.464381 s
19/02/23 09:30:05 INFO scheduler.JobScheduler: Finished job streaming job 1550885400000 ms.0 from job set of time 1550885400000 ms
19/02/23 09:30:05 INFO kafka.KafkaRDD: Removing RDD 26 from persistence list
新增订单数: 339
新增有效订单数: 58
有效订单总数:  1665
总订单数: 13895
新增订单营业额: 3449.0
订单总营业额: 193259.55
partition : 1 fromOffset:  4520 untilOffset: 4633
partition : 2 fromOffset:  4519 untilOffset: 4632
partition : 0 fromOffset:  4517 untilOffset: 4630
19/02/23 09:30:05 INFO scheduler.JobScheduler: Total delay: 5.777 s for time 1550885400000 ms (execution: 5.704 s)
19/02/23 09:30:05 INFO storage.BlockManager: Removing RDD 26
19/02/23 09:30:05 INFO scheduler.ReceivedBlockTracker: Deleting batches ArrayBuffer()
19/02/23 09:30:05 INFO scheduler.InputInfoTracker: remove old batch metadata: 1550881800000 ms
19/02/23 09:39:17 INFO storage.BlockManagerInfo: Removed broadcast_27_piece0 on 192.168.111.78:57788 in memory (size: 1841.0 B, free: 530.0 MB)
19/02/23 09:39:17 INFO storage.BlockManagerInfo: Removed broadcast_27_piece0 on node1:35624 in memory (size: 1841.0 B, free: 530.0 MB)
19/02/23 09:39:17 INFO storage.BlockManagerInfo: Removed broadcast_27_piece0 on node1:56857 in memory (size: 1841.0 B, free: 530.0 MB)
19/02/23 09:39:17 INFO storage.BlockManagerInfo: Removed broadcast_27_piece0 on node3:50329 in memory (size: 1841.0 B, free: 530.0 MB)
19/02/23 09:39:17 INFO spark.ContextCleaner: Cleaned accumulator 32
```

图 7-55　Spark Streaming 实时统计订单信息任务执行日志

从图 7-55 中可以看到，Spark Streaming 实时统计订单信息任务的执行时间为 2019-02-23 09:30:00。此时，观察 Redis 的存储情况，如图 7-56 所示。

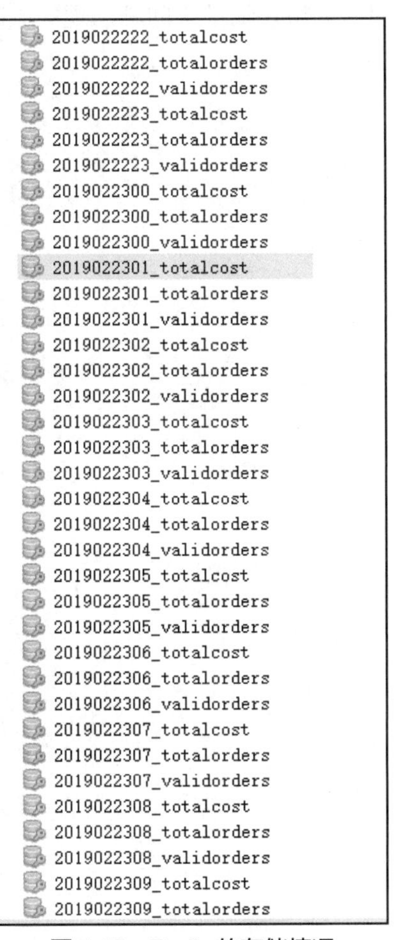

图 7-56　Redis 的存储情况

从图 7-56 中可以看到 Redis 的键的前缀时间已到 2019-02-23 09:00:00。

根据图 7-54、图 7-55 和图 7-56 的结果，可以证明实时统计订单信息任务的定时调度配置是正确的。

所有任务测试成功后，需要将每个任务的调度时间及参数根据调度策略修改回原状态，并等待任务执行，在下一个月的 3 日之后查看所有任务的执行情况。

至此，项目实施部署已完成，项目的维护人员需要定期检查及维护部署的系统，尤其需要注意检查每个月的定时任务是否按时并成功执行，如果出现问题，则需要立刻反馈给相关人员并及时进行解决。

小结

本章介绍了广电大数据用户画像项目的任务调度实现。项目的任务调度实现使用了 XXL-JOB 分布式调度平台模拟项目上线环境，先将 XXL-JOB 部署在节点上并设置了 Shell 执行器和 Java 执行器，再把通过 Shell 脚本实现的任务和通过 Java 调度的任务一一部署到

XXL-JOB 中，并设置了任务定时运行的时间。配置完成且任务执行成功后，只需在相应的时间点查看任务运行的状态、结果，确保项目正常运行即可。项目运行过程也伴随着维护过程，如果项目运行时出现问题，需仔细排查问题出现的原因，根据实际情况采取适当的措施。

第 8 章 基于 TipDM 大数据挖掘建模平台实现广电大数据用户画像

本书介绍了使用 Spark 来实现广电大数据用户画像，本章将介绍另一种工具——TipDM 大数据挖掘建模平台，通过该平台实现广电大数据用户画像。相较于传统的 Spark 解释器，TipDM 大数据挖掘建模平台具有流程化、去编程化、拖曳式等特点，可满足不懂编程的用户使用数据分析技术的需求。TipDM 大数据挖掘建模平台帮助读者更加便捷地掌握数据分析相关技术的操作，落实科教兴国战略、人才强国战略、创新驱动发展战略。

学习目标

（1）了解 TipDM 大数据挖掘建模平台的相关概念和特点。

（2）熟悉使用 TipDM 大数据挖掘建模平台配置广电大数据用户画像的总体流程。

（3）掌握使用 TipDM 大数据挖掘建模平台获取数据的方法。

（4）掌握使用 TipDM 大数据挖掘建模平台进行数据探索的操作方法。

（5）掌握使用 TipDM 大数据挖掘建模平台进行数据去重、数据筛选、表连接等操作的方法。

（6）掌握使用 TipDM 大数据挖掘建模平台构建模型的方法。

8.1 平台简介

TipDM 大数据挖掘建模平台是由广东泰迪智能科技股份有限公司自主研发的、面向大数据挖掘项目的工具。此平台使用 Java 语言开发，采用 B/S（Browser/Server，浏览器/服务器）结构，用户不需要下载客户端，可直接通过浏览器进行访问。TipDM 大数据挖掘建模平台具有支持多种语言、操作简单等特点，以流程化的方式对数据进行输入/输出、统计分析，与数据预处理、分析与建模等环节进行连接，从而实现大数据分析的目的。TipDM 大数据挖掘建模平台界面如图 8-1 所示。

本章将以广电大数据用户画像项目为例，介绍如何使用平台实现项目。在介绍之前，需要引入该平台的几个概念。

（1）组件：对建模过程涉及的输入/输出、数据探索及预处理、建模、模型评估等算法分别进行封装，每一个封装好的算法模块称为组件。

（2）工程：为实现某一数据分析目标，将各组件通过流程化的方式连接起来，整个数据分析流程称为一个工程。

（3）模板：用户可以将配置好的实训通过模板的方式分享给其他用户，其他用户可以使用该模板，创建一个无须配置算法便可运行的实训。

图 8-1　TipDM 大数据挖掘建模平台界面

TipDM 大数据挖掘建模平台主要有以下几个特点。

（1）平台算法基于 Python、R 以及 Hadoop/Spark 分布式引擎进行数据分析。Python、R 及 Hadoop/Spark 是目前十分流行的用于数据分析的语言，高度契合行业需求。

（2）用户可在没有 Python、R 或者 Hadoop/Spark 编程基础的情况下，使用直观的拖曳式图形界面构建数据分析流程，无须编程。

（3）平台提供公开可用的数据分析示例工程，可一键创建、快速运行工程；支持在线预览挖掘流程中每个节点的结果。

（4）平台包含 Python、Spark、R 三种编程语言的算法包，用户可以根据实际运用灵活选择不同的语言进行数据挖掘建模。

下面将对该平台的【模板】【数据空间】【我的项目】【系统组件】和【个人组件】5 个模块进行介绍，并对该平台的访问方式进行介绍。

8.1.1　模板

登录平台后，用户即可看到【模板】模块中系统提供的示例工程（模板），如图 8-2 所示。

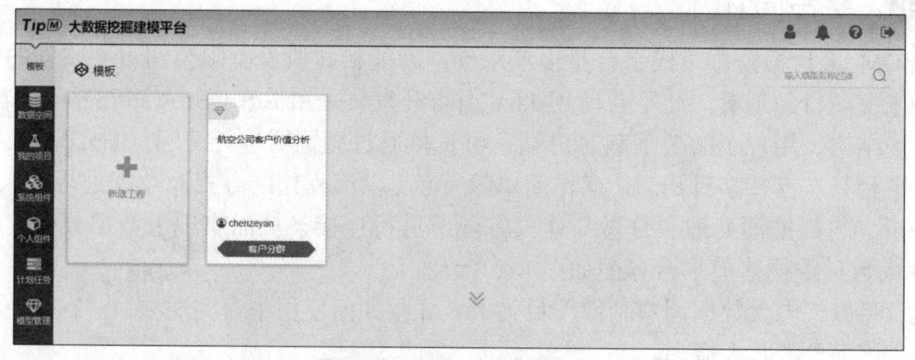

图 8-2　示例工程（模板）

【模板】模块主要用于标准大数据分析案例的快速创建和展示。通过【模板】模块，用户可以创建一个无须导入数据及配置参数即可快速运行的工程。同时，每一个模板的创建者都具有模板的所有权，能够对模板进行管理。用户可以将自己搭建的数据分析工程生成为模板，并使其显示在【模板】模块中，供其他用户一键创建。

8.1.2　数据空间

【数据空间】模块主要用于数据分析工程中的数据导入与管理，根据具体情况，用户可设置【数据来源】为【文件】或【数据库】。【文件】支持从本地导入任意类型的数据，如图 8-3 所示；【数据库】支持从 DB2、SQL Server、MySQL、Oracle、PostgreSQL 等常用关系数据库中导入数据，如图 8-4 所示。

新增数据集
* 名称
* 有效期（天）　到期后，系统将会自动清除关联的数据文件
* 描述　添加一些简短的说明　0/140
数据来源　文件　数据库　访问权限　私有　公开
数据文件　多个小文件　大数据集
将文件拖到此，或点击上传　可上传20个文件，总大小不超过500MB
暂无待上传文件
取消　确定

图 8-3　【数据来源】为【文件】

新增数据集
* 名称
* 有效期（天）　到期后，系统将会自动清除关联的数据文件
* 描述　添加一些简短的说明　0/140
数据来源　文件　数据库　访问权限　私有　公开
* URL　请输入数据库连接地址
* 用户名
* 密码
* SQL　请输入取数SQL
取消　确定

图 8-4　【数据来源】为【数据库】

8.1.3 我的项目

【我的项目】模块主要用于数据分析流程化的创建与管理，如图 8-5 所示。通过【工程】模块，用户可以创建空白工程，进行数据分析工程的配置，将数据输入/输出、数据预处理、建模、模型评估等环节通过流程化的方式连接起来，实现数据分析的目的。对于完成效果优秀的工程，可以将其保存为模板，让其他使用者学习和借鉴。

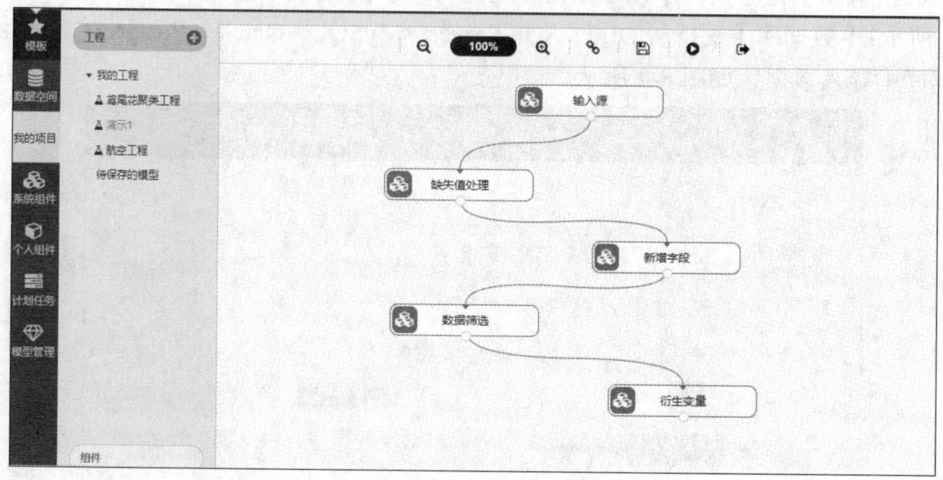

图 8-5 【我的项目】模块

8.1.4 系统组件

【系统组件】模块主要用于大数据分析内置常用算法组件的管理，提供 Python、R、Spark 算法包，如图 8-6 所示。

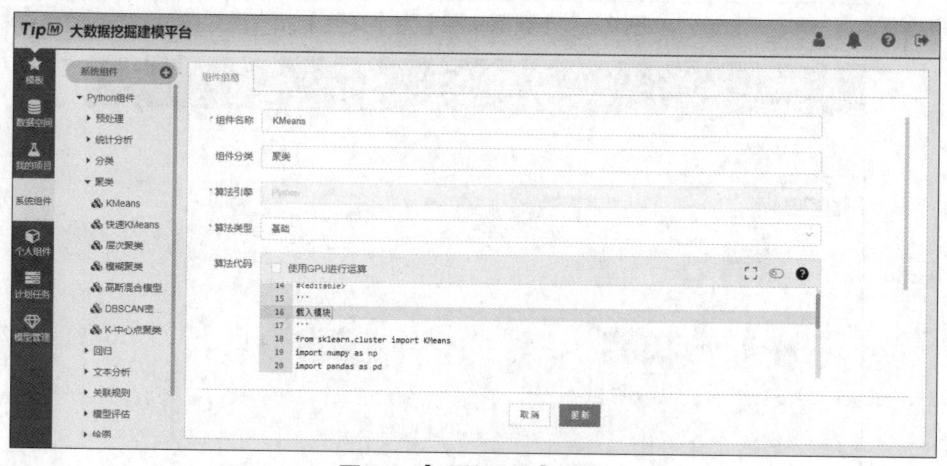

图 8-6 【系统组件】模块

Python 算法包提供 11 类算法，具体如下。

（1）【统计分析】类包括因子分析、全表统计、正态性检验、相关性分析、卡方检验、主成分分析、纯随机性检验和平稳性检验等。

（2）【数据预处理】类包括数据标准化、缺失值处理、表堆叠、数据筛选、类型转换、修改列名、特征构造、数据集划分、主键合并、数据离散化、排序、频数统计、记录去重和分组聚合等。

（3）【脚本】类包括 Python 脚本和 SQL 脚本等。

（4）【分类】类包括朴素贝叶斯、支持向量机、CART 分类树、逻辑回归、多层感知神经网络和最近邻分类等。

（5）【聚类】类包括层次聚类、DBSCAN 密度聚类和 K 均值聚类等。

（6）【回归】类包括 CART 回归树、线性回归、支持向量回归和最近邻回归等。

（7）【时间序列】类包括 ARIMA 模型等。

（8）【关联规则】类包括 Apriori 关联规则和 FP-Growth 关联规则等。

（9）【文本挖掘】类包括去除字符、分词与词性标注、TF-IDF（词向量化）、Hash Trick、word2vec（词向量化）、doc2vec（词向量化）、TF-IDF（关键词提取）、word2vec（关键词提取）、doc2vec（关键词提取）和主成分分析等。

（10）【深度学习】类包括 LeNet-5、AlexNet、VGG-16、LSTM 和 GRU 等。

（11）【画图】类包括柱状图、折线图、散点图、饼图和词云图等。

Spark 算法包提供 7 类算法，具体如下。

（1）【数据预处理】类包括记录去重、记录选择、数据映射、数据反映射、数据划分、SQL 脚本、缺失值处理、数据标准化、特征构造、表连接、表堆叠、独热编码和数据离散化等。

（2）【统计分析】类包括行列数目统计、频数统计、全表统计、相关性分析、卡方检验和主成分分析等。

（3）【分类】类包括逻辑回归、决策树分类、梯度提升树分类、朴素贝叶斯分类、随机森林分类、线性支持向量机和多层感知神经网络等。

（4）【聚类】类包括 K 均值、二分 K 均值聚类、混合高斯模型和文档主题生成模型等。

（5）【回归】类包括线性回归、广义线性回归、决策树回归、梯度提升树回归、随机森林回归和保序回归等。

（6）【智能推荐】类包括交替最小二乘法（Alternating Least Square，ALS）推荐等。

（7）【关联规则】类包括 FP-Growth 等。

R 语言算法包提供 8 类算法，具体如下。

（1）【统计分析】类包括卡方检验、因子分析、主成分分析、相关分析、正态性检验、全表统计、平稳性检验和纯随机性检验等。

（2）【数据预处理】类包括缺失值处理、异常值处理、表堆叠、主键合并、数据标准化、记录去重、数据离散化、排序、数据集划分、频数统计、新增序列、字符串拆分、字符串拼接、分组、修改类型、修改列名、特征构造、SQL 脚本和 R 脚本等。

（3）【分类】类包括朴素贝叶斯、CART 分类树、C4.5 分类树、反向传播（Back Propagation，BP）神经网络、最近邻分类、支持向量机和逻辑回归等。

（4）【聚类】类包括 K 均值聚类、DBSCAN 密度聚类和系统聚类等。

（5）【回归】类包括 CART 回归树、C4.5 回归树、线性回归、岭回归和最近邻回归等。

（6）【时间序列】类包括 ARIMA 模型、GM(1,1)灰度预测和指数平滑等。

（7）【关联规则】类包括 Apriori 关联规则等。

（8）【文本挖掘】类包括 Jieba 分词、去除停用词、余弦相似度、情感词定位、情感方向修正和 LDA 主题模型等。

8.1.5 个人组件

【个人组件】模块主要为了满足用户的个性化需求而设置。用户在使用过程中，可根据自己的需求定制算法，以便使用。目前，【个人组件】支持通过 Python 和 R 语言进行定制，如图 8-7 所示。

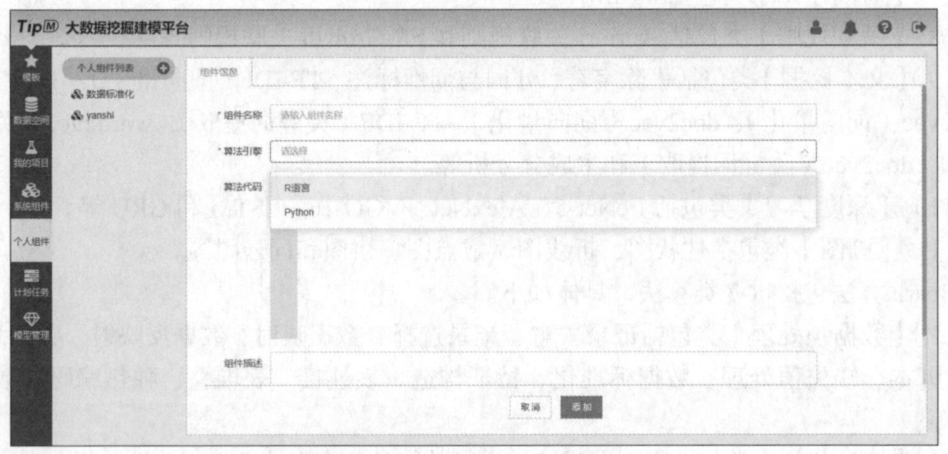

图 8-7 【个人组件】的定制

8.1.6 访问 TipDM 大数据挖掘建模平台的方式

本书读者可通过关注微信公众号获取平台的访问方式，具体步骤如下。

（1）微信搜索公众号“泰迪学社”或“Tip Data Mining”，关注公众号。

（2）关注公众号后，回复“建模平台”，获取 TipDM 大数据挖掘建模平台的访问方式。

8.2 广电大数据用户画像开发

本节以广电大数据用户画像为例，在 TipDM 大数据挖掘建模平台上配置相应工程，展示几个主要流程的配置过程。流程的详细配置过程可通过访问平台进行查看。

在 TipDM 大数据挖掘建模平台上配置广电大数据用户画像项目，主要包括以下 4 个步骤。

（1）将广电公司的数据导入 TipDM 大数据挖掘建模平台。

（2）对数据进行探索分析。

（3）对数据进行数据去重和数据筛选等操作。

（4）利用 SVM 算法建立分类模型，预测用户是否值得挽留，并实现用户挽留标签计算。

数据探索总流程如图 8-8 所示。

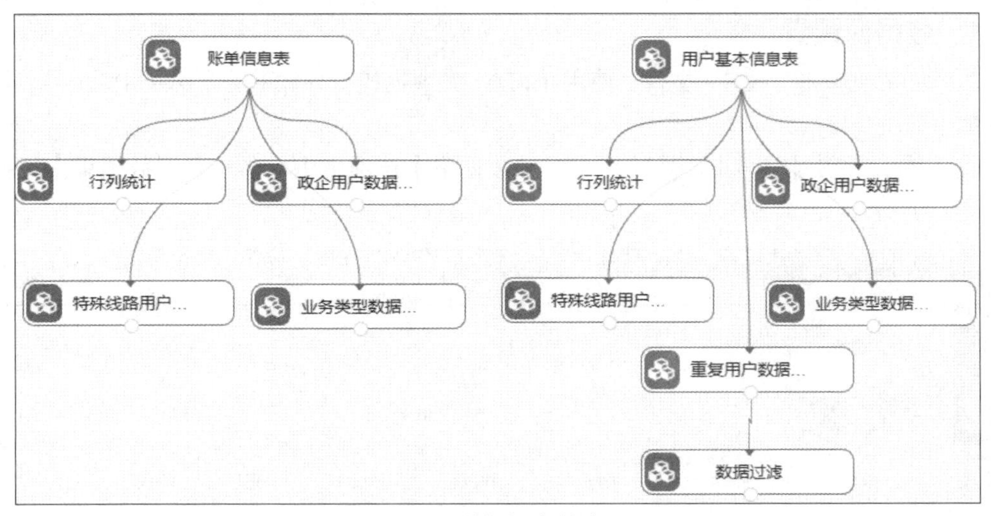

图 8-8　数据探索总流程

数据处理总流程如图 8-9 所示。

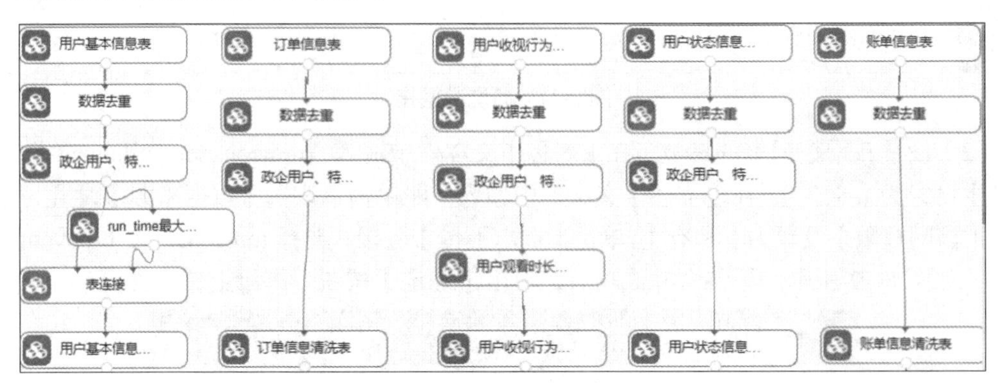

图 8-9　数据处理总流程

用户画像总流程如图 8-10 所示。

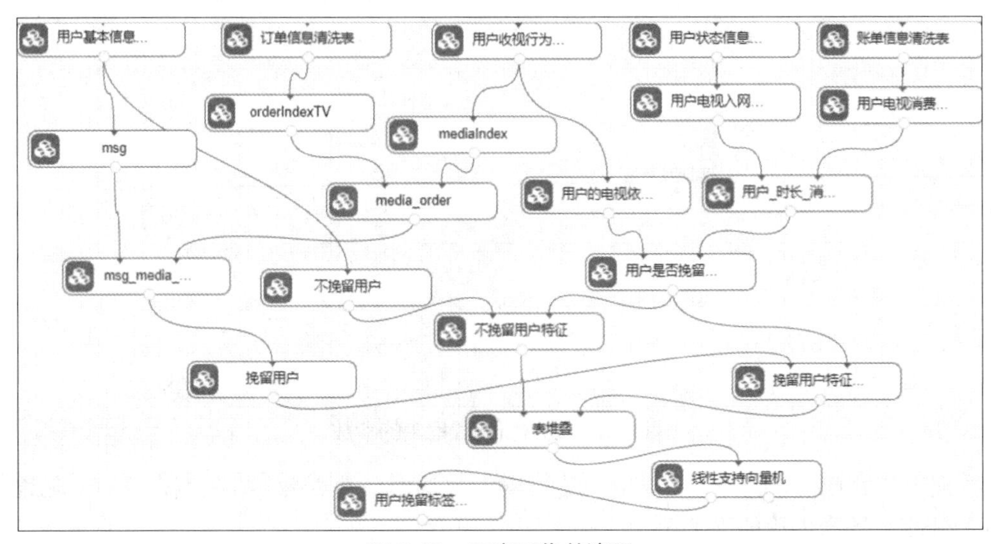

图 8-10　用户画像总流程

8.2.1　数据源配置

本章的数据来源于 CSV 文件，使用 TipDM 大数据挖掘建模平台导入 CSV 文件，具体步骤如下。

（1）单击【数据空间】模块，在【我的数据集】选项卡中单击【新增数据集】按钮，如图 8-11 所示。

图 8-11　新增数据集

（2）设置新增数据集的参数。在【名称】文本框中输入“mmconsume_billevents”，【有效期】设置为【永久】，在【描述】文本框中输入“账单信息表”，【数据来源】设置为【文件】，【访问权限】设置为【私有】，单击【点击上传】链接，选择 mmconsume_billevents.csv 文件，如图 8-12 所示。等待合并成功后，单击【确定】按钮，即可上传文件。

图 8-12　设置新增数据集的参数

数据上传完成后，新建一个名称为【广电大数据用户数据探索】的空白工程，配置【输入源】组件，具体步骤如下。

（1）在【工程】中的【组件】栏中找到【系统组件】→【内置组件】→【输入/输出】类。拖曳【输入/输出】→【输入源】组件到工程画布中。

（2）单击画布中的【输入源】组件，在工程画布右侧【参数配置】栏的【数据集】文本框中输入“mmconsume_billevents”，在弹出的下拉列表中选择【mmconsume_billevents】选项，如图 8-13 所示。右键单击【输入源】组件，选择【重命名】选项，输入“账单信息表”，如图 8-14 所示。

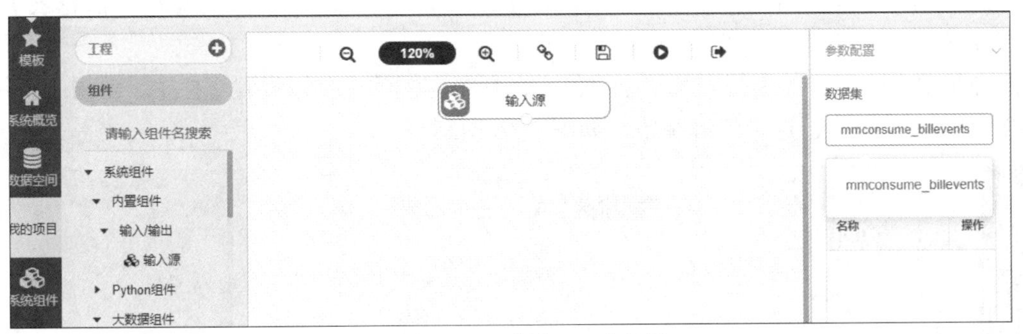

图 8-13　配置【输入源】组件

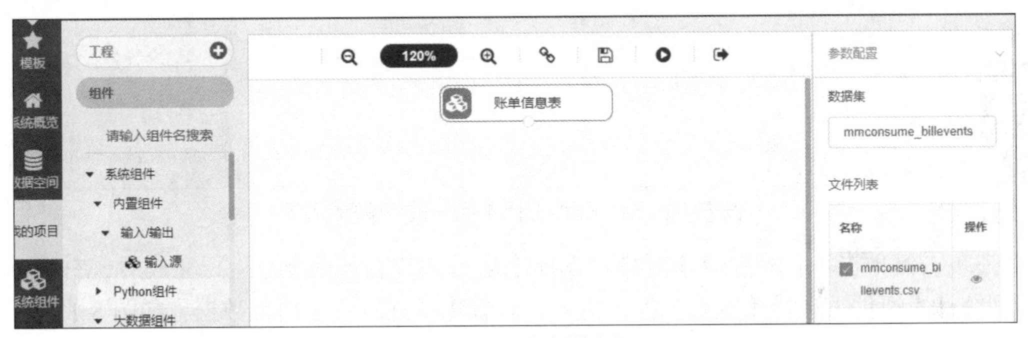

图 8-14　重命名输入源

（3）单击画布中的【账单信息表】组件，在工程画布右侧的【参数配置】栏中，单击【文件列表】选项组中的👁图标，查看数据集明细，如图 8-15 所示。

数据集明细

mmconsume_billevents.csv

35425485 Byte

文件内容　　　　最多显示前50列

terminal_no	phone_no	fee_code	year_month	owner_name	owner_code
1410005953	4477294	0B	2018-04-30 16:...	HC级	00
1410000342	4479074	0Y	2017-12-31 16:...	HC级	NULL
1410000342	4479074	0Y	2018-03-31 16:...	HC级	00
1410038590	4535578	0Y	2018-02-28 16:...	HC级	00
1410047737	4539813	0T	2017-12-31 16:...	HC级	NULL
1410047737	4539813	0Y	2018-06-30 16:...	HC级	00

图 8-15　查看数据集明细

8.2.2 数据探索

本小节介绍的数据探索主要是对账单信息表进行数据总体探索和异常数据探索。

1. 数据总体探索

对账单信息表进行基本的探索分析，查看数据中的记录数、字段数、缺失值个数以及重复值个数，具体步骤如下。

（1）拖曳【大数据组件】→【行列统计】组件到工程画布中，并与【账单信息表】组件相连接。

（2）单击画布中的【行列统计】组件，在画布右侧的【字段设置】栏中，单击【特征】选项组中的 图标，选中全部字段，如图 8-16 所示。

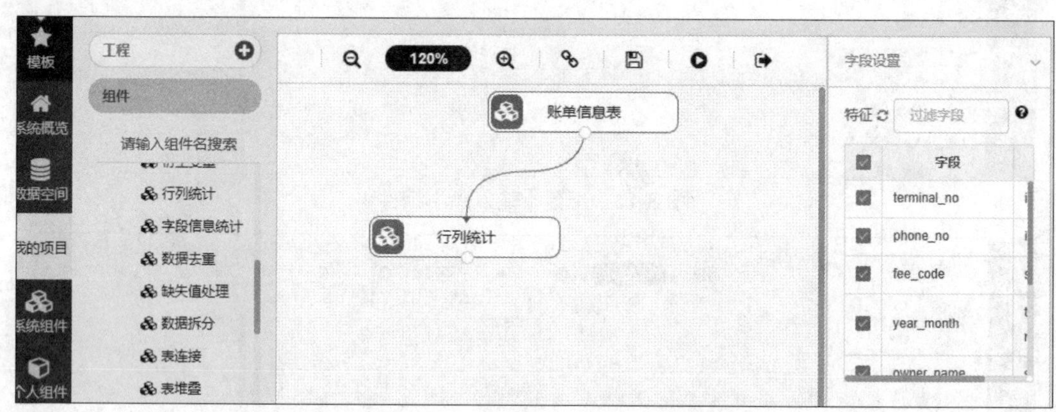

图 8-16　对【行列统计】组件进行字段设置

（3）右键单击【行列统计】组件，选择【运行该节点】选项。运行完成后，右键单击【行列统计】组件，选择【查看数据】选项，查看【行列统计】组件数据，如图 8-17 所示。

预览数据

RowAndCol	count
RowNumber	439158
ColNumber	9
NullNumber	108907
RepeatNumber	10683

图 8-17　查看【行列统计】组件数据

由图 8-17 可知，账单信息表中的记录数为 439158、字段数为 9、缺失值个数为 108907、重复值个数为 10683。

2. 异常数据探索

由于广电用户主要来自家庭用户，需要探索账单信息表中是否存在政企用户及其存在的数量，具体步骤如下。

（1）拖曳【大数据组件】→【预处理】→【分组聚合】组件到工程画布中，并与【账单信息表】组件相连接。

（2）单击【特征】选项组中的图标，选中全部字段。单击【分组主键】选项组中的图标，选中“owner_name”字段。

（3）右键单击【分组聚合】组件，选择【重命名】选项，输入“政企用户数据探索”。单击画布中的【政企用户数据探索】组件，在工程画布右侧的【字段设置】栏中，在【聚合函数】下拉列表框中选择【count】选项，如图 8-18 所示。

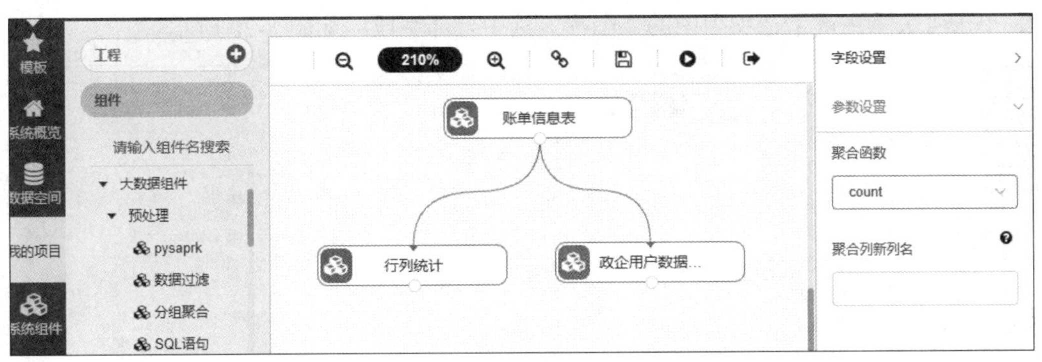

图 8-18　对【政企用户数据探索】组件进行字段设置

（4）右键单击【政企用户数据探索】组件，选择【运行该节点】选项。运行完成后，右键单击【政企用户数据探索】组件，选择【查看数据】选项，查看【政企用户数据探索】组件数据，如图 8-19 所示。

预览数据

owner_name	count
EA级	7863
HC级	16133563
HD级	5
HA级	1040
EE级	747042
HE级	4185291
HB级	5010
EB级	17559

图 8-19　查看【政企用户数据探索】组件数据

由图 8-19 可以看出，数据中有 owner_name 为“EA 级”“EE 级”“EB 级”的政企用户记录。政企用户属于异常数据，后续数据预处理中需要清洗 owner_name 为“EA 级”“EB 级”“EC 级”“ED 级”“EE 级”的政企用户数据。

8.2.3　数据处理

本小节介绍的数据处理主要是对账单信息表进行数据去重和数据筛选操作。

1. 数据去重

通过数据总体探索发现，账单信息表中存在重复记录数，需要对数据进行去重操作，具体步骤如下。

（1）拖曳【大数据组件】→【预处理】→【数据去重】组件到工程画布中，并与【账单信息表】组件相连接。

（2）在画布右侧的【字段设置】栏中，单击【特征】选项组中的图标，选中全部字段；单击【去重主键】选项组中的图标，选中全部字段，如图 8-20 所示。

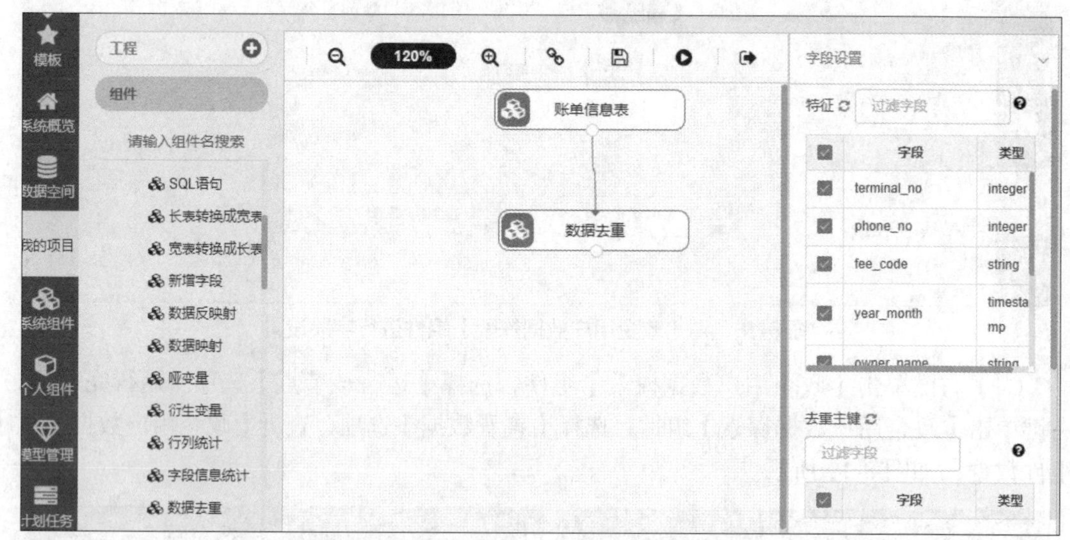

图 8-20　对【数据去重】组件进行字段设置

（3）右键单击【数据去重】组件，选择【运行该节点】选项，以运行该组件。

2. 数据筛选

通过异常数据探索发现，数据中存在异常数据，需要删除 owner_name 为 EA 级、EB 级、EC 级、ED 级、EE 级的数据，删除 owner_code 为 02、09、10 的数据，并保留 sm_name 为珠江宽频、数字电视、互动电视、甜果电视的数据，具体步骤如下。

（1）拖曳【大数据组件】→【预处理】→【数据过滤】组件到工程画布中，并与【数据去重】组件相连接。

（2）单击【特征】选项组中的图标，选中全部字段。

（3）右键单击【数据过滤】组件，选择【重命名】选项，输入“账单信息清洗表”。单击【账单信息清洗表】组件，在画布右侧的【过滤条件 1】栏中，单击【设置过滤条件列】选项组中的图标，选择【owner_name】选项，在【比较运算符】下拉列表框中选择【not in】选项，在【具体比较值】文本框中输入“EA 级,EB 级,EC 级,ED 级,EE 级”，如图 8-21 所示。

（4）单击画布右侧的【过滤条件 2】栏，在【逻辑运算符】下拉列表框中选择【and】选项，单击【设置过滤条件列】选项组中的图标，选择【owner_code】选项，在【比较运算符】下拉列表框中选择【not in】选项，在【具体比较值】文本框中输入“02,09,10”。

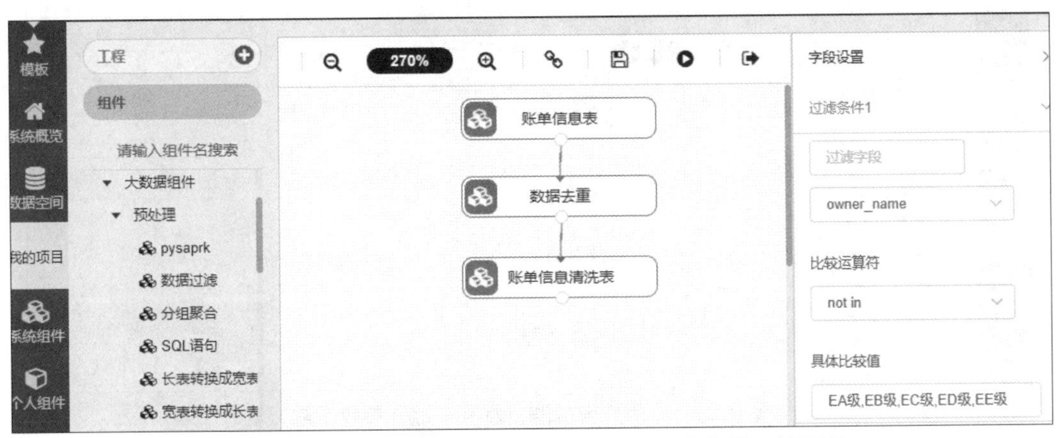

图 8-21 对【账单信息清洗表】组件进行过滤条件 1 的设置

（5）单击画布右侧的【过滤条件 3】栏，在【逻辑运算符】下拉列表框中选择【and】选项，单击【设置过滤条件列】选项组中的图标，选择【sm_name】选项，在【比较运算符】下拉列表框中选择【not in】选项，在【具体比较值】文本框中输入“珠江宽频,数字电视,互动电视,甜果电视”。

（6）运行【数据过滤】组件。

8.2.4 用户画像

本小节介绍的用户画像实现流程，包括利用 SVM 预测用户是否挽留和进行用户挽留标签计算。

1. 预测用户是否挽留

因为业务数据中没有一份可以直接用来预测用户是否挽留的数据，所以需要先利用业务数据构建一份适用于预测用户是否挽留的数据集，再利用 SVM 预测用户是否挽留。

（1）用户电视消费水平数据

用户电视消费水平用于计算用户平均每个月花费多少钱购买电视产品，这个特征值可根据 mmconsume_billevents 统计用户总的消费金额并除以 3 得到；之所以除以 3，是因为每次计算都是取 mmconsume_billevents 数据当前时间之前 3 个月的数据，具体步骤如下。

① 拖曳【大数据组件】→【预处理】→【SQL 语句】组件到工程画布中，并与【账单信息清洗表】组件相连接。

② 右键单击【SQL 语句】组件，选择【重命名】选项，输入“用户电视消费水平”。单击【用户电视消费水平】组件，在画布右侧【参数设置】栏的【sql 语句】文本框中输入“select phone_no, sum(should_pay)/3 consume from data where sm_name not like '%珠江宽频%' group by phone_no”，如图 8-22 所示。

③ 右键单击【用户电视消费水平】组件，选择【运行该节点】选项。运行完成后，右键单击【用户电视消费水平】组件，选择【查看数据】选项，查看【用户电视消费水平】组件数据，如图 8-23 所示。

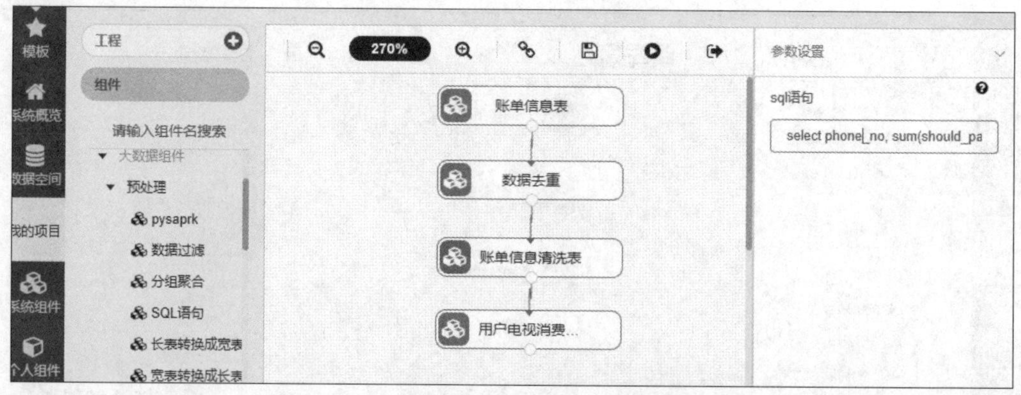

图 8-22 【用户电视消费水平】组件参数设置

预览数据

phone_no	consume
3496992	92.33333333333333
5085468	138.66666666666666
5362826	139.66666666666666
4001202	115.33333333333333
4002611	138.66666666666666
3365786	62.333333333333336
3336747	62.333333333333336
2413420	62.333333333333336
2330761	8.333333333333334

图 8-23 查看【用户电视消费水平】组件数据

由图 8-23 可知，字段“consume”表示用户平均每个月的消费金额。

（2）用户电视入网时长与用户电视消费水平数据

将用户的电视入网时长数据与用户电视消费水平数据进行表连接，以获得每个用户的电视入网时长与电视消费水平数据，具体步骤如下。

① 拖曳【大数据组件】→【预处理】→【表连接】组件到工程画布中，并分别与【用户电视入网时长】【用户电视消费水平】组件相连接。其中，【用户电视入网时长】组件的数据来源如下：在用户状态信息变更表中，计算当前时间与每个用户的 run_time 字段最大值的差值，将该差值作为每个用户的电视入网时长数据，该流程的详细配置过程可通过平台进行查看。

② 右键单击【表连接】组件，选择【重命名】选项，输入“用户_时长_消费_表连接”。单击【用户_时长_消费_表连接】组件，界面右侧将出现工程画布，在工程画布的【字段设置】栏中单击【左表特征】选项组中的 图标，选中全部字段；单击【右表特征】选项组中的 图标，选中全部字段；单击【左表主键】选项组中的 图标，选中“phone_no”字段；单击【右表主键】选项组中的 图标，选中“phone_no”字段。

③ 在工程画布右侧的【参数设置】的【选择连接方式】下拉列表框中选择【inner】选项，如图 8-24 所示。

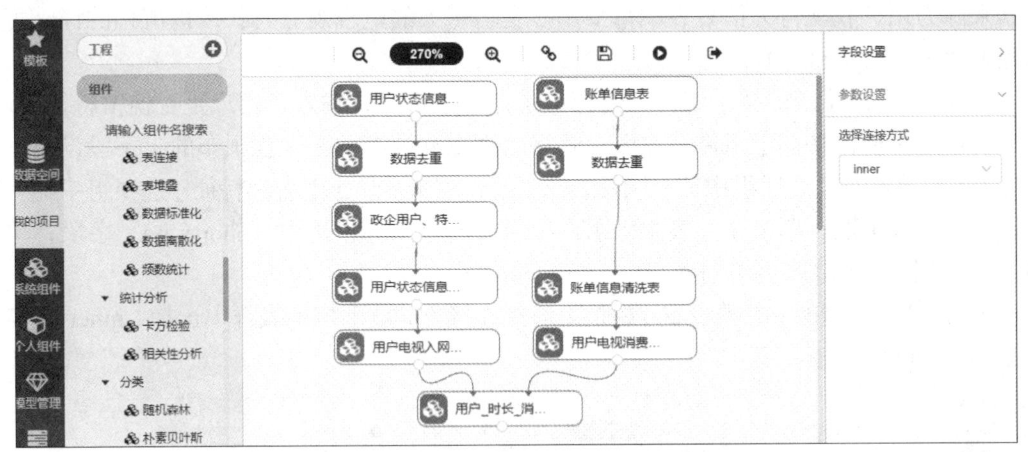

图 8-24　【用户_时长_消费_表连接】组件参数设置

④ 右键单击【用户_时长_消费_表连接】组件，选择【运行该节点】选项。运行完成后，右键单击【用户_时长_消费_表连接】组件，选择【查看数据】选项，查看【用户_时长_消费_表连接】组件数据，如图 8-25 所示。

预览数据

phone_no	join_time	consume
3262151	6.3455837441666665	62.333333333333336
5387763	2.5444209233333335	102.22666666666667
5106563	2.0968839916666666	62.333333333333336
5362826	3.0251897591666665	139.66666666666666
2751723	2.224044205833333	35.833333333333336
3071753	3.2742299816666667	74.83333333333333
2065841	5.777948744166667	98.17333333333333
5088571	8.761692055	111.33333333333333
2385685	11.333333333333334	74.83333333333333

图 8-25　查看【用户_时长_消费_表连接】组件数据

由图 8-25 可知，已对用户的电视入网时长与用户电视消费水平进行了表连接，其中，字段“join_time”表示用户从开户时间到当前时间的时长，字段“consume”表示用户平均每个月的消费金额。

（3）用户是否挽留特征数据

将用户的电视依赖度数据与用户的电视入网时长、电视消费水平数据进行表连接，获得每个用户的电视依赖度、电视入网时长与电视消费水平这 3 个特征，并将其作为判断用户是否挽留的特征，具体步骤如下。

① 拖曳【大数据组件】→【预处理】→【表连接】组件到工程画布中，并与【用户的电视依赖度】【用户_时长_消费_表连接】组件相连接。其中，【用户的电视依赖度】组件的数据来源为用户收视行为信息表中每个用户平均每天的电视观看时长，该流程的详细配置过程可通过平台进行查看。

② 右键单击【表连接】组件，选择【重命名】选项，输入“用户是否挽留特征”。单击【用户是否挽留特征】组件，界面右侧将出现工程画布，在工程画布的【字段设置】栏中单击【左表特征】选项组中的图标，选中全部字段；单击【右表特征】选项组中的图标，选中全部字段；单击【左表主键】选项组中的图标，选中“phone_no”字段；单击【右表主键】选项组中的图标，选中“phone_no”字段。

③ 在画布右侧的【参数设置】栏的【选择连接方式】下拉列表框中选择【inner】选项，如图 8-26 所示。

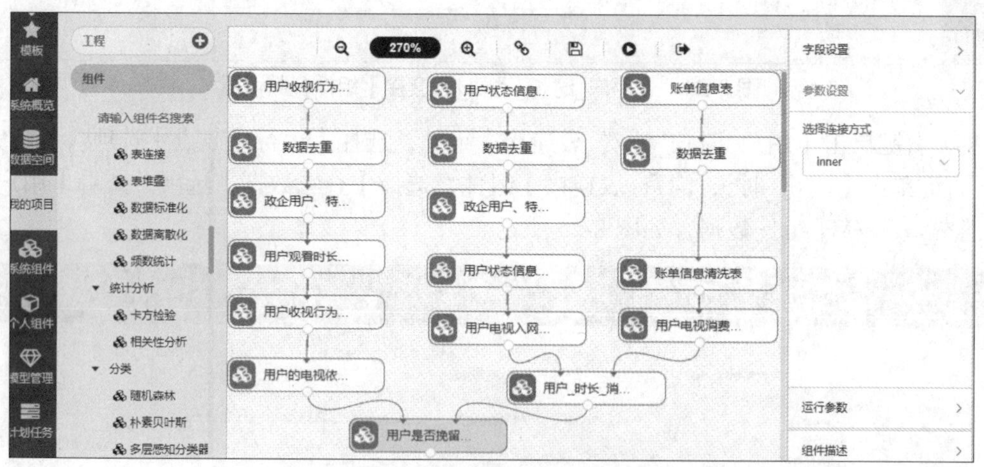

图 8-26 【用户是否挽留特征】组件参数设置

④ 右键单击【用户是否挽留特征】组件，选择【运行该节点】选项。运行完成后，右键单击【用户是否挽留特征】组件，选择【查看数据】选项，查看【用户是否挽留特征】组件数据，如图 8-27 所示。

预览数据

phone_no	count_duration	join_time	consume
3262151	0.1454166666666667	6.3455837441666665	62.333333333333336
2065841	0.3344753086419753	5.777948744166667	98.17333333333333
5088571	0.17320048309178745	8.761692055	111.33333333333333
3128042	0.31967171717171716	7.8236216275	62.333333333333336
3996303	0.1502777777777778	3.866943449166667	138.66666666666666
5085468	0.324260101010101	2.850623133333333	138.66666666666666
3496992	0.23606209150326798	6.7084010975	92.33333333333333
4062815	0.7722474747474748	3.01442475	120.66666666666667
4001202	0.38821637426900585	3.6666666666666665	115.33333333333333

图 8-27 查看【用户是否挽留特征】组件数据

由图 8-27 可知，字段“count_duration”为用户平均每天观看电视的时长，字段“join_time”为用户从开户时间到当前时间的时长，字段“consume”为用户平均每个月的消费金额。

（4）挽留用户特征数据

将挽留用户与用户是否挽留特征数据进行表连接，获得挽留用户的特征数据，具体步骤如下。

① 拖曳【大数据组件】→【预处理】→【表连接】组件到工程画布中，并与【挽留用户】【用户是否挽留特征】组件相连接。其中，【挽留用户】组件中的挽留用户数据来源如下：在用户收视行为信息表中，如果用户的观看时长大于 5.26h，且订单信息表和用户基本信息表中有该用户的记录，那么将这类用户数据归为挽留用户数据，该流程的详细配置过程可通过平台进行查看。

② 右键单击【表连接】组件，选择【重命名】选项，输入“挽留用户特征数据”。单击【挽留用户特征数据】组件，界面右侧将出现工程画布，在工程画布的【字段设置】栏中单击【左表特征】选项组中的图标，选中全部字段；单击【右表特征】选项组中的图标，选中全部字段；单击【左表主键】选项组中的图标，选中“phone_no”字段；单击【右表主键】选项组中的图标，选中“phone_no”字段。

③ 在画布右侧的【参数设置】栏的【选择连接方式】下拉列表框中选择【inner】选项，如图 8-28 所示。

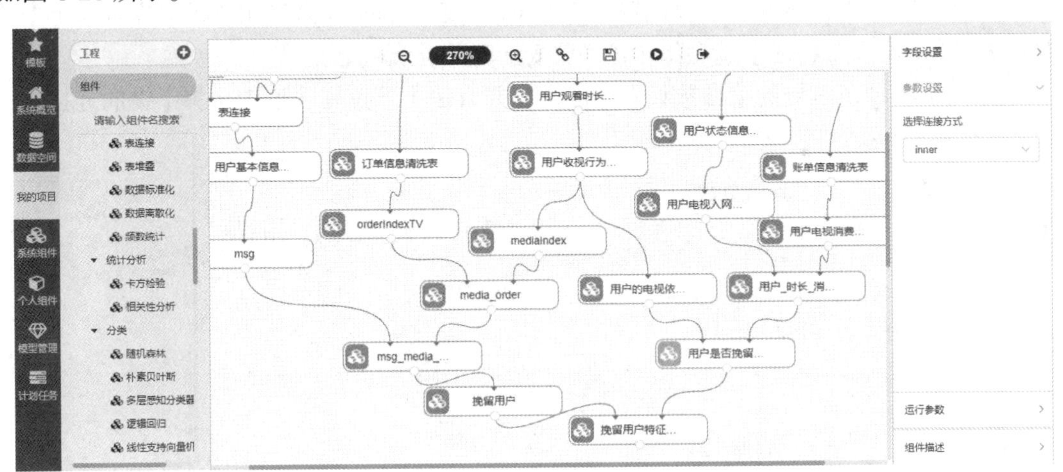

图 8-28 【挽留用户特征数据】组件参数设置

④ 右键单击【挽留用户特征数据】组件，选择【运行该节点】选项。运行完成后，右键单击【挽留用户特征数据】组件，选择【查看数据】选项，查看【挽留用户特征数据】组件数据，如图 8-29 所示。

由图 8-29 可知，字段“label”为 1 时表示该用户为挽留用户。

（5）表堆叠

将不挽留用户特征数据与用户挽留特征数据进行表堆叠，得到一份适用于预测用户是否挽留的数据集，具体步骤如下。

预览数据

phone_no	label	count_duration	join_time	consume
2158755	1	0.5206257631257631	2.0200162725	120.66666666666667
2261371	1	0.27707102672292544	2.5225405716666667	62.333333333333336
2290124	1	0.24181457431457432	11.333333333333334	62.333333333333336
2267876	1	0.24649425287356322	11.762266465000002	120.66666666666667
2255454	1	0.24437865497076022	5.7719966525	106.16666666666667
2149525	1	0.1297141316073355	5.775218475833333	74.0
2205765	1	0.1374179894179894	10.916666666666666	62.333333333333336
2081555	1	0.1711904761904762	11.327102779166667	115.33333333333333
2294549	1	0.19044946957007255	11.455006938333334	62.666666666666664

图 8-29　查看【挽留用户特征数据】组件数据

① 拖曳【大数据组件】→【预处理】→【表堆叠】组件到工程画布中，并与【不挽留用户特征】【挽留用户特征数据】组件相连接。其中，【不挽留用户特征】组件的数据来源如下：在用户基本信息表中提取“run_name”字段等于“主动销户”和“暂停”的不挽留用户，并统计不挽留用户的每月平均消费金额、电视入网时长和每天平均观看电视时长这3个特征，构成不挽留用户特征数据，该流程的详细配置过程可通过平台进行查看。

② 在右侧工程画布的【字段设置】栏中单击【表一特征】选项组中的图标，选中全部字段。单击【表二特征】选项组中的图标，选中全部字段。

③ 在工程画布右侧的【参数设置】的【合并方法】下拉列表框中选择【纵向连接】选项，如图 8-30 所示。

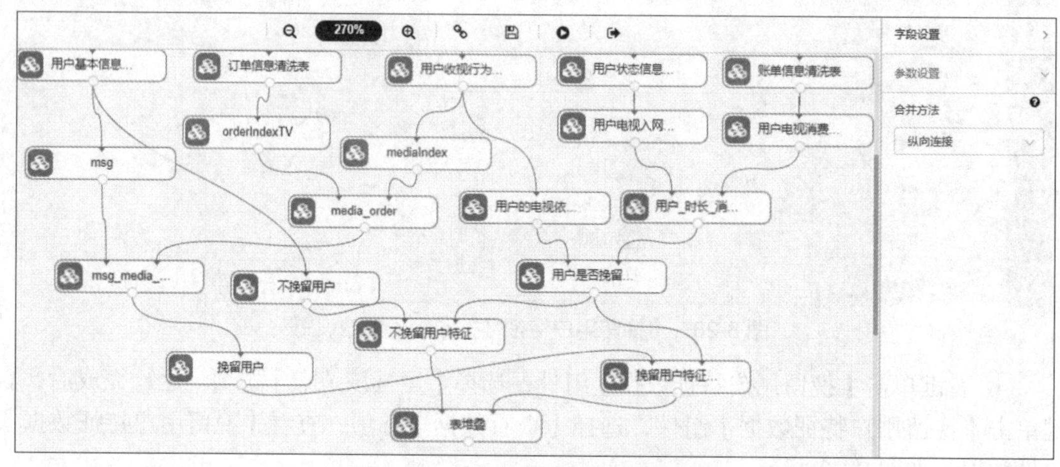

图 8-30　【表堆叠】组件参数设置

④ 右键单击【表堆叠】组件，选择【运行该节点】选项。运行完成后，右键单击【表堆叠】组件，选择【查看数据】选项，查看【表堆叠】组件数据，如图 8-31 所示。

预览数据

phone_no	label	count_duration	join_time	consume
2260888	0	0.2728864734299517	2.16818844	20.0
2158755	1	0.5206257631257631	2.0200162725	120.66666666666667
2261371	1	0.27707102672292544	2.5225405716666667	62.333333333333336
2290124	1	0.24181457431457432	11.333333333333334	62.333333333333336
2267876	1	0.24649425287356322	11.762266465000002	120.66666666666667
2255454	1	0.24437865497076022	5.7719966525	106.16666666666667
2149525	1	0.1297141316073355	5.775218475833333	74.0
2205765	1	0.1374179894179894	10.916666666666666	62.333333333333336
2081555	1	0.1711904761904762	11.327102779166667	115.33333333333333
2294549	1	0.19044946957007255	11.455006938333334	62.666666666666664

图 8-31　查看【表堆叠】组件数据

由图 8-31 可知，“label”字段为 0 的不挽留用户和“label” 字段为 1 的挽留用户数据已合并完成。

（6）模型训练

特征列和标签列数据构建完成后，使用 SVM 算法预测用户是否挽留，具体步骤如下。

① 拖曳【大数据组件】→【分类】→【线性支持向量机】组件到工程画布中，并与【表堆叠】组件相连接。

② 单击【线性支持向量机】组件，在右侧工程画布的【参数配置】栏中单击【特征】选项组中的图标，选中“phone_no”“count_duration”“join_time”“consume”字段。单击【标签】选项组中的图标，选中“label”字段。

③ 在工程画布右侧的【参数设置】栏的【是否设置截距】下拉列表框中选择【true】选项，设置【算法最大迭代次数】为【100】、【正则化参数】为【0】、【迭代时的收敛误差】为【0.0001】，在【是否标准化特征向量】下拉列表框中选择【true】选项，如图 8-32 所示。

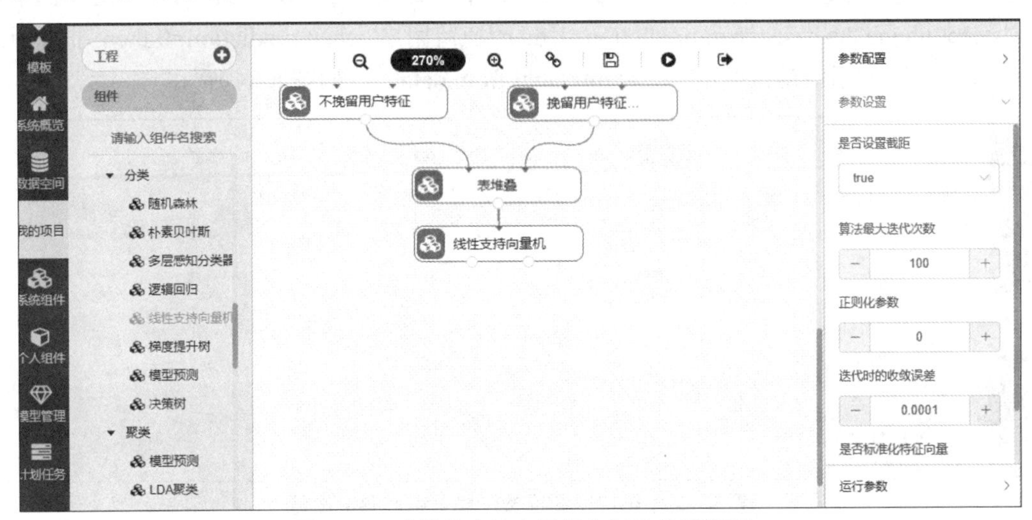

图 8-32　【线性支持向量机】组件参数设置

④ 右键单击【线性支持向量机】组件，选择【运行该节点】选项。运行完成后，右键单击【线性支持向量机】组件，选择【查看数据】选项，查看【线性支持向量机】组件数据，如图 8-33 所示。

预览数据

phone_no	label	count_duration	join_time	consume	prediction
2260888	0	0.2728864734299517	2.16818844	20.0	1.0
2158755	1	0.5206257631257631	2.0200162725	120.66666666666667	1.0
2261371	1	0.27707102672292544	2.5225405716666667	62.333333333333336	1.0
2290124	1	0.24181457431457432	11.333333333333334	62.333333333333336	1.0
2267876	1	0.24649425287356322	11.762266465000002	120.66666666666667	1.0
2255454	1	0.24437865497076022	5.7719966525	106.16666666666667	1.0
2149525	1	0.1297141316073355	5.775218475833333	74.0	1.0
2205765	1	0.1374179894179894	10.916666666666666	62.333333333333336	1.0
2081555	1	0.1711904761904762	11.327102779166667	115.33333333333333	1.0

图 8-33　查看【线性支持向量机】组件数据

由图 8-33 可知，“prediction”字段为【线性支持向量机】组件的预测结果。

2. 用户挽留标签计算

用户画像中有一个标签为用户是否挽留，该标签可根据 SVM 算法预测用户的挽留状态结果表 svm_prediction 中的 prediction 字段进行计算，prediction 为 1 的是挽留用户，prediction 为 0 的是不挽留用户，具体步骤如下。

（1）拖曳【大数据组件】→【预处理】→【SQL 语句】组件到工程画布中，并与【线性支持向量机】组件相连接。

（2）右键单击【SQL 语句】组件，选择【重命名】选项，输入“用户挽留标签计算”，单击【用户挽留标签计算】组件，在画布右侧的【参数设置】栏的【sql 语句】文本框中输入“select phone_no,case when prediction=1 then '挽留用户' when prediction=0 then '非挽留用户'　end as label,'用户是否挽留' as parent_label from data”，如图 8-34 所示。

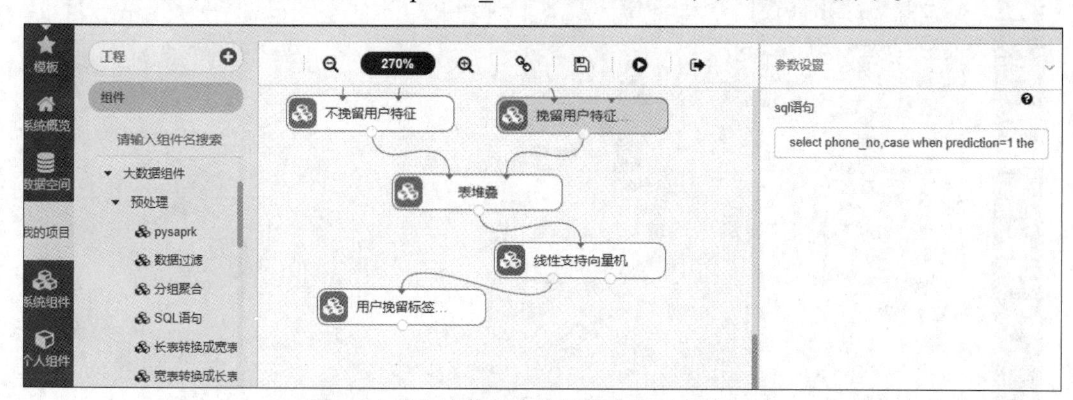

图 8-34 【用户挽留标签计算】组件参数设置

（3）右键单击【用户挽留标签计算】组件，选择【运行该节点】选项。运行完成后，右键单击【用户挽留标签计算】组件，选择【查看数据】选项，查看【用户挽留标签计算】组件数据，如图 8-35 所示。

预览数据

phone_no	label	parent_label
2260888	挽留用户	用户是否挽留
2158755	挽留用户	用户是否挽留
2261371	挽留用户	用户是否挽留
2290124	挽留用户	用户是否挽留
2267876	挽留用户	用户是否挽留
2255454	挽留用户	用户是否挽留
2149525	挽留用户	用户是否挽留
2205765	挽留用户	用户是否挽留
2081555	挽留用户	用户是否挽留

图 8-35　查看【用户挽留标签计算】组件数据

由图 8-35 可知，已进行用户挽留标签的计算。

小结

本章简单介绍了如何在 TipDM 大数据挖掘建模平台上配置广电大数据用户画像项目，从获取数据到数据探索分析，再到数据预处理，最后到数据建模，向读者展示了平台流程化的思维，使读者加深了对数据分析流程的理解。同时，TipDM 大数据挖掘建模平台去编程、拖曳式的操作，便于没有 Python、Spark 编程基础的读者轻松构建数据分析流程，从而实现数据分析的目的。